# STUDY GUIDE

To Accompany
## BIOLOGY

# STUDY GUIDE

To Accompany
## BIOLOGY

## Wessells
## Hopson

**Daniel B. Adams · Janet L. Hopson**

RANDOM HOUSE  NEW YORK

**First Edition**

9  8  7  6  5  4  3  2

Manufactured in the United States of America

Cover and text design by Victoria Vandeventer
Cover: Broad-billed hummingbird (*Cynanthus latirostris*). Photo by
Bob and Clara Calhoun/Bruce Coleman.
Project editor: Cathy de Heer
Copyeditor: Carol Dondrea
Compositor: Graphic Typesetting Service

ISBN 0-07-554561-6

9 780075 545613

# CONTENTS

# TO THE STUDENT

If you are reading this, you have embarked on a challenging voyage—the study of modern biology. This Study Guide is designed to help you complete that voyage successfully, and with the pleasure that comes from learning well. It is closely keyed to your text, Wessells/Hopson's BIOLOGY. If you use it effectively, you will learn the facts and concepts of biology more quickly, perform better on tests, and develop a thorough, lasting understanding of the fascinating "science of life."

Each chapter of the Guide has six parts: Chapter at a Glance, Chapter Preview, Learning Objectives, Concepts in Review, Key Terms, and a Self-Quiz. Most chapters also have Exercises that will reinforce your learning of important concepts and your understanding of basic biological processes.

**Chapter at a Glance**  Quickly recaps the text chapter's main sections.

**Chapter Preview**  An introductory survey of the text chapter, with hints about potential trouble areas or topics that may require extra study. This section also relates the chapter to others in the text by pointing out concepts that will be important in other chapters. If you read this section before you begin studying your text, you will get the most out of your hard work.

**Learning Objectives**  A list of specific learning goals to keep in mind as you study. If, after reading and working through the text, you can meet every objective, you can be confident that you have focused your time on key facts and concepts.

**Concepts in Review**  A concise summary of every main section of the text chapter. Although reading this material is no substitute for studying your text, you will find it extremely helpful in two ways: (1) as an integrated overview that emphasizes key concepts, and (2) as a focused chapter review before tests. If you wish to reread the more detailed text discussion of a particular point, references to text page numbers will enable you to do so quickly and easily.

**Key Terms**  All the most important vocabulary from the text chapter, with convenient page references.

**Self-Quiz**  A thorough combination of questions and problems to let you test your mastery of facts, concepts, and vocabulary. This section includes matching and completion exercises, true/false and short-answer questions, and a combined completion and multiple-choice review. Some questions involve simple recall, while others require you to analyze situations and integrate related facts and concepts.

You will find answers to self-quiz questions, including page references for the short-answer items, in a separate Answers section at the back of your Study Guide. In addition, Appendix A is a useful list of Latin and Greek roots for biological terms.

## Acknowledgments

The authors wish to express our sincerest gratitude to editor Beverly Fraknoi, whose outstanding organizational skills made possible the timely publication of this Study Guide. Ms. Fraknoi was instrumental in all phases of preparation, from planning and outlining to writing, editing, and production. Her vision and unstinting effort are evident throughout this book, as well as the entire ancillary package for BIOLOGY. We would also like to thank Beth Hatton for doing typing chores above and beyond the call of duty, Cathy de Heer and Cathy Miller for their high quality production support, and Ruth Veres for her editorial assistance under pressure. Finally, our thanks to Kathryn Gillam for many useful criticisms during the Study Guide's development. Without these people and many others, this project would not have been possible.

*Daniel B. Adams*
*Janet L. Hopson*

# 1
# THE STUDY OF LIFE

---

## CHAPTER AT A GLANCE

What Is Life?
Life on Earth: A Brief History
  *Early beliefs about the origin of life*
  *A modern view of the origin of life*
Evolution: A Theory That Changed Biology and Human Thought
  *The intellectual climate in Darwin's time*
  *Darwin's theory*
The Scientific Method: One Approach to Extending Knowledge
Biology, Society, and Your Future

---

## CHAPTER PREVIEW

The first chapter of any biology text tends to be short and simple—and this one is no exception to that rule. But do not assume that this chapter is simple*minded*. In fact, it sets the stage for much that you will learn later. In a sense, it is probably the most important chapter in the entire text.

The chapter begins by asking a crucial question: What, exactly, is life? Contrary to what you might think, the answer to this question is not at all obvious. Up until about 200 years ago, people in many cultures believed that inanimate entities such as fire and water were alive, and, as you will see in Chapter 20, today many biologists question whether the minute particles we call viruses are truly living organisms.

In exploring the boundaries between life and nonlife, biologists have devised seven tests, or characteristics, that mark a thing as being alive. These characteristics are outlined in the first section of Chapter 1. Next the chapter briefly considers the history of life on Earth (which will be elaborated in Chapter 19), and looks at some early and nonscientific beliefs about life's origins. In your reading you will see how the spontaneous generation hypothesis was laid to rest by Pasteur, and become acquainted with the concept of physical-biological coevolution—that is, how living organisms influence their nonliving environment and vice versa. This idea introduces a survey of early concepts of evolution, followed by a description of the life, travels, and evolutionary theories of Charles Darwin.

In the fourth major section, the chapter focuses on a biologist's most cherished tool—the scientific method. Be sure you understand the method's principles and how they are used—you will encounter them again and again as you proceed through this course.

# LEARNING OBJECTIVES

When you have mastered the concepts of this chapter, you will be able to:

1. List the seven characteristics of life.
2. Explain why the evolution of life on Earth can be pictured as a branching tree.
3. Identify Francesco Redi and Louis Pasteur, and describe their experiments.
4. Explain why the Earth and its life forms are said to have coevolved.
5. Briefly discuss the life of Charles Darwin and outline the two theories he developed.
6. Outline the scientific method and explain the differences between hypotheses, theories, and natural laws.
7. Explain the difference between inductive and deductive reasoning.
8. Explain what is meant by the term *cultural evolution*.
9. Explain how the scientific method is relevant to daily decision making, and how it applies to ethical and social problems.

# CONCEPTS IN REVIEW

## *Section I*   What Is Life?

Whether a single cell or a higher plant or animal, all living organisms have seven basic characteristics: They have a complex organization, the ability to take in and use energy, the ability to grow and develop, and the ability to reproduce; also they are responsive to the environment, show evolutionary adaptability, and show variation based on heredity. The phenomenon that we label life is simply a whole set of these processes that have resulted from the organization of nonliving, physical matter. In most species, each individual is unique. At the same time, it—and you—are the result of evolution. This nonreproducible biological history links each organism to a lineage of ancestors and descendants (text pp. 5–7).

## *Section II*   Life on Earth: A Brief History

No one really knows what the first organisms were like. They must have resulted from a "spontaneous generation" of very simple, self-replicating precursors of cells. This is a very different concept from that of the "spontaneous generation" of living things, such as maggots from rotten meat, which was in vogue during the days of Redi and Pasteur (text pp. 7–8).

The origin of life on Earth depended upon a very special set of physical and chemical conditions. However, although the physical Earth set the conditions for the formation of life, once that life came into existence, it in turn altered the physical nature of the world. As a result, the conditions that fostered the appearance of life on Earth no longer exist. This kind of reciprocal influence is termed *coevolution*. Living organisms have changed the Earth in numerous ways, including influencing the present oxygen-rich atmosphere, altering weather patterns, organizing and depositing huge amounts of materials in the Earth's crust, and so forth. These changes have affected how and where organisms may live (text p. 11).

As you will see in later chapters, all living things share a number of basic chemical traits. These shared traits are extremely strong evidence that all life evolved from a common origin (text p. 11).

## Section III    Evolution: A Theory That Changed Biology and Human Thought

The theory of evolution formulated by Charles Darwin is the most important unifying concept in biology. Simply stated, the theory says that all living things have evolved from a common ancestor that diverged into millions of species through a gradual process of change and variation. This change has taken place over the millions of years of the Earth's history, and continues today. Darwin proposed that the driving force of evolution was natural selection, a process in which natural events, or causes, tend to remove less-adapted individuals from a reproductive population so that their characteristics are not passed on to succeeding generations. Stated another way, in natural selection individuals that are better-adapted to their environment tend either to survive to reproductive age more successfully or to have more offspring survive to reproduce, and so pass on their characteristics with greater success than others in the population (text p. 13). Although a few people still reject Darwin's ideas, the concept of evolution has helped biologists understand and organize an incredible array of facts and observations about the living world (text p. 14).

In its most basic sense evolution simply means "change." As life forms evolve, descendants become different from their ancestors. In a later chapter you will see examples of cultural evolution, in which information is transferred from generation to generation in a nongenetic way.

## Section IV    The Scientific Method: One Approach to Extending Knowledge

The scientific method is a kind of "organized common sense" designed to extract facts from the external world. It begins with the observations of curious individuals, who then seek, through inductive reasoning, to arrive at an hypothesis—an educated guess—that will explain those observations. Once an hypothesis has withstood repeated attempts to falsify it, it can qualify as a theory. This testing process relies on control experiments, in which conditions are imposed that differ from the ones set forth in the hypothesis under study (text p. 16).

A theory is a general statement of how a process works. Usually, it is based on a number of thoroughly tested hypotheses. Once a theory exists, it can be used for deductive reasoning—that is, to make predictions on the basis of what is already known. For example, the gas carbon dioxide ($CO_2$) is formed under certain conditions by one atom of carbon and two atoms of oxygen. Thus, if we provide the right atoms and conditions in the future, then we expect (predict), as we have seen many times in the past, that $CO_2$ will be formed. When a theory is repeatedly tested and corroborated in this way—as has happened with the theory of evolution—it becomes a "natural law," accepted as scientific fact (text p. 16).

## KEY TERMS

| | | |
|---|---|---|
| coevolution    *text page 11* | inductive reasoning    *15* | theory    *16* |
| control experiment    *16* | life    *4* | theory of natural selection    *13* |
| cultural evolution    *19* | natural law    *16* | |
| deductive reasoning    *16* | scientific method    *16* | |
| hypothesis    *15* | spontaneous generation    *7* | |

# SELF-QUIZ: TESTING WHAT YOU HAVE LEARNED

## Matching Key Terms

Match each term on the left with the most appropriate description on the right.

| | |
|---|---|
| 1. coevolution | a. organized common sense |
| 2. inductive reasoning | b. predicting new facts |
| 3. theory | c. educated guess |
| 4. life | d. grows and reproduces |
| 5. control experiment | e. Charles Darwin |
| 6. theory of evolution | f. interrelated change |
| 7. cultural evolution | g. nongenetic transformation |
| 8. natural selection | h. general statement |
| 9. natural law | i. scientific fact |
| 10. spontaneous generation | j. life from nonlife |
| 11. deductive reasoning | k. conditions different from hypothesis |
| 12. scientific method | l. better-adapted survive |
| 13. hypothesis | m. specific to general |

## True or False?

1. _____ The scientific method is useful in daily decision making.

2. _____ Evolution is essentially a form of history.

3. _____ Francesco Redi convinced scientists of his time that spontaneous generation was false.

4. _____ The secret of Pasteur's success was his use of a straight-necked flask.

5. _____ Cells are the fundamental units of life.

6. _____ Buffon believed in special creation and denied that all life is related.

7. _____ Darwin sailed on the HMS *Hornblower* in 1831.

8. _____ Darwin based his theory of evolution by natural selection in part on the results of animal domestication.

9. _____ Thomas Malthus believed that organisms never reproduce fast enough.

10. _____ Scientific tests are designed to prove hypotheses.

## Completion

1. The fundamental units of all living things are _____.
2. The idea that an organism could acquire a characteristic durings its lifetime, and then pass that characteristic on to its offspring, was formulated by _____.
3. Answers to scientific problems, or explanations of a certain set of phenomena, are formulated as an _____.
4. Organisms interact and influence one another in a process called _____.
5. When a theory is shown to withstand numerous tests, it may be elevated to the status of a _____.

6. The form of reasoning wherein predictions are made based on observed and tested theories is called _____.

7. Darwin proposed his theory of _____ as a mechanism for evolution.

## Short Answer

1. Define life by describing the seven characteristics of living things.

2. How did the work of Redi, and later Pasteur, weaken the notion of spontaneous generation?

3. What were Darwin's two major theories?

4. Compare and contrast inductive and deductive reasoning.

5. What is a control experiment?

6. Is there any place in science for morality?

## Multiple-Choice Review

In the following sentences, fill in the blanks. Complete each statement by circling the correct response.

1.  Charles Darwin's theory of _____ offered a reasonable mechanism whereby life could change by virtue of the fact that there is genetic variation. This theory depends upon:

    a.  spontaneous generation of the first life.
    b.  a control experiment.
    c.  the presence of water and an atmosphere.
    d.  different rates of reproduction and survival.
    e.  the survival of acquired characteristics.

2.  Testing of a scientific hypothesis is best accomplished by attempting to _____ the hypothesis because:

    a.  that's the proper way according to the rules.
    b.  it's easy to prove anything just by selecting the good evidence.
    c.  control experiments do not apply.
    d.  inductive reasoning has to be used.
    e.  other scientists can duplicate the work.

3.  Ironic as it may seem, the scientific work of Francesco Redi was actually hampered by the

    invention of the _____ because:

    a.  it allowed his critics to falsify his experiments.
    b.  it allowed even more organisms to be seen, apparently arising from nowhere.
    c.  it showed that Pasteur was wrong.
    d.  it showed that his flasks actually held small cells.
    e.  none of the above

4.  One of the most important reasons that Earth can harbor life is that _____ exists

    here as a _____. This would not be possible if Earth's temperature ranged:

    a.  much below 0°C or above 100°C.
    b.  below 32°C or above 100°C.
    c.  below 32°C or above 98.8°C.
    d.  much below 0°C or above 98.8°C.
    e.  none of the above

5.  Once a scientific _____ has withstood numerous tests and predicted observations, it may be elevated to the level of:

    a.  theory.
    b.  hypothesis.
    c.  a natural law.
    d.  an example of deductive reasoning.
    e.  none of the above

# Exercise

Fill in the following table, using your text as a guide.

| The Seven Characteristics of Life | |
|---|---|
| **Characteristic** | **Example** |
|  |  |
|  |  |
|  |  |
|  |  |
|  |  |
|  |  |
|  |  |

# 2
# ATOMS, MOLECULES, AND LIFE

## CHAPTER AT A GLANCE

## CHAPTER PREVIEW

In this chapter you will draw on your knowledge of basic chemistry. You will begin by learning that the elements important to living systems are no different in their makeup from those in nonliving entities such as rocks and coal. However, the subset of elements found in organisms—including hydrogen, carbon, nitrogen, and oxygen—is quite different from the subset commonly seen in the nonliving world. A fundamental concept is the fact that the atoms that make up all elements interact in specific ways. As you will see throughout this text, such interactions are literally the stuff of which life is made.

To help you understand the chemical interactions that make life possible, this chapter surveys the structure of atoms, including the characteristics of protons, neutrons, and electrons. It then focuses on the various types of chemical bonds that create molecules and compounds, and considers the basic steps of chemical reactions. These concepts will provide a framework for much of what you will study in later chapters. In particular, be sure you completely understand the principles that underlie ionic, covalent, and polar bonding before you proceed to the next section on the physical properties of water. This section on water is vitally important, because the properties of water affect virtually all biological processes.

The last section of the chapter considers life as a continuum of organization of physical matter, from atoms to molecules to cells and organisms. With this perspective, you will be ready to forge ahead, in Chapter 3, into the realm of interacting biological molecules.

# LEARNING OBJECTIVES

When you have mastered the concepts of this chapter, you will be able to:

1. Define the terms *element* and *atom,* and distinguish between the two.
2. List the types of subatomic particles, and explain how they interact to form atoms.
3. Define the term *atom* in terms of atomic orbitals and their energy levels.
4. Explain the four rules of electron orbitals and their corresponding energy levels.
5. Explain how atoms bond to form elements and compounds.
6. Name the three basic types of bonds, explain how they are formed, and compare their strengths.
7. Explain the meaning of chemical formulas and equations, and what they tell us about reactants and products.
8. List the physical properties of water and explain their causes.
9. Describe and explain the dissociation of water.

# CONCEPTS IN REVIEW

## *Section I*   Elements and Atoms: Building Blocks of All Matter

Elements are the fundamental substances of the universe. This means they are composed of atoms of only one kind. In general, those atoms cannot be changed in any way—either added to or subtracted from—without changing the nature of the element.

Atoms have structure: a central nucleus composed of protons and neutrons, and areas around the nucleus that are occupied by electrons. The important characteristics of atoms are largely determined by these three major types of particles. An atom's atomic number is equal to the number of protons, while the atomic weight or mass is simply the number of protons added to the number of neutrons (text p. 27). Isotopes of elements have the same atomic number but different atomic weights.

Atoms are electrically neutral because the negatively charged electrons are equal in number to the positively charged protons in the nucleus. The space in which an electron moves about the nucleus is called an orbital, and electrons are distributed throughout an atom's orbitals according to a basic set of rules which states that simplest, lowest energy level orbitals must be filled before any higher orbitals are filled (text pp. 27–28). Only two electrons can occupy an orbital at any one time.

## *Section II*   Molecules and Compounds: Aggregates of Atoms

Two or more atoms bound together make up a molecule. Chemical bonding takes place when unpaired (valence) electrons in the outermost orbitals of atoms fuse into a shared molecular orbital. An atom can form as many bonds as there are unpaired electrons in its outermost orbital (text p. 32). The pairing of electrons in bonds forms an energetically stable unit.

Atoms may be held together by chemical bonds of three major types, and the kind of bond that holds together a molecule or compound has a direct bearing on the substance's properties. The first type, a covalent bond, forms when electrons are shared by two atoms. In a single covalent bond, only one pair of electrons is shared; in double and triple bonds, two or three pairs, respectively, are shared (text p. 33).

An ionic bond forms when one atom gives up a valence electron and another adds the free electron to its outermost orbital. Atoms that have lost or gained electrons bear a charge and are called ions (text p. 34). In a third type of bond, called a polar bond, electrons are shared but tend to spend more time orbiting one nucleus than the other. The ability to attract electrons from other atoms in a molecule is called electronegativity (text p. 34). Water molecules ($H_2O$) are polar: the electrons spend more time orbiting the single oxygen nucleus than the two hydrogen nuclei.

The strength of a chemical bond, called bond energy, is the amount of energy (measured in kilocalories) needed to break it. All covalent and ionic bonds are strong bonds, because a great deal of energy is needed to break them. Weak bonds, including hydrogen bonds and van der Waals forces, are easily broken (text p. 35).

The numbers of different atoms that make up a molecule are expressed in the shorthand of a molecular formula. Molecules can also be represented by a structural formula that shows the rough arrangement of its various atoms in space and the number of bonds between them (text p. 35). The process in which molecules and ions (reactants) interact to form new substances (products) is called a chemical reaction.

## *Section III*   Water: Life's Precious Nectar

Life on Earth could not exist were it not for the presence of water in a liquid state. The physical properties of water include a high melting and boiling point, a high specific heat and heat of vaporization, cohesion, tensile strength, adhesion, capillarity, and surface tension (text pp. 36–37). These properties derive from the structure of water molecules, as well as from the weak hydrogen bonds that water molecules tend to form with one another (text pp. 39–40).

Water is Earth's most widespread solvent—a substance capable of forming a homogeneous mixture with molecules of another substance. Compounds that dissolve readily in water are termed hydrophilic; those that tend to be insoluble in water—because they contain atoms linked by nonpolar covalent bonds—are called hydrophobic compounds (text p. 41). Such compounds are basic to life on Earth: a hydrophobic layer of lipids (a type of fat) covers the surface of every living cell.

A final important property of water is dissociation—its slight tendency to fall apart, separating into hydrogen ions ($H^+$) and hydroxyl ions ($OH^-$). A solute in water that gives up (donates) $H^+$ ions is an acid, while a compound that decreases the number of $H^+$ ions in solution is a base. The pH scale expresses the concentration of $H^+$ in acid and base solutions (text p. 42). A buffer is any chemical substance that binds $H^+$ ions when their concentration is high, and releases $H^+$ ions when their concentration is low. Buffers are extremely important to cells and organisms because they help them resist changes in pH when acids or bases are produced or added (text p. 43).

## *Section IV*   Atoms to Organisms: A Continuum of Organization

The main point of this final section is that there is physical and chemical evolution as well as biological evolution. The various elements that make up the Earth and its life forms were formed shortly after the beginning of the universe, and can be ordered into a sort of hierarchical classification, from subatomic particles to atoms to molecules and so on. The same kind of increase in complexity is seen in biological evolution. Thus we can view the world around us as a continuum of organization, from the simplest element to the most complex organism.

## KEY TERMS

| | | |
|---|---|---|
| acid *text page 42* | electronegativity  34 | neutron  27 |
| adhesion  37 | element  25 | pH  42 |
| atom  25 | gram molecular weight  35 | polar bond  34 |
| atomic number  27 | heat of vaporization  36 | product  36 |
| atomic orbital  30 | hydrogen bond  39 | proton  27 |
| atomic weight  27 | hydrophilic  40 | reactant  36 |
| base  42 | hydrophobic  41 | solute  40 |
| bond energy  35 | ion  34 | solvent  40 |
| buffer  43 | ionic bond  34 | specific heat  36 |
| capillarity  37 | isotope  27 | strong bond  35 |
| chemical reaction  35 | kilocalorie  35 | structural formula  35 |
| cohesion  37 | mole  35 | surface tension  38 |
| compound  32 | molecular formula  35 | tensile strength  37 |
| covalent bond  33 | molecular orbital  32 | weak bond  35 |
| electron  27 | molecule  32 | |

# SELF-QUIZ: TESTING WHAT YOU HAVE LEARNED

## Matching Key Terms

Match each term on the left with the most appropriate description on the right.

| | |
|---|---|
| 1. atom | a. composed of identical atoms |
| 2. covalent bonds | b. atoms bound together |
| 3. isotope | c. contains atoms of more than one element |
| 4. element | d. water hating |
| 5. proton | e. Avogadro's number |
| 6. hydrophobic | f. measure of bond strength |
| 7. kilocalorie | g. like molecules cling together |
| 8. compound | h. tendency to move up in narrow space |
| 9. solvent | i. unlike molecules cling together |
| 10. capillarity | j. water loving |
| 11. cohesion | k. matter's smallest characteristic units |
| 12. molecule | l. shared electrons |
| 13. mole | m. dissolved solutes |
| 14. adhesion | n. same atomic number, different atomic weights |
| 15. hydrophilic | o. add them to get atomic number |

## True or False?

1. _____ Oxygen (O), sulfur (S), and gold (Au) are elements.

2. _____ Atoms are easily seen with an electron microscope.

3. _____ Carbon and its isotopes have the same atomic weights.

4. _____ Atomic orbitals may hold more than two electrons.

5. _____ Covalent bonds share electrons.

6. _____ Electronegativity involves attracting electrons.

7. _____ Water may exist in two physical states.

8. _____ Surface tension is a liquid's tendency to minimize its surface area.

9. _____ Water is a polar molecule.

10. _____ The pH scale is logarithmic.

## Completion

1. Atoms that have the same atomic number but different _____ are called

   _____ .

2. Protons have a _____ charge, while neutrons have _____, and electrons
   have a _____ charge.

3. _____ are formed when there is a partial transfer of electrons in the bonding.

4. It requires exactly _____ of heat energy to raise the temperature of 2 kilograms of
   water by 1°C.

5. Weak chemical bonds are _____ by chance molecular collisions, or require few

   _____ per _____ to break.

6. Two of the significant characteristics of water are its high _____ and also its high

   _____ .

7. _____ is a measure of the resistance molecules exert against being pulled apart.

8. A substance that gives up $H^+$ ions to a solution is an _____ .

## Short Answer

1. What are the four basic rules for filling orbitals?

2. What are Bohr models, and how are they used?

3. How do covalent bonds differ from ionic bonds?

4. Why is water such a good solvent?

5. What is the difference between a molecular and structural formula?

6. What are the properties of water that make it so useful (and critical) for life?

## Multiple-Choice Review

In the following sentences, fill in the blanks. Complete each statement by circling the correct response.

1. An element with four _____ will have how many occupied $2p$ orbitals?

   a. 1
   b. 2
   c. 3
   d. 4
   e. none

2. Atoms that share three pairs of _____ electrons are said to be:

   a. double-bonded.
   b. triple-bonded.
   c. single-bonded.
   d. bonded.
   e. none of the above

3. _____ is the ability to attract electrons from other atoms. This creates relatively weak bonds that are:

   a. electrical.
   b. double.
   c. polar.
   d. covalent.
   e. carbonized.

4. _____ bonds are broken easily by chance collisions, or require very little energy to separate, and include:

   a. van der Waals forces.
   b. covalent bonds.
   c. structural bonds.
   d. triple bonds.
   e. reaction forces.

5. Ionically bonded _____ share their electrons:

   a. polarly.
   b. unequally.
   c. equally.
   d. subequally.
   e. They do not share electrons.

## Exercise

Study the figure below and answer the questions on page 16.

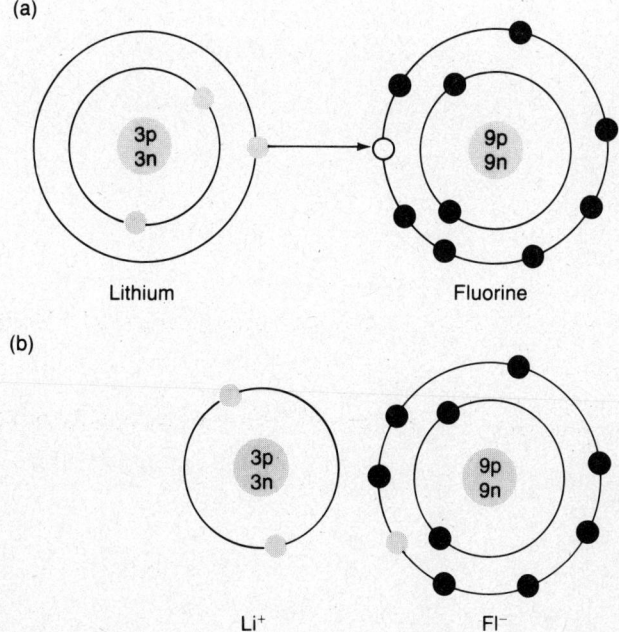

(a)

Lithium    Fluorine

(b)

Li⁺    Fl⁻

What is happening in part (a)?

What kind of bonding does it represent?

Are the arrangements shown in (b) stable or unstable?

# 3

# THE MOLECULES OF LIVING THINGS

---

## CHAPTER AT A GLANCE

---

## CHAPTER PREVIEW

This chapter deals with the four major classes of large molecules that are unique to living things: carbohydrates, lipids, proteins, and nucleic acids. This chemistry background will be useful to you as you study later chapters, and most of the concepts are fairly simple. In the first section you will encounter some basic chemistry terminology, including *functional groups, monomers, polymers,* and *isomers*. In the sections that follow you will see how these entities are used to construct the major groups of large biological molecules.

Next the chapter considers the four main molecule classes, starting with a survey of the carbohydrates. You will learn how chemical energy is stored and how it is used to build simple and complex sugars that serve as biological fuels. Following this section you will look at the three main types of lipids. The first type is fats and oils, which are especially rich in energy (and are used for long-term storage) because of the huge numbers of C–H covalent bonds they contain. Phospholipids—the chemical result of adding nitrogen and phosphorus to the carbon, hydrogen, and oxygen of fats and oils—come next, followed by steroids. You may be familiar with steroids because of references to them in the popular press as "drugs" that "cheating" athletes abuse. So you may be surprised to learn that steroids are really types of hormones and vitamins. The following section on proteins is a traditional problem area for students, but it should not be. The way that protein structure progresses in complexity from primary to secondary, tertiary, and quaternary is actually commonsensical—use the chapter figures for help, and try drawing a few yourself. The final two sections focus on nucleic acids (the building blocks of DNA), and the important concepts of macromolecule specificity and complementarity.

# LEARNING OBJECTIVES

When you have mastered the concepts of this chapter, you will be able to:

1. List the four main types of compounds found in living organisms, and explain how they are structured.
2. Explain how energy is stored in the bonds of carbohydrates, and compare and contrast the various carbohydrates used by organisms.
3. Explain how energy is stored in lipids, how it can be used by organisms, and how lipids are critical elements of biological membranes.
4. Describe the basic structure of proteins, amino acids, and polypeptides.
5. Outline the four basic structural organization patterns of proteins.
6. Explain why protein shape is important in biological processes.
7. Describe the basic chemical structure of nucleic acids.

# CONCEPTS IN REVIEW

## *Section I*   The Fundamental Components of Biological Molecules

Most of the compounds that make up living things are of four main types: nucleic acids, proteins, carbohydrates, and lipids. All four contain carbon, and are thus organic molecules; many of the molecules in these four classes are also macromolecules, with molecular weights of about 10,000 daltons or more. The structure of carbon permits it to form as many as four covalent bonds with other atoms. As a result, carbon atoms make up the linear or ringlike "backbones" of more than one million compounds (text pp. 47–48).

It is important and useful to learn to think of molecules as three-dimensional structures. A molecule composed of one carbon atom with four different atoms attached to it can be arranged in two ways, each the mirror image of the other—so-called stereoisomers. All stereoisomers of a compound have the same chemical properties. Two compounds with identical formulas but different arrangements of atoms are called isomers (text p. 48). And compounds that have the same molecular formula but different properties are known as structural isomers.

Each cluster of atoms called a functional group imparts similar chemical behaviors to all the molecules to which it is attached. Functional groups may be attached to carbon atoms; the biologically important groups include the hydroxyl and carboxyl groups, and the aldehyde, methyl, ketone, and disulfide groups (text pp. 49–50).

All macromolecules are polymers, which are made up of monomers. The specific order of these constituent monomers is a key feature of these substances. The polymers are formed by means of condensation reactions, and are broken down by way of hydrolysis (text p. 50).

## Section II    Carbohydrates: Sources of Stored Energy

Carbohydrates, including sugars and starches, are the most abundant carbon compounds in living organisms. All are composed of carbon, hydrogen, and oxygen in a ratio of 1:2:1. The basic carbohydrate subunits are simple sugar molecules, or monosaccharides, such as glucose. These function as monomers that can be joined together via covalent glycosidic bonds to form more complex disaccharides and polysaccharides (text pp. 51–52). Carbohydrates that are especially important include starch, the main nutrient reserve of plants; glycogen, the carbohydrate storage form in animals; and cellulose, the fibrous structural material of plant cells and wood.

## Section III    Lipids: Energy, Interfaces, and Signals

The three types of lipids—fats and oils, phospholipids, and steroids—all share one property: they are insoluble in water. The basic units of fats and oils are glycerol and fatty acids; these are called triglycerides because the usual structure is three fatty acids linked to one glycerol molecule. The linkage takes place via ester bonds, which form between alcohol and carboxylic acid. Fats and oils are energy-rich because they contain many C–H covalent bonds. In saturated fat and oil molecules, each carbon in the molecular backbone is linked to a maximum number of hydrogen atoms. In unsaturated molecules, some adjacent carbon atoms are joined by double covalent bonds (text p. 56).

Phospholipids contain atoms of nitrogen and phosphorus in addition to C, O, and H atoms. Their basic structure consists of two fatty acid chains joined to glycerol (again, by ester bonds) and a phosphate group. The charged phosphate group constitutes a water-soluble "head," while the fatty acid chains form two hydrophobic "tails." This arrangement is vital to the structure and function of cell membranes (text p. 57).

Unlike other lipids, steroids do not consist of fatty acids and glycerols. Instead, they consist of interconnected rings of carbon atoms with functional groups attached. Some act as vitamins and others are hormones (text p. 58).

## Section IV    Proteins: The Basis of Life's Diversity

Proteins make up a varied class of macromolecules that are classified according to function. Structural proteins form major portions of diverse plant and animal cells. Other proteins serve as enzymes, while still others act as chemical messengers such as antibodies. All proteins are built from amino acids, in which a central carbon is bound to an hydrogen, an amino group, a carboxyl group, and a side chain (text p. 59).

The chains of amino acids that make up proteins are linked by peptide bonds in a condensation reaction. In addition, many proteins include a disulfide bond. Two joined amino acid units (residues) are called a dipeptide. Repeated condensation reactions can create long amino acid chains called polypeptides (text p. 61).

There are four categories of protein structure. Primary structure refers to the linear sequence of amino acids in a protein, which determines that protein's function. Secondary structure refers to the patterns in which specific subunits occur that make up a particular protein. Configurations of secondary structure include the α-helix (alpha-helix, helical coiling of a polypeptide chain) and the β-pleated (beta-pleated) sheet (an accordionlike sheet of connected polypeptide molecules) (text pp. 63–64). Tertiary structure encompasses the precise folding patterns that give proteins characteristic shapes. Some proteins also have quaternary structure, the arrangement of separate polypeptides in a three-dimensional shape held together by weak bonds (text pp. 63–66).

Proteins assume their complex three-dimensional shapes automatically. This structure is determined by each protein's amino acid sequence, and is held and stabilized by chemical bonds, van der Waals forces, and hydrophobic interaction (text pp. 67–69).

## Section V    Nucleic Acids: The Code of Life

Like proteins, nucleic acids have a specific linear sequence of subunits. These subunits are called nucleotides, and each consists of a nitrogen-containing base, a five-carbon sugar, and a phosphate. The two types of nucleic acids are DNA (deoxyribonucleic acid), which contains the sugar deoxyribose, and RNA (ribonucleic acid), which contains the sugar ribose. The four nitrogen-containing bases commonly found in nucleotides are adenine, guanine, cytosine, and thymine. RNA contains uracil instead of thymine. In its native state, DNA consists of two chains of nucleotides twisted around each other in a double-helix configuration and held together by hydrogen bonds (text pp. 69–70).

## Section VI    The Specificity and Complementarity of Macromolecules

Biological molecules act in distinctly characteristic ways. This is what is meant by specificity. Complementarity refers to the tendency of groups of atoms in molecules in three-dimensional conformations to form lock-and-key type bonds. Such complementary fit is a critical component of many essential biological processes.

## KEY TERMS

| | | |
|---|---|---|
| α-helix    *text page 63* | glycerol   55 | peptide bond   61 |
| amino acid   59 | glycogen   53 | phospholipid   57 |
| β-pleated sheet   64 | glycosidic bond   52 | polymer   50 |
| carbohydrate   51 | hydrolysis   50 | polypeptide   61 |
| cellulose   53 | isomer   48 | polysaccharide   51 |
| condensation reaction   50 | lipid   55 | protein   58 |
| disaccharide   51 | macromolecule   47 | starch   52 |
| enzyme   59 | monomer   50 | stereoisomer   48 |
| ester bond   56 | monosaccharide   51 | steroid   57 |
| fatty acid   56 | nucleic acid   69 | structural isomer   49 |
| functional group   49 | nucleotide   69 | triglyceride   56 |
| glucose   51 | organic compound   47 | |

# SELF-QUIZ: TESTING WHAT YOU HAVE LEARNED

## Matching Key Terms

Match each term on the left with the most appropriate description on the right.

| | | | |
|---|---|---|---|
| 1. cellulose | | a. | splitting with water |
| 2. glycogen | | b. | form polymers |
| 3. amino acids | | c. | long carbon chains + carboxyl group |
| 4. functional group | | d. | many parts |
| 5. condensation reaction | | e. | make up proteins |
| 6. polymers | | f. | grape sugar |
| 7. organic | | g. | plant nutrient reserve |
| 8. glucose | | h. | single parts |
| 9. hydrolysis | | i. | animal energy storage |
| 10. monomer | | j. | compounds with carbon |
| 11. macromolecule | | k. | specially arranged atoms |
| 12. ester bonds | | l. | fibrous structural material |
| 13. fatty acid | | m. | insoluble in water |
| 14. starch | | n. | 10,000 daltons or more |
| 15. lipid | | o. | form between alcohol and carboxylic acid |

## True or False?

1. _____ Inorganic compounds always contain carbon.

2. _____ Isomers are mirror images of each other.

3. _____ Glucose molecules contain six carbon atoms.

4. _____ Enzymes speed up chemical reactions.

5. _____ Steroids are hydrophobic.

6. _____ Proteins are made from the twenty-two amino acids.

7. _____ Breaking disulfide bonds denatures a protein.

8. _____ Proteins are stable under a wide range of temperatures.

9. _____ There are about 500,000 possible nucleic acids.

10. _____ Proteins have the property of self-assembly.

## Completion

1. Chemical compounds that have identical molecular formulas but possess different properties are _____ of each other.

2. Carbohydrates are composed of sugar molecules called _____ , which can be joined into more complex _____ or _____ .

3. _____ is a six-carbon sugar that is a structural isomer of glucose.

4. _____ is the major component of insect shells (exoskeletons).

5. Fats and oils are composed of _____ and _____ .

6. Lipids are insoluble in water because they are non_____ and _____ .

7. Amino acids are linked together covalently by _____ .

8. When several (two or more) polypeptides are cross-linked by hydrogen bonds, they form a _____ .

## Short Answer

1. What is the difference between organic and inorganic compounds? Are all organic compounds made by living things?

2. What is a stereoisomer?

3. What is a functional group? Name the ten functional groups listed in the text.

4. Why is cellulose usually indigestible by most living things? Do any animals use it for food?

5. Why are fats and oils such high energy foods?

6. Why are phospholipids so important to life?

## Multiple-Choice Review

In the following sentences, fill in the blanks. Complete each statement by circling the correct response.

1. Compounds that have the same molecular formula but different _____ are:

   a. functional isomers.
   b. radical isomers.
   c. stereoisomers.
   d. structural isomers.
   e. none of the above

2. Fats and oils are also called _____ because fatty acids usually join to one glycerol molecule in:

   a. twos.
   b. threes.
   c. fours.
   d. fives.
   e. none of the above

3. The most common polysaccharide storage material in plants is _____ , while in animals the liver stores:

   a. glycogen.
   b. chitin.
   c. cellulose.
   d. glycerol.
   e. none of the above

4. Nucleic acids always have a _____ group at one end and on the other end:

   a. an aldehyde group.
   b. an hydroxyl group.
   c. a carboxyl group.
   d. a methyl group.
   e. none of the above

5. Glycerol and fatty acids are linked by _____ bonds, which form between carboxylic acids and:

   a. water.
   b. alcohols.
   c. sulphur.
   d. phosphates.
   e. none of the above

## Exercise

Write the formulas for the reaction products of reactions (a) and (b).

(a)

$$\text{H—N—C—C—OH} \quad + \quad \text{H—N—C—C—OH} \quad \rightarrow$$

(b)

What type of bond has formed in (a)? in (b)?

# 4
# CHEMICAL REACTIONS, ENZYMES, AND METABOLISM

---

## CHAPTER AT A GLANCE

The Energetics of Chemical Reactions
*Energy transformations and the laws of nature*
  The first law of thermodynamics
  The second law of thermodynamics
Free-Energy Changes
*Dependence of equilibrium between reactants and products on free energy*
Rates of Chemical Reactions
*Temperature and reaction rates*
*Concentration and reaction rates*
*Catalysts and reaction rates*
Enzymes and How They Work
*Enzyme function*
*Enzymes and reaction rates*
Metabolic Pathways
*Control of enzymes and metabolic pathways*
Looking Ahead

---

## CHAPTER PREVIEW

The central principle of Chapter 4 is that chemical reactions are the basis of all change in biological systems, and that they all involve changes in energy. The first section introduces the three types of energy—kinetic, potential, and chemical—and describes the first and second laws of thermodynamics. The text then turns to the important concept of "free" energy, or, as it is defined, the energy free and available to do work as a result of a chemical reaction. This subject sometimes brings with it the "fear of the triangle" for those students who have problems with math. But the triangle symbol ($\Delta$) in energy equations is simply a shorthand way of writing "change." Once you understand this, you'll be able to proceed without difficulty to consider the two types of chemical reactions: exergonic (energy out) reactions in which $\Delta G$ is a negative number, and endergonic (energy in) reactions in which $\Delta G$ is a positive number.

The study of endergonic and exergonic reactions leads to the idea that the number of productive collisions between molecules is what determines the rate of chemical reactions. Accordingly, the next section of the chapter describes the factors that can alter reaction rates, including temperature, concentration of reactants, and action of enzymes. Next you will consider how enzymes (biological catalysts) work; take care to study the text figures on the enzyme–substrate complex.

The last section of this chapter focuses on the metabolic pathways that make up nearly all of what a cell "does." Metabolism has two main components—building up molecules and tearing them down. You should pay special attention to the section on negative feedback because this is the most important mechanism controlling metabolism.

# LEARNING OBJECTIVES

When you have mastered the concepts of this chapter, you will be able to:

1. State the first law of thermodynamics and explain why it is important in biological systems.
2. State the second law of thermodynamics and explain why it is important in biological systems.
3. Discuss the energy relationships that occur between products and reactants in a chemical reaction.
4. Describe the effects of temperature, concentration, and catalysts on chemical reactions.
5. Define enzymes and explain how they work.
6. List the kinds of reactions and processes that make up metabolism, and explain how each works.
7. Outline the role of enzymes in metabolic pathways.

# CONCEPTS IN REVIEW

## *Section I*   The Energetics of Chemical Reactions

Chemical reactions underlie every change in the living world. To bring about a reaction, energy is required. In addition to light, heat, and electrical energy, other important energy forms are kinetic energy (motion), potential energy (stored), and chemical energy (that stored in atoms and molecules and their bonds) (text p. 75).

The first law of thermodynamics states that energy can change from one form to another but can never be created or destroyed. No energy transformation is ever 100 percent efficient; some is always given off as heat and is no longer available to do useful work. This fundamental fact is reflected in the second law of thermodynamics. This law states that the total energy in a system decreases as conversions take place and heat dissipates. Hence the universe is becoming increasingly disordered. The amount of disorder in a system is known as entropy. For a cell in a living system to stay alive and counteract the tendency of complex molecules to degrade, the cell must continuously carry out energy-producing chemical reactions (text p. 77).

## *Section II*   Free-energy Changes

An important concept in the study of cell energetics is the idea of free energy—the energy available to do work as a result of a chemical reaction. An exergonic reaction is one in which less total and free energy remain in the products after the reaction. In effect, energy is released. In an endergonic reaction the products have more total and free energy than did the reactants. Thus, for this type of reaction, there must be an input of energy from an outside source (text p. 77). Free-energy changes underlie all life processes.

Chemical reactions are said to be at equilibrium when the combined free energies of reactants and products are equal. At equilibrium, net conversions of reactants to products, or vice versa, no longer take place, even though the reactions may continue in both directions (text p. 78). However, true chemical equilibrium is rarely achieved in living cells. In a sense, life requires such an imbalance, because in biological systems the energy freed during an exergonic reaction is frequently used to drive an endergonic one. Such coupling of reactions is the basis for many interrelated reactions in living cells.

## Section III    Rates of Chemical Reactions

Life depends on the fact that covalent bonds hold molecules together stably: it also requires that those bonds sometimes be broken. In order for a molecule to break its bonds and participate in a reaction, it must be in an energetically activated state (text p. 80). Activation energy is derived from collisions between molecules. Hence, the number of productive collisions in a system determines the rate of chemical reactions. In organisms, the rates of many reactions—slow under other circumstances—must be very fast. Reactions in cells may be speeded by increases in temperature or changes in the concentrations of reactants or products: however, the most important influence on reaction rates is the presence or absence of chemical catalysts (text p. 81). Called enzymes, biological catalysts work not through an input of energy, but rather by lowering the activation energy necessary for a reaction to proceed.

## Section IV    Enzymes and How They Work

Enzymes have two characteristics crucial to their functioning. First, they are specific: a given enzyme can act only on a particular substrate, and can usually catalyze only one type of reaction. Second, enzymes can be controlled by the presence or absence of critical compounds. Most enzymes are globular proteins. On the surface of each is a small active site that is reciprocal to the shape of a specific substrate. This reciprocal arrangement is the key to enzyme specificity (text p. 83). The active site may contain a prosthetic group that is essential to the enzyme's activity. Enzymes that lack permanently bound prosthetic groups depend instead on the temporary binding of cofactors or coenzymes.

An enzyme carries out four functions in lowering the activation energy of molecules: (1) it forms a complex (called an ES complex) with the reacting molecules; (2) it increases the local concentration of the molecules; (3) the enzyme then orients the molecules in space so that the reaction can proceed efficiently; and (4) it somewhat distorts the shape of the molecules, thereby raising their free energies (text pp. 84–87).

An enzyme cannot bring about a reaction that is energetically unfavorable—that is, it cannot change an endergonic reaction into an exergonic one. Reactions that are enzyme-mediated can be speeded up by a rise in temperature. Also, the reaction rate is proportional to the amount of enzyme present. Finally, the rate of a reaction will increase as the concentration of the enzyme substrate increases, allowing more active sites on enzyme molecules to bind. This is true only up to a point, however. Eventually the enzyme will be saturated, with all active sites engaged (text p. 89).

## Section V    Metabolic Pathways

The combination of simultaneous, interrelated chemical reactions taking place at any given time in a cell is referred to as cellular metabolism. Reactions that use energy to synthesize biological molecules are collectively termed *anabolic*, while those that break down molecules and yield energy are collectively termed *catabolic* (text p. 90). All anabolic and catabolic reactions in cells are coupled, with the energy released in catabolism used to drive the reactions of anabolism.

Many of the enzyme-catalyzed chemical events in cells are part of a metabolic pathway—a series of reactions in which compounds are progressively built up or degraded. At each step of such a pathway, an enzyme catalyzes a reaction that alters the starting material (text p. 91). The sets of enzymes involved in a pathway are segregated in distinctive compartments in cells, separated from enzymes that catalyze other reactions. This association allows the enzymes to work much more efficiently.

The controllability of enzymes is a crucial feature that sets them apart from other catalysts. One type of enzyme control is negative feedback, in which the concentration of a reaction product feeds back to control the rate of the reaction (text p. 92). Allosteric enzymes are those in which the binding of unused product molecules to a specific site on the enzyme causes it to change to a shape that is incompatible with the functioning of the enzyme's active site (text p. 92). In another type of enzyme control, called competitive inhibition, a compound other than the normal enzyme substrate competes for the enzyme's active site, preventing binding of the substrate.

## KEY TERMS

activation energy
*text page 80*

active site   83

allosteric enzyme   92

anabolism   90

catabolism   90

catalyst   91

chemical energy   75

chemical reaction   75

cofactor   83

competitive inhibition   92

endergonic reaction   77

energetics   75

entropy   77

enzyme saturation   89

enzyme–substrate complex
  (ES)   84

equilibrium (chemical)   78

exergonic reaction   77

first law of thermodynamics
  76

free energy   77

induced fit   85

kinetic energy   75

metabolic pathway   91

metabolism   90

negative feedback   92

noncompetitive feedback   92

potential energy   75

second law of
  thermodynamics   76

substrate   83

# SELF-QUIZ: TESTING WHAT YOU HAVE LEARNED

## Matching Key Terms

Match each term on the left with the most appropriate description on the right.

1. anabolism
2. free energy
3. first law of thermodynamics
4. exergonic
5. potential energy
6. allosteric enzyme
7. kinetic energy
8. catabolism
9. catalyst

a.  breakdown
b.  stored
c.  stored in chemical bonds
d.  disorder
e.  build up
f.  energy in
g.  regulation
h.  energy conservation
i.  available energy

10. induced fit
11. negative feedback
12. chemical energy
13. activation energy
14. endergonic
15. entropy

j.  needed to break bonds
k.  energy of motion
l.  other shape
m.  increases rate
n.  change in shape
o.  energy out

## True or False?

1. _____ A considerable amount of cell energy is utilized by enzymes.

2. _____ Enzyme–substrate complexes are held together with strong bonds.

3. _____ Equilibrium refers to the fact that the free energies of reactants and products are equal.

4. _____ ($\Delta G > 0$) is an exergonic reaction.

5. _____ Enzymes are usually very specific.

6. _____ Anabolism makes bigger molecules out of smaller ones.

7. _____ Enzyme–substrate complexes take relatively long periods of time to form.

8. _____ Enzyme saturation occurs in noncatalyzed reactions.

9. _____ Negative feedback usually assumes a noncompetitive inhibition format.

10. _____ Catalysts raise activation-energy barriers.

## Completion

1. The gunpowder in a rifle shell has a certain amount of _____, or _____, energy. When fired, the traveling bullet has _____ energy.

2. When an exergonic reaction yields the energy to drive an endergonic reaction, the two reactions are said to be _____ .

3. The metabolic process of _____ makes larger molecules out of smaller ones, whereas the process of _____ makes smaller molecules out of larger ones.

4. The type of enzyme control wherein the active site is occupied by other than the normal substrate is called _____ .

5. In _____ reactions $\Delta G$ is a positive number.

6. In _____ reactions $\Delta G$ is a negative number.

7. _____ are substances that increase reaction rates without being used up in the reaction.

8. _____ is a form of enzyme control wherein the build-up level of amino acid controls the rate of its synthesis.

## Short Answer

1. What is the first law of thermodynamics?

2. What is the second law of thermodynamics? What aspect of this law does evolution "defy," and how?

3. What are the two free-energy changes, and how are they symbolized?

4. How does heat speed up a reaction?

5. How do catalysts work?

6. How does an enzyme–substrate complex work?

## Multiple-Choice Review

In the following sentences, fill in any blanks. Complete each statement by circling the correct response.

1. The amount of disorder in a system is known as _____. In general, disorder in the known universe is:

    a. decreasing.
    b. increasing.
    c. staying about the same.
    d. oscillating back and forth.
    e. none of the above

2. In an _____ reaction the products have more total and free energy than the reactants, and ΔG is:

   a. less than the free energy of the reactants.
   b. a positive number.
   c. less than the free energy of the products.
   d. a negative number.
   e. none of the above

3. The amount of energy required to break chemical bonds for a reaction is:

   a. ΔG.
   b. kinetic energy.
   c. activation energy.
   d. reaction energy.
   e. none of the above

4. Catalysts accomplish their work by reducing a reaction's

   a. chemical energy.
   b. kinetic energy.
   c. potential energy.
   d. activation energy.
   e. none of the above

5. The production of amino acids, carbohydrates, and other biological molecules is accomplished

   in _____ reactions, or:

   a. anabolism.
   b. catabolism.
   c. metabolism.
   d. metabolic pathways.
   e. none of the above

## Exercise

In the figure, label the enzymes, substrate(s), and reaction products.

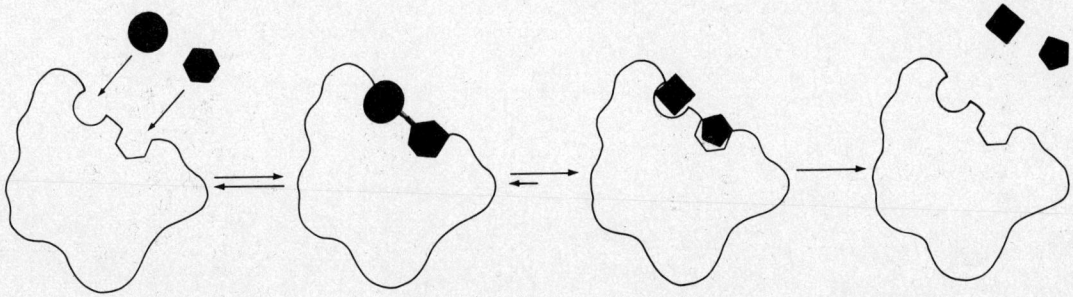

In a short paragraph, describe the process that is depicted here.

# 5
# CELLS: THEIR PROPERTIES, SURFACES, AND INTERCONNECTIONS

---

## CHAPTER AT A GLANCE

How Cells Are Studied
  *Microscopy*
  *Means of studying cell function*
Characteristics of Cells
  *The nature and diversity of cells*
  *Limits on cell size*
Box: Sorting Cells with a Glowing Technique
The Cell Surface
  *The plasma membrane*
  *Movement of materials into and out of cells: Role of the plasma membrane*
  *Osmosis and cell integrity*
Box: The Secrets of Winter Wheat
Cell Walls and the Glycocalyx
  *Cell walls*
  *Glycocalyx*
Linkage and Communication Between Cells
Multicellular Organization
  *Tissues, organs, and systems*
  *The price of multicellularity*

---

## CHAPTER PREVIEW

Chapter 5 is one of the most important you will study in your biology course. Why? Because it lays the foundation for your knowledge of the most fundamental unit of life: the cell. As you will see, the remainder of your text elaborates what goes on in cells or examines the interactions among organisms that are made from cells. The chapter begins with a look at Anton van Leeuwenhoek's invention and use of a light microscope. Some of the objects he saw and recorded are truly amazing, even by today's standards. Next the text describes modern transmission and scanning electron microscopes, which permit higher resolution of small objects but can examine only dead cells.

The following two sections present vital concepts that you will encounter again and again in the chapters that follow. First, the cell theory lays out the characteristics of the basic units of life. Also extremely important is the explanation of surface-to-volume ratio and why it places definite limits on cell size. As the discussion of this concept proceeds, you will look closely at the cell surface and its properties. The means by which molecules travel across this barrier are not difficult to learn; take time to fully understand the workings of osmosis and the anatomy of the lipid bilayer.

Next, the chapter probes the composition and structure of the cell wall and glycocalyx, and describes the five classes of linkage and communication between and among adjacent cells. Chapter 5 closes by introducing the cellular hierarchy—tissues, organs, and organ systems—that you will study in detail in the chapters on development and physiology.

# LEARNING OBJECTIVES

When you have mastered the concepts of this chapter, you will be able to:

1. State the three major and two auxiliary tenets of the cell theory, and explain why the theory is a cornerstone of modern biology.
2. Describe the organizational differences between prokaryotic and eukaryotic cells.
3. Explain the relationship of surface area to volume, and tell why it is so important to living systems.
4. Describe the structure of the plasma membrane, and explain how materials may pass through it.
5. List and describe the various junctions, linkages, and connections that occur between cells.
6. Describe the features that distinguish plant cells and animal cells.
7. Explain how biologists organize multicellular organisms into hierarchies of tissues, organs, and organ systems.

# CONCEPTS IN REVIEW

## *Section I*    How Cells Are Studied

Most cells are very small, and it is therefore no surprise that they remained unknown to scientifically minded people until the 17th century. It was not until the discovery of the light microscope that people started looking in close detail at living organisms and tissue and noticing, as Robert Hooke did, that all living things are constructed of one or more compartments or "cells."

Unfortunately, the light microscope cannot resolve objects that are as small as cells because of limits to resolution set by the wavelength of visible light. Also, at very high magnifications, there is a limited "depth of field" (the area in which objects in the field of view are in clear focus) that is often less than the thickness of the cell or tissue being studied. The magnification limits of the light microscope are overcome by transmission electron microscopes and scanning electron microscopes, in which streams of electrons, rather than light, irradiate specimens. SEMs also provide great depth of field. Both types of electron microscopes have a serious drawback, however: They cannot be used to examine living cells. Specimens must often go through elaborate preparation before they can be viewed (text pp. 98–99).

## *Section II*    Characteristics of Cells

Modern cell theory consists of six statements of basic facts about cells. Some of the statements are little more than common sense. For example, the first two statements say that cells are the basic units of life and that all organisms are composed of cells. These statements are possible because no one has ever discovered an organism that does not consist of one or more cells. Statement number three says that all cells arise from preexisting cells. This means that all cells have parents—there is no spontaneous generation of life (you will study this question in detail in Chapter 19). A fourth statement declares that cells are the *functional* units of life, in which the chemical reactions necessary for maintaining and reproducing life take place. Finally, the last two statements describe features of multicellular species: populations of cells may behave as functional units—in other words, they may act in an organized fashion, travel as a unit, show a division of labor, specialize, and so on; and cells of multicellular animals must adhere to solid substrates in order to divide, move, and assume specialized shapes (text p. 100).

Cells are classified into two groups, based on their anatomy. A prokaryotic cell such as a bacterium is very small, lacks a nucleus (so that its hereditary material is located throughout the cell), and has no specialized cell organelles. Eukaryotic cells have organelles, a membrane-bound nucleus full of hereditary material, and a cytoskeleton, and can form living organisms of very many cells. The differences between prokaryotes and eukaryotes will be elaborated in future chapters (text p. 102).

In addition to sorting cells based on their anatomy, biologists also like to classify cells based on the *source of the cell's energy.* Autotrophs ("auto" = self; "trophic" = feeding) can feed themselves by making their own food either from sunlight or from nonliving chemical compounds. Heterotrophs ("hetero" = other; "trophic" = feeding) cannot make their own food, and so use other cells—either autotrophs or other heterotrophs—as food sources (text p. 102).

Both plant and animal cells are enclosed by a thin plasma membrane. Plant cells are differentiated by an outer rigid cell wall and some characteristic organelles (Chapter 6); they are usually interconnected by thin bridges of cytoplasm, or plasmodesmata. Animal cells need to be mobile, so they lack the rigid cell wall but contain many of the same organelles that plant cells have. In both plants and animals the ratio between a cell's surface and its volume is crucial to virtually all life processes, from respiration to digestion; in addition, it sets limits on the absolute size of cells (text p. 105). The only way cells can become really large is if their surface area is vastly increased while the volume is kept low (long narrow cells), or if they contain a large amount of stored substances (like eggs) that do not need ongoing active support.

## *Section III*    The Cell Surface

In all cells a plasma membrane surrounds the cytoplasm. The membrane has a specific construction—that of a lipid bilayer—that allows materials to pass into and out of the cell. The fluid-mosaic model describes the lipid bilayer, in which proteins float in a fluid phospholipid "sea." The outer portion of the bilayer, which consists of the hydrophilic heads of the phospholipid molecules, serves as a barrier between the cell and its watery environment (text pp. 107–108).

Molecules can pass into and out of the cell through the membrane in a number of ways. Lipid-soluble substances may cross the membrane through simple diffusion—the tendency of substances to move from regions of high concentration to those of low concentration. In passive transport, a carrier molecule facilitates diffusion, while in active transport a carrier must expend energy to actually move substances against a concentration gradient. Osmosis involves the diffusion of water through a selectively permeable membrane from solutions of low solute concentration (hypotonic) to an area of higher solute concentration (hypertonic) (text pp. 110–114).

## Section IV    Cell Walls and the Glycocalyx

Cell walls are protective barriers found around the outside of the plasma membrane in bacteria, fungi, and plants. Cell walls are composed largely of cellulose in plants and fungi, and of polysaccharides, sugars, and lipids in bacteria (text pp. 115–116). Animal cells have a coating of sugar polymers and proteins, the glycocalyx, outside their plasma membranes. These groups of sugar molecules are involved in cell recognition and other functions (text p. 117).

## Section V    Linkage and Communication Between Cells

Several types of junctions hold cells together and provide channels for intercellular communication. Zonulae adherens and desmosomes provide structural linkage between animal cells; tight junctions prevent leakage across plasma membranes. The actual sites (holes and pores) through which materials pass into and out of cells are called gap junctions. Plasmodesmata serve as large connective bridges between cells in plants (text pp. 117–120).

## Section VI    Multicellular Organization

The first statement of the cell theory says that all organisms are composed of cells. Moving up the ladder of complexity, cells are organized into a hierarchy of functions as follows: (1) individual cells; (2) tissues—groups of cells that perform the same or closely related functions; (3) organs—groups of several tissues that act together to perform specific functions; and (4) organ systems—a group of organs that carry out the components of an activity (text pp. 120–121).

## KEY TERMS

active transport   *text page 111*

anchorage dependence   *117*

autotroph   *102*

basal lamina   *117*

bilayer (lipid)   *106*

carrier-facilitated diffusion   *110*

cell   *97*

cell theory   *100*

cellulose   *115*

cell wall   *115*

chitin   *116*

concentration gradient   *111*

connective tissue   *120*

cotransport   *112*

desmosome   *117*

diffusion   *108*

epithelium   *117*

eukaryotic cell   *102*

extrinsic protein   *107*

fluid-mosaic model   *107*

gap junction   *118*

glycocalyx   *116*

heterotroph   *102*

hypertonic solution   *113*

hypotonic solution   *113*

intrinsic protein   *107*

isotonic solution   *113*

organ   *120*

organelle   *102*

organ system   *120*

osmosis   *112*

osmotic pressure   *113*

plasma membrane   *106*

plasmodesmata   *119*

prokaryotic cell   *102*

surface-to-volume ratio   *105*

tight junction   *118*

zonulae adherens   *117*

# SELF-QUIZ: TESTING WHAT YOU HAVE LEARNED

## Matching Key Terms

Match each term on the left with the most appropriate description on the right.

| | | |
|---|---|---|
| 1. prokaryote | a. | thousands of kinds |
| 2. organelle | b. | extracellular fluid |
| 3. diffusion | c. | form of active transport |
| 4. isotonic | d. | "sugar coat" |
| 5. cells | e. | firm physical contact |
| 6. glycocalyx | f. | water movement |
| 7. zonula adherens | g. | membrane-bound internal structure |
| 8. cotransport | h. | higher concentration to lower concentration |
| 9. osmosis | i. | contains intrinsic and extrinsic proteins |
| 10. lipid bilayer | j. | no nucleus |

## True or False?

1. _____ In an SEM, electrons emitted by gold form the image of the specimen.

2. _____ The magnification limit of the light microscope is 1200X.

3. _____ Surface-to-volume ratio limits cell size only in certain rare cases.

4. _____ The lipid bilayer is a characteristic of plant cells.

5. _____ The glycocalyx of animal cells is not involved in cell–cell recognition.

6. _____ In simple diffusion, large protein molecules may traverse the plasma membrane.

7. _____ Eukaryotic cells have numerous structures not found in prokaryotic cells.

8. _____ The fluid-mosaic model describes the structure of the animal cell nucleus.

9. _____ Osmosis requires the input of energy.

10. _____ Bacteria have an outer cell wall.

## Completion

1. In _____, sodium ions and an amino acid or sugar molecule bind to a carrier, which transports them into cells.

2. The cell walls of plants and some fungi are composed largely of _____.

3. _____ is the need to adhere to a solid surface.

4. An animal cell is enclosed by the _____, which in turn is surrounded by the

_____.

5. _____ seal plasma membranes together at ridges and so prevent leakage.

_____ are sites of pores that permit ions, small molecules, and electric currents to pass.

6. The energy-requiring process of _____ takes place against a _____.

## Short Answer

1. What is a major drawback in the use of electron microscopes to study cells?

2. What are the two features used to classify cells?

3. List the three most general statements of the cell theory.

4. Why does the surface-to-volume ratio limit cell size?

5. What structural feature of a lipid bilayer forms a watertight barrier between a cell and its environment?

6. What are the roles of intrinsic and extrinsic proteins in the lipid bilayer?

7. What is the difference between active transport and carrier-facilitated diffusion?

## Multiple-Choice Review

Complete each of the following statements by circling the correct response.

1. Plants are considered to be:

   a. heterotrophs.
   b. auxotrophs.
   c. autotrophs.
   d. zerotrophs.
   e. none of the above

2. The most obvious difference between a prokaryote and a eukaryote is:

   a. the presence of a cell wall in prokaryotes.
   b. the presence of proteins in the cytoplasm of eukaryotes.
   c. the presence of a cytoskeleton in eukaryotes.
   d. the absence of a nucleus in prokaryotes.
   e. the presence of plasmodesmata in prokaryotes.

3. The organizational hierarchy in organisms from simplest to most complex is:

   a. organs–tissues–cells.
   b. tissues–organs–cells.
   c. cells–organs–tissues.
   d. tissues–organs–cells.
   e. cells–tissues–organs.

4. When the extracellular fluid is hypertonic relative to the cell contents:

   a. water tends to leave the cell.
   b. water tends to enter the cell.
   c. there is little or no water movement.
   d. only ions tend to enter the cell.
   e. none of the above

5. Desmosomes are analogous to:

   a. belts or straps around the cell.
   b. tunnels through the plasma membrane.
   c. spot welds or buttons.
   d. antennae.
   e. structural support girders.

# Exercises

## Exercise 1

Fill in the missing labels on this drawing of a typical animal cell. If you have colored pencils, make the glycocalyx brown, the mitochondria blue, the Golgi complex orange, and the ER yellow.

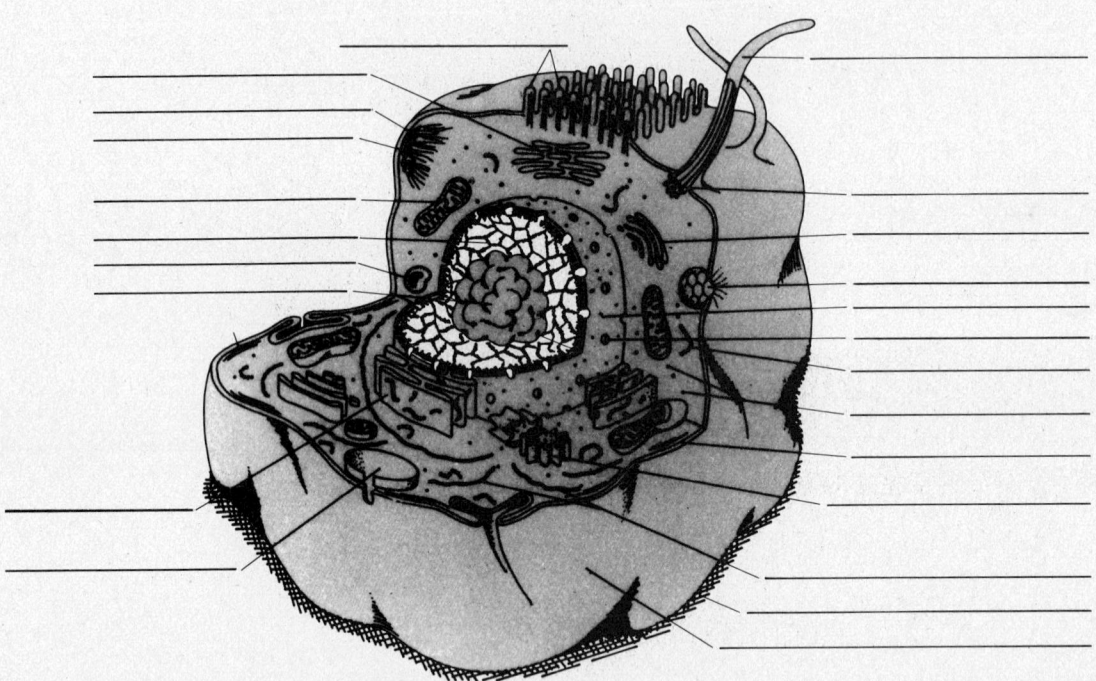

*Exercise 2*

Fill in the missing labels on this drawing of a typical plant cell. Use colored pencils to color the chloroplasts and cell wall green, the vacuole blue, and the nucleolus purple.

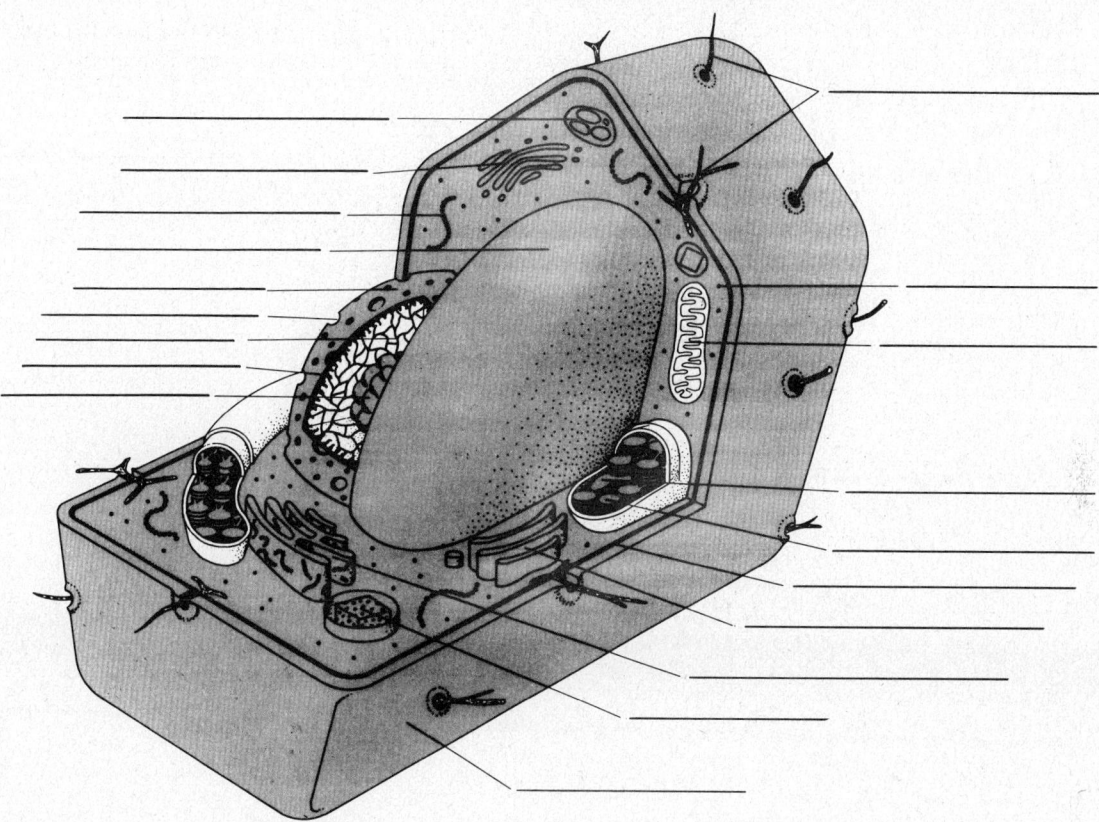

# 6

# INSIDE THE LIVING CELL: STRUCTURE AND FUNCTION OF INTERNAL CELL PARTS

## CHAPTER AT A GLANCE

## CHAPTER PREVIEW

In a sense this chapter is a continuation of Chapter 5. It provides the framework of facts and concepts about cells that you will need for much of the remainder of the course. The differences between prokaryotic and eukaryotic cells and plant and animal cells are especially important—you'll be wise to pay close attention to Table 6-1, where these differences are clearly summarized.

In the chapter's opening section you'll learn about the nature and properties of the semifluid cytoplasm. This section includes an interesting historical account of early debates over whether the "living state" was a property of cytoplasm or of the nucleus. Next, the focus shifts to the nucleus—the site of the chromosomes, and usually itself the largest cell organelle.

The remainder of the chapter surveys the other organelles and their functional anatomy. In studying the cytoskeleton, you will find it useful to refer to Figure 6-15 to help in sorting out the differences between microfilaments, microtubules, and intermediate filaments. The next section, on cellular movement, describes three kinds of movements seen in cells—creeping or gliding, swimming, and internal movements—and the structures and molecules that make them possible. This section is all the more meaningful when you consider that movement is a fundamental property of all living cells.

# LEARNING OBJECTIVES

When you have mastered the concepts of this chapter, you will be able to:

1. Describe the role of the cytoplasm, and how various organelles are distributed in it.
2. List the organelles of typical plant and animal cells, and tell where they are distributed.
3. Explain the role of the nucleus, what materials are found within it, and what processes the nucleus controls.
4. Describe the anatomy of a ribosome, and list its major function in cell protein synthesis.
5. Describe the anatomy of the two types of endoplasmic reticula, and explain the function of each.
6. Explain how the Golgi complex acts to package cell products.
7. Discuss the role of vacuoles, and compare the various plant and animal vacuole types.
8. Describe the anatomy of a mitochondrion, and explain how that anatomy is linked to the generation of ATP.
9. Describe the anatomy of the components of the cytoskeleton.
10. Explain how cells move.
11. Describe the anatomy of cilia and flagella.

# CONCEPTS IN REVIEW

## *Section I*   Cytoplasm: The Dynamic, Mobile Factory

Most of the properties we associate with life are properties of the cytoplasm. Much of the mass of a cell consists of this semifluid substance, which is bounded on the outside by the plasma membrane. Organelles are suspended within it, supported by the filamentous network of the cytoskeleton. Dissolved in the cytoplasmic fluid are nutrients, ions, soluble proteins, and other materials needed for cell functioning (text p. 125).

## Section II    The Nucleus: Information Central

The eukaryotic cell nucleus is the largest organelle, and it houses the genetic material (DNA) on chromosomes. (In prokaryotes the hereditary material is found in the nucleoid). The nucleus also contains one or two organelles, called nucleoli, that play a role in cell division. A pore-perforated sac, the nuclear envelope, separates the nucleus and its contents from the cytoplasm (text p. 127).

## Section III    Organelles: Specialized Work Units

All eukaryotic cells contain most of the various kinds of organelles, each of which has a specialized function in the cell. Organelles described in this section include ribosomes, the endoplasmic reticulum, the Golgi complex, vacuoles, lysosomes, mitochondria, and the plastids of plant cells.

There may be a few hundred or many thousands of ribosomes within a cell, a reflection of the fact that ribosomes are the sites at which amino acids are assembled into proteins for export or for use in cell processes. A complete ribosome is composed of one larger and one smaller subunit. During protein synthesis the two subunits move along a strand of mRNA, "reading" the genetic sequence coded in it and translating that sequence into protein (text pp. 127–128). Several ribosomes may become attached to a single mRNA; the combination is called a polysome.

The endoplasmic reticulum, a lacy array of membranous sacs, tubules, and vesicles, may be either rough (RER) or smooth (SER). Both play roles in the synthesis and transport of proteins (text pp. 129–130). The RER, which is studded with polysomes, also apparently gives rise to the nuclear envelope after a cell divides. SER lacks polysomes and is active in the synthesis of fats and steroids, and in the oxidation of toxic substances in the cell. Both types of ER serve as compartments within the cell in which specific products can be isolated and shunted to particular areas within or outside the cell.

Transport vesicles may carry exportable molecules from the ER to another membranous organelle, the Golgi complex. Within the Golgi complex molecules are modified and packaged for export or for delivery elsewhere in the cytoplasm (text p. 132).

Vacuoles in cells, which appear to be hollow sacs, are actually filled with fluid and soluble molecules. The most prominent vacuoles appear in plant cells and serve as water reservoirs and storage sites for sugars and other molecules. Vacuoles in animal cells carry out phagocytosis and pinocytosis (text p. 135).

A subset of vacuoles is organelles known as lysosomes, which contain digestive enzymes (packaged in lysosomes in the Golgi complex) that can break down most biological macromolecules. Their role is to digest food particles and degrade damaged cell parts.

Mitochondria are the sites of energy-yielding chemical reactions in all cells. In addition, plant cells contain plastids, which utilize light energy to manufacture carbohydrates in the process of photosynthesis. It is on the large surface area provided by the inner cristae of mitochondria that ATP-generating enzymes are located (text pp. 137–138). Mitochondria are self-replicating, and it is likely that they are descendants of once-freeliving prokaryotes.

There are two types of plastids: (1) leucoplasts, which lack pigments and serve as storage sites; and (2) chromoplasts, which contain pigments. The most important chromoplasts are chloroplasts, organelles that contain the chlorophyll used in photosynthesis (text p. 139). The internal structure of chloroplasts includes stacks of membranes called grana, which are embedded in a matrix called the stroma. You will study chloroplasts in detail in Chapter 8.

## Section IV    The Cytoskeleton

All eukaryotic cells have a cytoskeleton: a convoluted latticework of filaments and tubules that appears to fill all available space in the cell and provides support for various other organelles. A large portion of the cytoskeleton consists of threadlike microfilaments composed mainly of the contractile protein actin. These microfilaments are involved in many types of intracellular movements in plant and animal cells. A main structural component of the cytoskeleton is microtubules, which are composed of the globular protein tubulin and which together act as a scaffold that provides a stable cell shape (text p. 140). Cytoskeletal intermediate filaments appear to impart tensile strength to the cell cytoplasm (text p. 141). Mechanoenzymes such as myosin, dynein, and kinesin interact with the cytoskeletal filaments and tubules to generate forces that cause movements.

## Section V    Cellular Movements

Although the cytoskeleton provides some stability to cells, it is its microtubules and filaments and their associated proteins that enable cells to move by creeping or gliding (text pp. 142–143). Such movements require a solid substrate to which the cell can adhere, and its movement is guided by the geometry of the substrate's surface. Some cells also exhibit chemotaxis, the ability to move toward or away from the source of a diffusing chemical.

Certain eukaryotic cells can swim freely in liquid environments, propelled by whiplike cilia or flagella (text p. 143). Both cilia and flagella have the same internal structure: nine doublets (pairs of microtubules) are arranged in a ring and extend the length of the cilium or flagellum. Two more microtubules run down the center of the ring. Also, every cilium or flagellum grows from the cell surface only where a basal body is located. Movement is based on the activities of tiny dynein side arms that extend from one of the microtubules of each doublet (text p. 144).

Within most plant cells nutrients, proteins, and other materials are moved about via cytoplasmic streaming. The process occurs as myosin proteins attached to organelles push against microfilaments arrayed throughout the cell (text p. 145). Microfilaments and microtubules are responsible for almost all major cytoplasmic movements. During cell division, microtubules of the spindle—assembled from tubulin subunits near organelles called centrioles—move the chromosomes. You will study more about this process in Chapter 9.

## KEY TERMS

actin    *text page 139*

basal body    *144*

centriole    *146*

chemotaxis    *143*

chloroplast    *139*

chromosome    *125*

cilium    *143*

cytoplasm    *125*

cytoplasmic streaming    *145*

cytoskeleton    *135*

dynein    *144*

endocytosis    *135*

endoplasmic reticulum
    (ER)    *128*

exocytosis    *135*

flagellum    *143*

gene    *125*

Golgi complex    *131*

intermediate filament
    *141*

lipofusion granule    *136*

lysosome    *136*

matrix    *138*

mechanoenzyme    *141*

microbody    *137*

microfilament    *139*

microtubule    *140*

mitochondrion    *137*

myosin    *139*

nuclear envelope    *126*

nucleoid    *126*

nucleolus    *126*

nucleus    *125*

phagocytosis    *132*

pinocytosis    *135*

plastid    *137*

polysome    *128*

ribosome    *127*

stroma    *139*

tubulin    *140*

vacuole    *132*

# SELF-QUIZ: TESTING WHAT YOU HAVE LEARNED

## Matching Key Terms

Match each term on the left with the most appropriate description on the right.

| | | | |
|---|---|---|---|
| 1. polysome | | a. | protein synthesis |
| 2. pinocytosis | | b. | baglike structure |
| 3. exocytosis | | c. | power generator |
| 4. plastid | | d. | where flagella grow |
| 5. Golgi complex | | e. | toward or away from a chemical stimulus |
| 6. flagella | | f. | engulfment |
| 7. phagocytosis | | g. | RNA and ribosomes |
| 8. lysosome | | h. | weblike |
| 9. basal body | | i. | in plants only |
| 10. chemotactic | | j. | control room |
| 11. nucleus | | k. | expel |
| 12. vacuole | | l. | vacant |
| 13. ribosome | | m. | whiplike |
| 14. cytoskeleton | | n. | cell drinking |
| 15. mitochondrion | | o. | packaging |

## True or False?

1. _____ Ribosomes are derived from the nucleoli.

2. _____ Unlike other cell membranes, the nuclear envelope has no pores.

3. _____ The smooth endoplasmic reticulum is held in place by the cytoskeleton.

4. _____ Structural proteins are exportable.

5. _____ The nuclear envelope is produced by the rough endoplasmic reticulum.

6. _____ White blood cells work by phagocytosis.

7. _____ Prokaryotic cells have microbodies.

8. _____ Mitochondria are self-replicating.

9. _____ Carotenoids are colorless molecules.

10. _____ Grana are surrounded by stomata.

## Completion

1. Phagocytosis is a method of cell feeding that first requires that the food be _____.

2. The _____ packages some fifty hydrolytic enzymes in _____.

3. _____ are lysosomelike vesicles containing waste products. They are thought to be involved with cell _____.

4. Both _____ and _____ are thought to have arisen from endosymbiosis.

5. The cytoskeleton is composed of very fine _____, medium _____, and larger _____.

6. Creeping and gliding cell movements are usually _____ dependent.

7. _____ behavior is shown when a cell moves toward or away from a chemical substance.

8. Flagella grow from the cell surface only at the site of a _____.

## Short Answer

1. Why is the nucleus often described as the "control room" of the cell?

2. Name eight organelles of a eukaryotic cell.

3. What are the two kinds of endoplasmic reticulum, and how do they differ?

4. What process takes place in the Golgi complex?

5. How do plant and animal vacuoles differ?

6. What are lipofuscin granules? How are they thought to play a part in aging?

7. What are plastids? What are their two basic types?

8. What are the three types of filaments that make up the cytoskeleton?

9. What are the three basic types of cell movement, and how does each work?

## Multiple-Choice Review

In the following questions, fill in any blanks. Finish each sentence by circling the letter of the correct response.

1. Most of the properties associated with the processes of life are properties of:

    a. the nucleus.
    b. DNA.
    c. the cytoplasm.
    d. endosymbionts.
    e. none of the above

2. Oxidation of toxic materials is a major function of the:

    a. cytoplasm.
    b. ribosomes.
    c. vacuoles.
    d. smooth endoplasmic reticulum.
    e. rough endoplasmic reticulum.

3. Lysosomes contain:

    a. hydrolytic enzymes.
    b. genetic material.
    c. stored fats.
    d. proteins.
    e. carbohydrates.

4. Chromoplasts are a type of:

    a. pigment.
    b. storage bin.
    c. nutrient tank.
    d. plastid.
    e. none of the above

5. An mRNA molecule and its _____ make up a:

    a. multisome.
    b. polysome.
    c. lysosome.
    d. monosome.
    e. none of the above

# Exercises

## Exercise 1

Not all the cell components shown in the table below are found in all cell types. For each basic cell type listed—prokaryote, plant, and animal—mark "X" if a component is always or often present and "O" if a component is rarely or never present.

**Components of Prokaryotic, Plant, and Animal Cells**

| Component | Prokaryote | Plant Cell | Animal Cell |
|---|---|---|---|
| Cell wall | _____ | _____ | _____ |
| Glycocalyx | _____ | _____ | _____ |
| Plasma membrane | _____ | _____ | _____ |
| Cytoskeleton | _____ | _____ | _____ |
| Nucleus | _____ | _____ | _____ |
| Chromosomes | _____ | _____ | _____ |
| Mitochondria | _____ | _____ | _____ |
| Plastids | _____ | _____ | _____ |
| Ribosomes | _____ | _____ | _____ |
| Endoplasmic reticulum | _____ | _____ | _____ |
| Golgi complex | _____ | _____ | _____ |
| Vacuoles | _____ | _____ | _____ |
| Lysosomes | _____ | _____ | _____ |
| Cilia (9 + 2) | _____ | _____ | _____ |
| Flagellum | _____ | _____ | _____ |
| Centrioles | _____ | _____ | _____ |

## *Exercise 2*

Identify the two structures shown below and label the parts indicated in each.

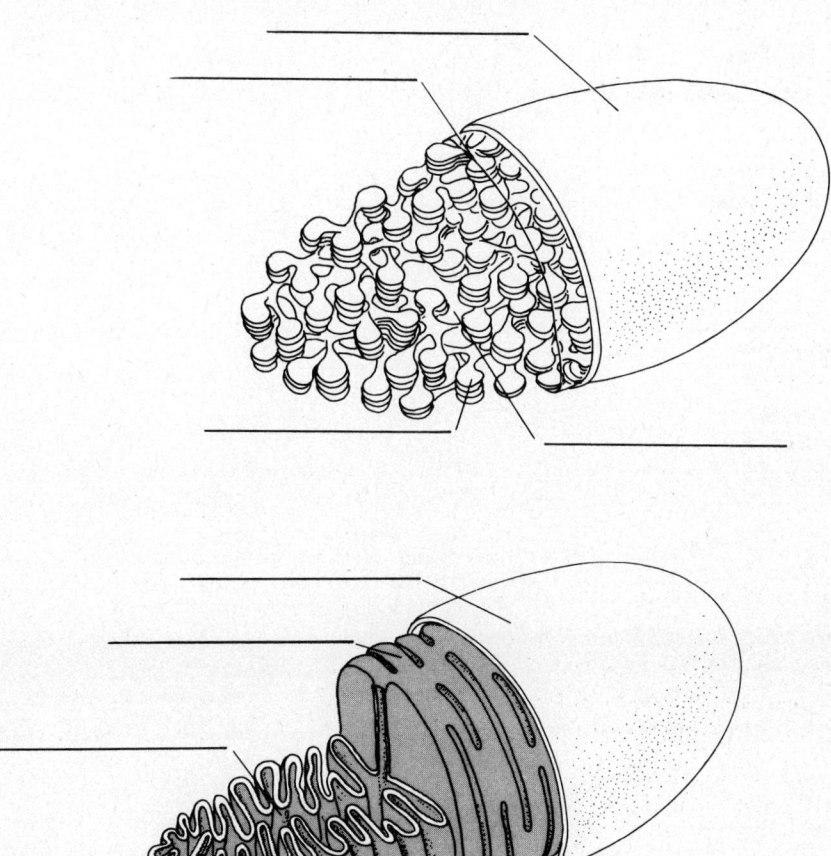

# 7

# HARVESTING THE ENERGY STORED IN NUTRIENTS: FERMENTATION AND CELLULAR RESPIRATION

---

## CHAPTER AT A GLANCE

---

## CHAPTER PREVIEW

In this chapter and in Chapter 8 you will encounter the mechanisms through which cells and whole organisms obtain the energy they need to live. Here, the focus is on fermentation, glycolysis, and cell respiration—processes that go on in all cells and that produce energy, in the form of the "energy currency" molecule ATP, to fuel cell activities. In Chapter 8 you will consider the chemical opposite of cellular respiration: the specialized process of photosynthesis in which energy-containing molecules are created.

This is a challenging chapter, because the energy-releasing chemical reactions in cells involve many steps. Fortunately, your text outlines the steps with care and points out underlying principles that will help you organize the material. In the first section you will look at the role of ATP as an energy storage molecule and learn its structure. Next the text describes oxidation-reduction reactions and explains how energy is actually stored in molecules of ATP. In this section you will want to take extra care to understand the differences between oxidation (removal of electrons) and reduction (addition of electrons).

The subject of glycolysis is treated next, and once again you will need to allow plenty of time to absorb and organize details. As you will see, the nine steps of glycolysis begin the slow liberation of the energy contained in glucose bonds, and produce a small amount of ATP. Once you have mastered this section, you will be ready to look at cellular respiration, the process that produces many molecules of ATP and that involves an "energy cascade" known as the electron transport chain. The information in Chart 7-1 and Table 7-1 is particularly helpful.

After introducing the mechanism of the electron transport chain, the text next probes the Mitchell hypothesis, or the chemiosmotic coupling hypothesis—a detailed model of the events in mitochondria that generate ATP. This material will require careful reading, but it won't be difficult if you have understood the previous two sections. Finally, the chapter closes with another look at the mechanism of negative feedback—the means by which cellular metabolism is controlled so that an appropriate amount of energy is stored as ATP, and raw materials are available to the cell as needed.

# LEARNING OBJECTIVES

When you have mastered the concepts of this chapter, you will be able to:

1. Describe the molecular structure of ATP, ADP, and AMP, and explain how they are used in cellular energy processes.
2. Explain what is meant by oxidation and reduction, and name the molecules that are important in each.
3. Outline glycolysis and fermentation, and show how much ATP is produced by each process.
4. Outline the reactions of fermentation, and show how much ATP the process produces.
5. Describe the events of cellular respiration, and explain how much ATP is generated in the Krebs cycle and electron transport chain.
6. Describe in detail the anatomy of the mitochondrial membranes, and explain the role of the membranes in metabolism.
7. Compare and contrast the metabolism of fats and proteins with that of carbohydrates.

# CONCEPTS IN REVIEW

## Section I    ATP: The Cell's Energy Currency

The prime energy-storage molecule of cells is adenosine triphosphate (ATP). Energy stored in its phosphate groups (added during coupled exergonic reactions) is released during coupled endergonic reactions, during which the groups are removed. The energy-storage and transfer function of ATP is always associated with its phosphate groups and with the interconversion of ATP, ADP, and AMP (text pp. 152–153). The transfer of a phosphate group is termed phosphorylation. ATP's first two phosphate groups release much free energy when cleaved, and are known as high-energy groups. The bonds linking them, which are broken during phosphorylation, are called high-energy phosphate bonds.

The energy released by the cleavage of ATP to ADP drives the vast majority of endergonic biological reactions. All such reactions involve a common intermediate: a product of an initial reaction (in this case, phosphorylation) that becomes a reactant in a subsequent reaction (text p. 154). Thus, by way of the phosphorylated common intermediate, the energy to power biosynthesis becomes available.

## Section II    Oxidation-Reduction Reactions

Energy from nutrients is channeled into ATP via oxidation-reduction reactions. Oxidation is the removal of electrons from an atom or compound, and reduction is the addition of electrons (text p. 154). The reactions always occur simultaneously and take place in all key metabolic reactions. The change in free energy that occurs is expressed in kilocalories. Electrons may be transferred in the form of hydrogen atoms. Two major electron carriers—or energy intermediates—are the coenzymes $NAD^+$ and FAD (text p. 155).

## Section III    Glycolysis: The First Phase of Energy Metabolism

Glycolysis is the initial sequence of reactions used by most cells to derive energy from glucose. During the nine reaction steps of glycolysis, glucose is broken down to two molecules of pyruvate. This process nets two ATP molecules via the phosphorylation of two ADP molecules; at the same time, two molecules of $NAD^+$ are reduced to NADH (text pp. 157–160).

The pyruvate molecules generated by glycolysis can be further acted on during the anaerobic reaction sequences of fermentation and during cellular respiration. In fermentation, pyruvate is converted to lactate, ethanol, or other products; there is no net gain of ATP. However, $NAD^+$ is regenerated to again participate in basic steps of the glycolytic pathway (text pp. 160–161). Both glycolysis and fermentation take place in the cell cytoplasm.

## Section IV    Cellular Respiration

Because the cells and tissues of large organisms require large amounts of energy, they rely on the large amounts of ATP generated during cellular respiration. Cellular respiration takes place in mitochondria in eukaryotes. As it proceeds, the pyruvate derived from glycolysis is shunted into the Krebs cycle while other products go to the electron transport chain, where oxygen accepts electrons in the form of hydrogen atoms and is reduced to $H_2O$. The end result of cellular respiration is thirty-six molecules of ATP formed for every molecule of glucose consumed (text p. 161).

Before the Krebs cycle begins, pyruvate generated by glycolysis is oxidized. The result is a high-energy acetate compound, acetyl CoA (plus $CO_2$) (text p. 162). Acetyl CoA drives the first step of the Krebs cycle (citric acid cycle), in which citric acid is formed. In the remaining nine steps of the cycle a series of enzymes transfers electrons to the acceptors $NAD^+$ and FAD. During the oxidation of one intermediate compound, ATP is formed. In all, two ATPs result from the Krebs cycle for each original molecule of glucose. NADH has also been generated during the preceding glycolysis. During the next phase of cellular respiration, the electron transport chain, the energy in NADH and $FADH_2$ is retrieved and stored in large amounts of ATP (text pp. 163–164).

The electron transport chain refers to a series of pigment-containing cytochrome compounds that serve as electron carriers. Cytochromes are embedded in the inner membranes of mitochondria.

They pass along the electrons lost by NADH and FADH$_2$ as those compounds are oxidized during the third phase of cellular respiration. At three points in the chain, ATP is formed from ADP during the process of oxidative phosphorylation (text p. 164).

## Section V   Mitochondrial Membranes and the Mitchell Hypothesis

A widely accepted model for the mechanism through which ATP is formed during the electron transport chain is the chemiosmotic coupling hypothesis of Peter Mitchell. According to this model, the movement of electrons through the chain is accompanied by a protein-pumping mechanism that sets up an energy gradient, consisting of hydrogen ions (protons), across the inner mitochondrial membrane. These protons are pumped from the inner mitochondrial matrix to the outer compartment. As they flow back through the complex molecules known as coupling factor, free energy is released and conserved as ATP (text p. 167).

## Section VI   Metabolism of Fats and Proteins

Like glucose, proteins and fats are broken down during metabolism and converted to compounds that serve as intermediates in the Krebs cycle. Fragments of protein molecules may be converted to pyruvate, acetate, or acetyl CoA. A gram of protein yields about as much ATP as a gram of glucose. Fats (triglycerides) are first hydrolyzed to glycerol and free fatty acids. These components then go on to participate in glycolysis and the Krebs cycle. One gram of fatty acids provides 2.5 times the ATP of a gram of either sugar or protein. Intermediates in the Krebs cycle and their precursors, such as pyruvate and acetyl CoA, also serve as starting points for biosynthesis (text p. 169).

## Section VII   Control of Metabolism

The amount of energy stored as ATP is regulated via negative feedback, which controls the rate at which metabolic pathways function. The two mechanisms involved in this negative feedback are allosteric enzymes (Chapter 4) and the covalent modification of enzymes, which generates a temporary change in the chemical structure of the catalyst (text p. 171).

## KEY TERMS

adenosine triphosphate (ATP)
  *text page 52*

biosynthesis   *169*

cellular respiration   *161*

chemiosomatic coupling
  hypothesis   *166*

coupling factor   *167*

cytochrome   *164*

electron transport chain   *161*

FAD (flavin adenine
  dinucleotide)   *155*

fermentation   *160*

glycolysis   *156*

Krebs cycle   *161*

NAD$^+$ (nicotinamide adenine
  dinucleotide)   *155*

oxidation   *154*

oxidation-reduction
  reaction   *154*

oxidative
  phosphorylation   *164*

phosphorylation   *153*

pyruvate   *156*

reduction   *154*

# SELF-QUIZ: TESTING WHAT YOU HAVE LEARNED

## Matching Key Terms

Match each term on the left with the most appropriate description on the right.

| | | | |
|---|---|---|---|
| 1. | reduction | a. | energy storage |
| 2. | electron transport chain | b. | two protons, two electrons |
| 3. | coupling factor | c. | initial sequence |
| 4. | fermentation | d. | Mitchell |
| 5. | pyruvate | e. | add electrons |
| 6. | FAD | f. | flow through membrane |
| 7. | NAD$^+$ | g. | remove electrons |
| 8. | chemiosmotic coupling | h. | manufacture biological molecules |
| 9. | ATP | i. | pigment protein |
| 10. | biosynthesis | j. | ADP to ATP |
| 11. | Krebs cycle | k. | no $O_2$ electron acceptors |
| 12. | phosphorylation | l. | final phase of respiration |
| 13. | cytochrome | m. | three-carbon compound |
| 14. | glycolysis | n. | citric acid cycle |
| 15. | oxidation | o. | add two electrons, one proton |

## True or False?

1. \_\_\_\_\_ Fermentation requires moderate levels of oxygen.

2. \_\_\_\_\_ Removing phosphate groups from ATP requires energy.

3. \_\_\_\_\_ Removing hydrogen atoms is termed *hydrogenation*.

4. \_\_\_\_\_ Pyruvate is a three-carbon compound.

5. \_\_\_\_\_ The tricarboxylic acid cycle oxidizes pyruvate to $CO_2$.

6. \_\_\_\_\_ Electron movement through the electron transport chain takes place via proton pumping.

7. \_\_\_\_\_ Biosynthesis is a catabolic process.

8. \_\_\_\_\_ All cytochromes contain a heme group.

9. \_\_\_\_\_ Common intermediates are reactants and products.

10. \_\_\_\_\_ Anaerobic cells cannot carry out fermentation.

## Completion

1. The most fundamental metabolic chain of reactions is _____. Its product may be broken down further in the absence of oxygen during _____, or in the presence of oxygen during _____.

2. No biological process is 100 percent efficient. Only about half of all possible energy is ever harvested; the remainder is wasted in the form of _____.

3. The removal of electrons is _____, while the addition of electrons is _____.

4. Once pyruvate is in the mitochondrial matrix, it is oxidized to _____.

5. ADP is used to form ATP during the electron transport chain in a process of

_____.

6. Another term for anabolism is _____.

7. In _____ an enzyme is inactivated by attaching a chemical group that reduces that enzyme's catalytic function.

8. Glycolysis yields _____ ATP, and the Krebs cycle yields _____ ATP per molecule of glucose.

## Short Answer

1. What is phosphorylation? Why is it such an important reaction for cells?

2. What is a hydrolysis reaction?

3. What types of reactions involve the addition or subtraction of electrons or hydrogen atoms?

4. Briefly outline the nine steps of glycolysis.

5. What types of cells carry out fermentation reactions?

6. What is the Mitchell hypothesis?

7. How much energy (in terms of ATP per molecule of glucose) is generated from: glycolysis, the Krebs cycle, the electron transport chain?

8. How do enzymes control metabolism?

## Multiple-Choice Review

In the following questions, fill in any blanks. Finish each sentence by circling the letter of the correct response.

1. In the majority of anabolic (biosynthetic) reactions the stored energy in _____ is released when that molecule is cleaved to form:

   a. $AMP + 2 P_i$.
   b. ATP.
   c. $ADP + 1 P_i$.
   d. glycogen.
   e. none of the above

2. Aerobic _____ have no specialized organelles, so respiration takes place:

   a. on the inner plasma membrane.
   b. in DNA.
   c. in mitochondria.
   d. in the cytoplasm.
   e. around the chromosome.

3. In humans the Krebs cycle takes place in:

   a. the cytoplasm.
   b. the liver.
   c. the plasma membrane.
   d. the matrix of the mitochondria.
   e. none of the above

4. Proton pumping in electron transport is associated with the term:

   a. *glycolytic pathway*.
   b. *chemiosmotic coupling*.
   c. *heme group*.
   d. *yield*.
   e. none of the above

5. Enzymes may be inactivated by picking up chemical groups that serve to reduce that enzyme's function in a process of:

   a. allosteric concentration.
   b. covalent modification.
   c. becoming degraded.
   d. coupling.
   e. none of the above

## Exercise

Fill in the energy yield for each of the respiration steps listed below.

**Approximate Maximal ATP Yield from the Complete Respiration of 1 Mole of Glucose**

| Process | ATP Gain per Glucose |
|---|---|
| Glycolysis | |
| Krebs cycle | |
| Electron transport chain from 2 NADH generated from glycolysis | |
| Electron transport chain from 8 NADH generated from Krebs cycle | |
| Electron transport chain from 2 FADH$_2$ generated from Krebs cycle | |
| Total | |

# 8

# PHOTOSYNTHESIS: HARNESSING SOLAR ENERGY TO PRODUCE CARBOHYDRATES

---

## CHAPTER AT A GLANCE

---

## CHAPTER PREVIEW

Just as cellular respiration releases energy, photosynthesis captures it from the sun and ultimately "packages" it in carbohydrates. In a real sense photosynthesis is the most important chemical process on the modern Earth. After all, it is the source of the chemical energy that fuels the life processes of nearly all organisms! As with the material on cellular respiration in Chapter 7, you will be asked to apply your knowledge of basic chemistry in this chapter. You can begin by learning the simple (but profound) equations for photosynthesis presented in the chapter's introductory section. The section also includes details on chloroplast anatomy (shown in Figure 8-2).

Next the chapter looks at the exact nature of light and how it is absorbed by the photosynthetic pigments of plants. Chlorophyll, the prime photosynthetic pigment, is what makes plants green. The following two sections explain the chemical steps involved in the two separate sets of photosynthetic reactions—called light and dark reactions, respectively. As you'll discover, energy in sunlight is converted to chemical-bond energy in the light reactions, and then used to build carbohydrates during the dark reactions. Important mechanisms include the two types of photophosphorylation, in which electrons excited by light energy zigzag down an electron transport chain, which leads to the generation of ATP. In studying the dark reactions you will want to pay attention to the roles of two compounds, ribulose bisphosphate (RuBP) and the enzyme ribulose bisphosphate carboxylase.

The following sections survey chloroplast anatomy in even more detail, linking the functions of photosynthesis to specific physical sites, and examining the various conditions wherein oxygen can actually cause problems in photosynthesis, including the inefficient process of photorespiration. The chapter then briefly considers $C_4$ plant species, with their special leaf anatomy and biochemical pathway that enable them to thrive under arid conditions. A final section describes how one of the most important chemical components of photosynthesis, $CO_2$, is cycled through the environment.

# LEARNING OBJECTIVES

When you have mastered the concepts of this chapter, you will be able to:

1. Name the two basic reactions of photosynthesis, and state them in equation form.
2. Describe the anatomy of a chloroplast, and tell where each reaction of photosynthesis takes place.
3. Explain the physical properties of light, and how energy is transferred in terms of electrons and orbital energy.
4. List the various photosynthetic pigments and the plant groups in which they are found.
5. Outline the light and dark reactions of photosynthesis, and define cyclic and noncyclic photophosphorylation.
6. Explain how oxygen can inhibit photosynthesis.
7. Compare and contrast $C_3$ and $C_4$ plants in terms of their anatomy, metabolism, and adaptations to arid climates.
8. Explain the importance of carbon dioxide cycling.

# CONCEPTS IN REVIEW

## *Section I*   An Overview of Photosynthesis

Photosynthesis occurs only in the chlorophyll-containing cells of green plants, algae, and certain protists and bacteria. Overall, it is a process that converts light energy to chemical energy stored in the form of molecular bonds. From the point of view of chemistry and energetics, it is the opposite of cellular respiration (text p. 175). Whereas cellular respiration is highly exergonic and releases energy, photosynthesis requires energy and is highly endergonic.

Photosynthesis starts with $CO_2$ and $H_2O$ as raw materials, and proceeds through two sets of partial reactions. In the first, called the light reactions, water molecules are split (oxidized), and $O_2$ is released. These reactions must take place in the presence of light energy. In the second, called dark reactions, $CO_2$ is reduced (via addition of H atoms) to carbohydrate. These chemical events rely on the electron carrier NADPH and on ATP, but do not use light directly (text p. 176).

Both light and dark reactions take place in chloroplasts. As you learned in Chapter 6, the most prominent structures in chloroplasts are stacks of grana, flattened sacs that are embedded in the gel-like matrix of the stroma. Each flattened sac in a granum is called a thylakoid, and the internal space of each sac is surrounded by the thylakoid membrane, where most of the enzymes and pigments for the light reactions are embedded. The dark reactions take place in the stroma.

## Section II    How Light Energy Reaches Photosynthetic Cells

The energy in light photons in the visible part of the spectrum can be captured by biological molecules to do constructive work. The pigment chlorophyll in plant cells absorbs photons within a particular absorption spectrum—a statement of the amount of light absorbed by chlorophyll at different wavelengths (text p. 179). When light is absorbed, it alters the arrangement of electrons in the absorbing molecule. The added energy of the photon boosts the energy condition of the molecule from its stable ground state to a less-stable excited state. During the light reactions of photosynthesis, as the absorbing molecule returns to the ground state, the "excess" excitation energy is transmitted to other molecules and stored as chemical energy (text p. 179).

All photosynthetic organisms contain various classes of chlorophylls and one or more carotenoid (accessory) pigments that also contribute to photosynthesis (text p. 180). Groups of pigment molecules, called antenna complexes, are present on thylakoids. Light striking any one of the pigment molecules is funneled to a special chlorophyll *a* molecule, termed a reaction-center chlorophyll, which directly participates in photosynthesis. In most photosynthetic organisms there are two types of reaction-center chlorophylls, P680 and P700, each associated with an electron acceptor molecule and an electron donor. These aggregations are known, respectively, as photosystem I (P700) and photosystem II (P680) (text pp. 181–182).

## Section III    The Light Reactions: Converting Solar Energy to Chemical-bond Energy

The photosystems of the light reactions are responsible for packaging light energy in the chemical compounds ATP and NADPH. This packaging takes place through a series of oxidation-reduction reactions, set in motion when light strikes the P680 reaction center in photosystem II. In this initial event, water molecules are cleaved, oxygen is released, and electrons are donated. These electrons are accepted first by plastiquinone and then by a series of carriers as they descend an electron transport chain. For each four electrons that pass down the chain, two ATPs are formed. The last acceptor in the chain is the P700 reaction center of photosystem I. At this point, incoming photons boost the energy of the electrons and they are accepted by ferredoxin. Ferredoxin is then reoxidized, and the coenzyme $NADP^+$ is reduced to NADPH. The ATP generated previously and the NADPH then take part in the dark reactions (text pp. 182–183).

The production of ATP from the transport of electrons excited by light energy down an electron transport chain is termed photophosphorylation. The process is similar to oxidative phosphorylation in mitochondria, which you studied in Chapter 7. The one-way flow of electrons through photosystems II and I is called noncyclic photophosphorylation; plants derive additional ATP through cyclic photophosphorylation, in which some electrons are shunted back through the electron transport chain between photosystems II and I (text p. 184).

## Section IV    The Dark Reactions: Building Carbohydrates

In the dark reactions of photosynthesis, driven by ATP and NADPH, $CO_2$ is converted to carbohydrate. The reactions are also known as the Calvin-Benson cycle, in honor of the biologists who worked out the steps of the pathway. Atmospheric $CO_2$ is fixed as it reacts with ribulose bisphosphate (RuBP), a reaction that is catalyzed by the enzyme ribulose bisphosphate carboxylase. The reduction of $CO_2$ to carbohydrate (fructose diphosphate) is completed via several more steps of the cycle. Finally, RuBP is regenerated so that the cycle may continue (text p. 186).

## Section V    The Key Role of Chloroplast Architecture in Photosynthesis

In chloroplasts, the pigments and enzymes that participate in the steps of the light reactions are highly ordered elements within the thylakoid membrane; this arrangement ensures functional efficiency. Arrangements of other molecules also ensure that the chemical products of the light reactions are present in the stroma, where they drive the production of carbohydrates (text p. 187).

## Section VI    Oxygen: An Inhibitor of Photosynthesis

High levels of oxygen in plant cells can disrupt photosynthesis, and can also cause photorespiration—an inefficient form of the dark reactions in which $O_2$, rather than $CO_2$, is fixed and no carbohydrate is produced (text p. 188).

## Section VII    Reprieve from Photorespiration: The $C_4$ Pathway

Most plants are $C_3$ plants; that is, they experience decreased carbohydrate production under hot, dry conditions due to the effects of photorespiration. Among $C_4$ plants, however, a special leaf anatomy and a unique biochemical pathway enable them to thrive in arid conditions. Thus, $C_4$ plants both lessen photorespiration by carrying out photosynthesis only in cells that are insulated from high levels of $CO_2$, and also possess a novel mechanism for carbon fixation (text pp. 189–190).

## Section VIII    The Carbon Cycle

The simultaneous activities of photosynthesis and cellular respiration establish the carbon cycle, in which $CO_2$ is fixed by photosynthesizers into carbohydrates, $O_2$ is released, and carbohydrates are taken in by all organisms and used to fuel cellular processes. During aerobic respiration, $O_2$ is fixed and $CO_2$ is released back into the environment (text p. 191).

## KEY TERMS

# SELF-QUIZ: TESTING WHAT YOU HAVE LEARNED

## Matching Key Terms

Match each term on the left with the most appropriate description on the right.

| | | | |
|---|---|---|---|
| 1. $C_3$ plant | | a. | surrounds a lumen |
| 2. pigment | | b. | principal pigment |
| 3. light reactions | | c. | dark reactions |
| 4. thylakoid | | d. | electron acceptor for $CO_2$ |
| 5. ground state | | e. | moist climates |
| 6. reaction center | | f. | light optional |
| 7. photorespiration | | g. | water oxidized |
| 8. RuBP | | h. | wave and particle |
| 9. Calvin-Benson cycle | | i. | greenhouse effect |
| 10. $C_4$ plant | | j. | a specific site |
| 11. photon | | k. | banana shape |
| 12. chlorophyll | | l. | inefficient dark reaction |
| 13. chloroplast | | m. | dry climates |
| 14. carbon cycle | | n. | absorbs |
| 15. dark reactions | | o. | most stable |

## True or False?

1. _____ Dark reactions cannot proceed in light.

2. _____ Photon energy is inversely proportional to wavelength.

3. _____ Chlorophyll $b$ is not found in any prokaryote.

4. _____ The reaction center in photosystem II is P700.

5. _____ Cyclic photophosphorylation produces additional ATP in plants.

6. _____ The chemiosmotic theory applies to mitochondria but not to chloroplasts.

7. _____ Light reactions take place in the chloroplast stroma.

8. _____ The Calvin-Benson cycle may take place either in light or dark.

9. _____ Light reactions take place in the thylakoid membrane.

10. _____ $C_3$ plants grow slowly in hot, dry weather.

## Completion

1. Complete photosynthesis includes two reactions: a first step where _____ is _____, and a second where _____ is _____.

2. All photosynthetic organisms contain, in addition to chlorophyll, one or more _____.

3. A group of chlorophyll molecules, proteins, and various pigments form _____, which form aggregates on the thylakoid disc.

4. The last electron acceptor in the electron transport chain of the first series of light reactions is _____, the reaction-center chlorophyll in photosystem _____.

5. During noncyclic photophosphorylation, electrons flow in a _____ pattern.

6. Closure of the stomata prevents _____ from leaving the plant, and _____ from entering.

7. $C_4$ plants have veins surrounded by an inner layer of _____ cells.

8. High levels of atmospheric $CO_2$ can act to trap heat near the Earth's surface in a phenomenon called the _____.

## Short Answer

1. Write the chemical reactions for photosynthesis and respiration.

2. What are the two basic subreactions of photosynthesis?

3. Where in the cell do the two reactions of photosynthesis take place?

4. What is the physical nature of light? How is light energy transferred to a photosynthetic cell?

5. How are photosynthetic pigments organized into complexes?

6. What is meant by the "zigzag" scheme of electron flow?

7. What happens in noncyclic photophosphorylation?

8. How can oxygen inhibit photosynthesis?

## Multiple-Choice Review

In the following questions, fill in any blanks. Finish each sentence by circling the letter of the correct response.

1. In the first stage of photosynthesis water is:

   a. oxidized.
   b. phosphorylated.
   c. reduced.
   d. heated.
   e. none of the above

2. The distance a photon travels during a complete vibration is:

   a. its particle property.
   b. radiation.
   c. wavelength.
   d. strength.
   e. none of the above

3. Chlorophyll is green because it does not absorb light from the _____ portion of the spectrum.

    a. red
    b. green
    c. blue
    d. ultraviolet
    e. infrared

4. The last electron acceptor in the first electron transport system of the light reactions is:

    a. $O_2$.
    b. P700.
    c. photosystem II.
    d. water.
    e. none of the above

5. The Calvin-Benson cycle takes place in:

    a. ribosomes
    b. inner plasma membrane
    c. cytoplasm
    d. stroma
    e. none of the above

## Exercise

Label the parts of a $C_3$ leaf and a $C_4$ leaf.

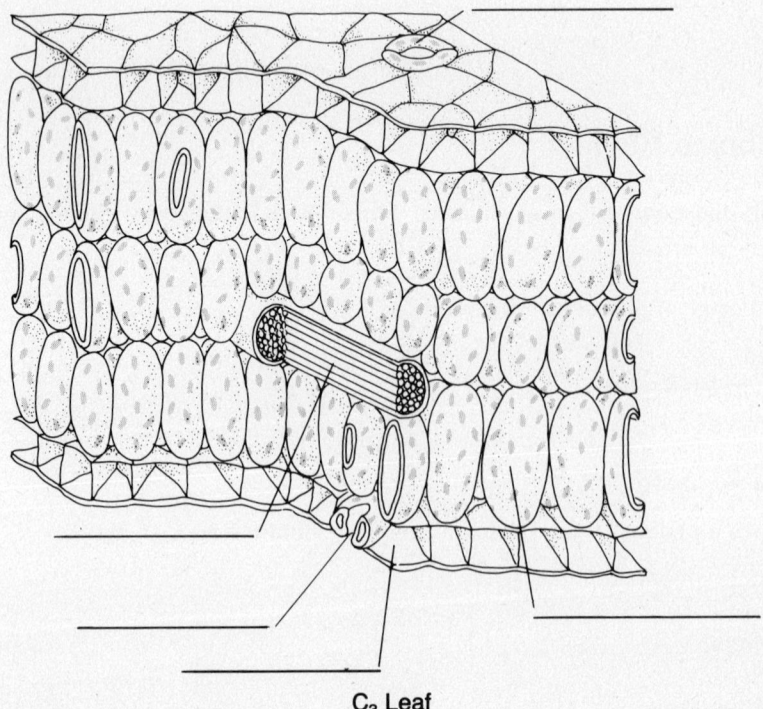

$C_3$ Leaf

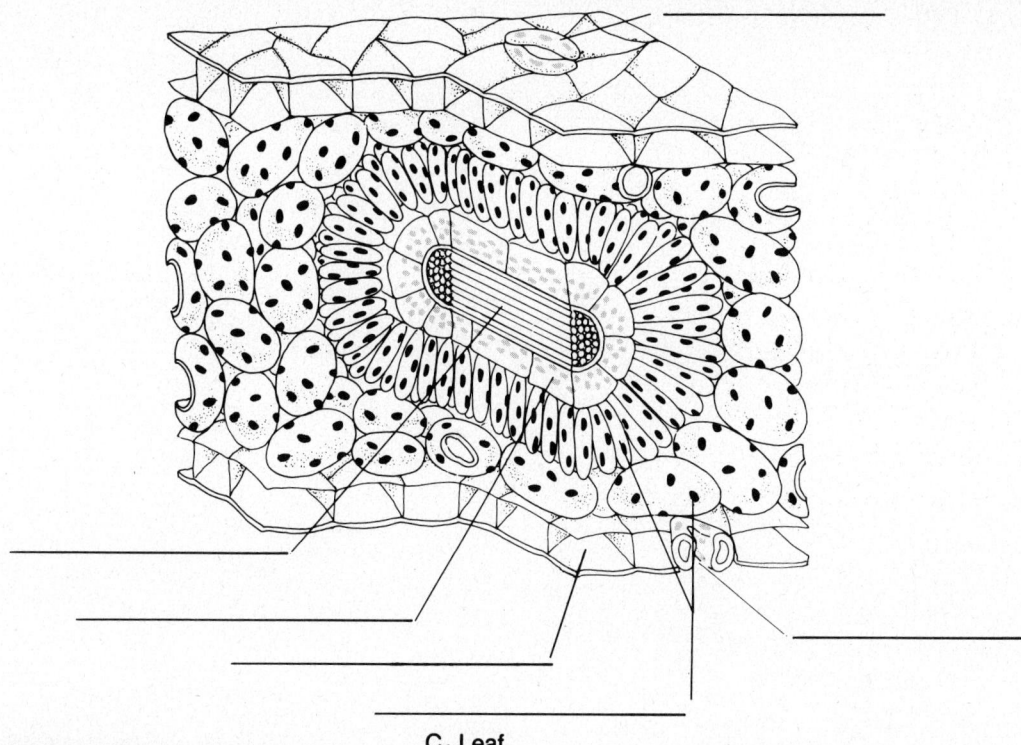

**C$_4$ Leaf**

Briefly describe the special biochemical pathway of C$_4$ plant photosynthesis (begins with a stable four-carbon intermediate).

# 9
# CELLULAR REPRODUCTION: MITOSIS AND MEIOSIS

---

## CHAPTER AT A GLANCE

---

## CHAPTER PREVIEW

This chapter introduces one of the most basic and remarkable of biological events—cellular reproduction. As you read and study, you will see how chromosomes in the nucleus of a dividing cell, which contain the genetic blueprint of the entire organism, are duplicated and distributed in a beautifully orchestrated "dance" so that each daughter cell receives the proper number. This is the process called mitosis. And you will see how, in meiosis, the process takes place in cells of the reproductive system to produce eggs and sperm having half the normal number of chromosomes.

The chapter begins with a look at the history of research on the cell nucleus, and the discovery that the nucleus is the repository of an organism's genetic material—DNA—organized into chromosomes. After a survey of the four phases of the cell cycle, during which growth, metabolism, and division occur, the focus narrows to one phase, mitosis, during which chromosome duplication and cell division (cytokinesis) take place. Mitosis in turn is subdivided into four phases. Because mitosis is a critical and fundamental biological event, you should carefully study the details of the mitotic phases. This preparation will be useful in the later section on meiosis.

Next, the text describes cytokinesis, the actual physical process of division of the cell cytoplasm, which generates two separate cells from one. In this section, be sure to have a clear idea of the differences between cytokinesis in plant and animal cells.

Meiosis is similar to mitosis in important respects, and different in major ways, too. This combination of similarities and differences often presents difficulty for students. Plan to master mitosis before you take up meiosis, and spend time reviewing the comparison of mitosis and meiosis in Figure 9-13. Also, take note of one of meiosis' most significant outcomes: the random distribution of parental chromosomes. This idea will play a major role in some of the genetic mechanisms you will encounter in the next few chapters.

The final section of Chapter 9 compares two biological extremes—asexual and sexual reproduction.

# LEARNING OBJECTIVES

When you have mastered the concepts of this chapter, you will be able to:

1. Describe the chemical and physical makeup of chromosomes, explain how they differ between sexes, and tell how cells are classified by chromosome number.
2. Name the four phases of mitosis and explain what happens to the chromosomes and associated cell structures in each phase.
3. Describe the process of cytokinesis and list the major differences between cytokinesis in plants and animals.
4. Name and describe the sequence of events during meiosis, and explain how meiosis differs from mitosis.
5. Compare and contrast cell reproduction in sexual and asexual organisms, and list the advantages and disadvantages of each.

# CONCEPTS IN REVIEW

## Section I    The Nucleus and Chromosomes

The cell nucleus is the main repository of genetic information. Within the nucleus are the chromosomes—tightly coiled strands of DNA and clusters of associated proteins. Long stretches of the continuous DNA molecule wind around these clusters of positively charged proteins, or histones, forming beadlike complexes known as nucleosomes. More coiling and supercoiling produces a dense chromosome structure (text p. 199). Each long strand of DNA combines with histones and non-histone proteins to make up the substance chromatin.

A pictorial display of an organism's chromosomes in the coiled, condensed state is known as a karyotype. Karyotypes reveal that, in most cells, two copies of all but the sex chromosomes are present; these two copies are referred to as homologous pairs. Non–sex chromosomes are called autosomes. Organisms whose cells contain two sets of parental chromosomes are called diploid; those with cells containing a single set of parental chromosomes are called haploid (text p. 201).

## Section II    The Cell Cycle

The cell cycle is a regular sequence in which the cell grows, prepares for division, and divides to form two daughter cells, which each then repeat the cycle. Such cycling in effect makes single-celled organisms immortal. Many cells in multicellular organisms, including animal muscle and nerve cells, either slow the cycle or break out of it altogether (text p. 201).

The normal (mitotic) cell cycle consists of four phases. The first three are: $G_1$, the period of normal metabolism; S phase, during which normal synthesis of biological molecules continues and, in addition, DNA is replicated and histones are synthesized; and $G_2$, a brief period of metabolism and additional growth. Together the $G_1$, S, and $G_2$ phases are called interphase. The fourth phase of the cell cycle is M phase, the period of mitosis during which the replicated chromosomes condense and move, and the cell divides (text p. 202). It is believed that properties of the cell cytoplasm control the cell cycle, along with external stimulators and inhibitors such as chalones (text p. 203).

## Section III    Mitosis: Partitioning the Hereditary Material

Biologists divide the mitotic cycle into four phases. At the beginning of prophase, each chromosome has replicated itself and now consists of two highly condensed, identical chromatids attached to each other at a centromere (text p. 204). As prophase ends and metaphase begins, the condensed chromosomes become associated with the spindle. Eventually, the chromosomes become arranged in a plane at a right angle to the spindle fibers. This plane is called the metaphase plate. Next, during anaphase, the two sister chromatids of each chromosome split, and one from each pair is drawn toward each pole of the cell. During telophase, nuclear envelopes begin to form around each set of chromosomes and division of the cytoplasm takes place.

As mitosis proceeds, the spindle microtubules play a crucial role in ensuring that both paired and separated chromatids move in the right direction at the proper times (text pp. 205–206). Each half of the spindle forms as microtubules extend from each pole of a dividing cell to the region of the metaphase plate. During prophase, other microtubules, the centromeric fibers, extend outward from the spindle poles to structures on the chromosomes called kinetochores. During anaphase, the fibers begin to shorten, and the chromatids begin to move apart (text p. 206).

The spindle forms differently in plant and animal cells. In animals it is associated with the centriole (Chapter 6), while in plant and fungal cells spindle formation is associated with regions called microtubule-organizing centers (text p. 206).

## Section IV    Cytokinesis: Partitioning the Cytoplasm

The division of the cell cytoplasm at the end of mitosis is called cytokinesis. In animal cells it takes place as a ring of actin filaments contracts around the cell equator, pinching the cell in two. In plant cells, which are bounded by a cell wall, cytokinesis involves the building of a new cell plate across the dividing cell at its equator. Cell wall material is then deposited in the region of the cell plate (text p. 207).

In certain developmental events, cytokinesis can occur without the duplication of chromosomes and division of the nucleus. Likewise, chromosomes can be duplicated and nuclei can divide without subsequent cytokinesis (text p. 207).

## Section V    Meiosis: The Basis of Sexual Reproduction

Meiosis is a special form of cell division that takes place in the reproductive organs that produce sex cells. As with mitosis, it takes place after DNA replication has occurred, and involves two sequential nuclear divisions (meiosis I and meiosis II). These divisions result in four daughter cells, each with half the number of chromosomes of the parent cell. The phenomenon of crossing over during meiosis results in exchanges of genetic information between chromosomes. Hence, the homologous chromosomes distributed to different progeny cells are not identical (text p. 208).

As in mitosis, there are two chromatids for each chromosome at the beginning of prophase I. During this phase, the homologous chromosomes undergo synapsis, or pairing, which is brought about by a bridging structure of proteins and RNA called the synaptinemal complex (text p. 208). The homologous pairs stay together when they align on the metaphase plate. Unlike the anaphase of mitosis, during anaphase I the two chromatids of each chromosome stay joined at the centromere and move together to one of the two poles of the cell. It is this event that results in the halving of the chromosome number in the four daughter cells that result from meiosis (text p. 210).

During telophase I nuclear envelopes enclose the chromosomes in nuclei, and in most species cytokinesis (the first nuclear division) follows. The second nuclear division begins with metaphase II, in which the chromosomes in each daughter cell again align on a metaphase plate. The centromeres finally divide, and each sister chromatid moves to one of the poles of the spindle. The next phase is telophase II, followed again by cytokinesis. The result of the entire process of meiosis is four haploid cells in which parental chromosomes are randomly distributed (text pp. 210–212).

## Section VI    Asexual Versus Sexual Reproduction

Mitosis and meiosis make possible simple cell division and sexual reproduction, respectively. Each means of passing on hereditary information has advantages. In asexual reproduction the parent organism gives rise to offspring that are genetic clones of the parent. The advantages of this type of reproduction are that it preserves the parent's successful genetic complement, requires little or no specialization of reproductive organs, and is more rapid than sexual reproduction. A major disadvantage of the asexual mode is that a single catastrophic event or disease may readily destroy an entire population of genetically identical organisms. A prime benefit of sexual reproduction is that it provides genetic variability and a ready mechanism for the elimination of deleterious mutations. It also allows "new" gene forms to arise and spread through populations (text p. 214).

## KEY TERMS

| | | |
|---|---|---|
| anaphase *text page 205* | chromatin *200* | G$_2$ phase *202* |
| autosome *200* | chromosome *199* | haploid *201* |
| cell cycle *201* | clone *214* | histone *199* |
| cell plate *207* | crossing over *212* | homologous *200* |
| centromere *204* | cytokinesis *206* | interphase *202* |
| chalone *203* | diploid *201* | karyotype *200* |
| chromatid *202* | G$_1$ phase *201* | kinetochore *206* |

M phase   *202*                nucleosome   *199*           synapsis   *208*

meiosis   *208*                prophase   *204*             synaptinemal complex   *208*

metaphase   *205*              S phase   *202*              telophase   *205*

metaphase plate   *205*        sex chromosome   *200*

mitosis   *197*                spindle   *205*

# SELF-QUIZ: TESTING WHAT YOU HAVE LEARNED

## Matching Key Terms

Match each term on the left with the most appropriate description on the right.

| | | | |
|---|---|---|---|
| 1. cytokinesis | a. | DNA + histones |
| 2. synapsis | b. | two sets |
| 3. histone | c. | X and Y |
| 4. mitosis | d. | one set |
| 5. clone | e. | chromosome display |
| 6. chalone | f. | set of microtubules |
| 7. spindle | g. | plant division |
| 8. chromatid | h. | gamete production |
| 9. nucleosome | i. | genetically identical |
| 10. diploid | j. | cell division |
| 11. meiosis | k. | crossing over |
| 12. cell plate | l. | inhibits cell division |
| 13. sex chromosome | m. | division of cytoplasm |
| 14. karyotype | n. | single chromosome copy |
| 15. haploid | o. | + charged protein |

## True or False?

1. _____ Autosomes include X and Y chromosomes.

2. _____ Cells enter $G_2$ at the end of the S phase.

3. _____ Chalones promote cell division.

4. _____ The nuclear envelope forms in telophase.

5. _____ Plant cells have the most prominent centrioles.

6. _____ Nuclear division may occur without cytokinesis.

7. _____ Homologous chromosomes pair in synapsis.

8. _____ Chromosomes break at chiasmata.

9. _____ Sexual organisms cannot be cloned.

10. _____ All daughter cells are haploid.

## Completion

1.  If human chromosomes are stained on a slide, the resulting display, which is called a
    _____, should contain twenty-two pairs of _____, and a pair of
    _____.

2.  The "S" in S *phase* stands for _____, because during this time DNA is
    _____, and histones are _____.

3.  Instead of centrioles, most plants and fungi have regions called _____ centers.

4.  During meiosis, but not during mitosis, homologous chromosomes pair up in a process of
    _____.

5.  Mitosis produces _____ progeny cells, each with a _____ set of
    chromosomes.

6.  Meiosis produces _____ daughter cells, each with a _____ set of
    chromosomes.

7.  Microtubules called centromeric fibers attach to the chromosomes at the _____.

8.  Two chromatids attach to each other at the _____.

## Short Answer

1.  What is meant by "haploid" and "diploid"? Where are cells of these types found?

2.  What are the phases that make up the cell cycle?

3.  What happens in cytokinesis? How do plants and animals differ in this process?

4. What are chalones?

5. How are chromatids moved toward the poles?

6. How can cytokinesis take place independently of chromosome duplication and nuclear division?

7. What are the basic differences between mitosis and meiosis?

8. What are the advantages of sexual reproduction in terms of variation?

## Multiple-Choice Review

In the following questions, finish each sentence by circling the letter of the correct response.

1. DNA is replicated during the:
   a. S phase.
   b. M phase.
   c. $G_2$.
   d. $G_1$.
   e. $G_0$.

2. In a typical vertebrate the longest phase is:

   a. S.
   b. M.
   c. $G_2$.
   d. $G_1$.
   e. none of the above

3. The spindle forms during the mitotic phase of:

   a. metaphase.
   b. anaphase.
   c. prophase.
   d. synapsis.
   e. none of the above

4. Male sex chromosomes can never be:

   a. haploid.
   b. homologous.
   c. diploid.
   d. analogous.
   e. duplicated.

5. Crossing over occurs during:

   a. $G_0$.
   b. $G_1$.
   c. synapsis.
   d. cytokinesis.
   e. M phase.

# Exercise

Supply the missing labels in these diagrams of mitosis and meiosis.

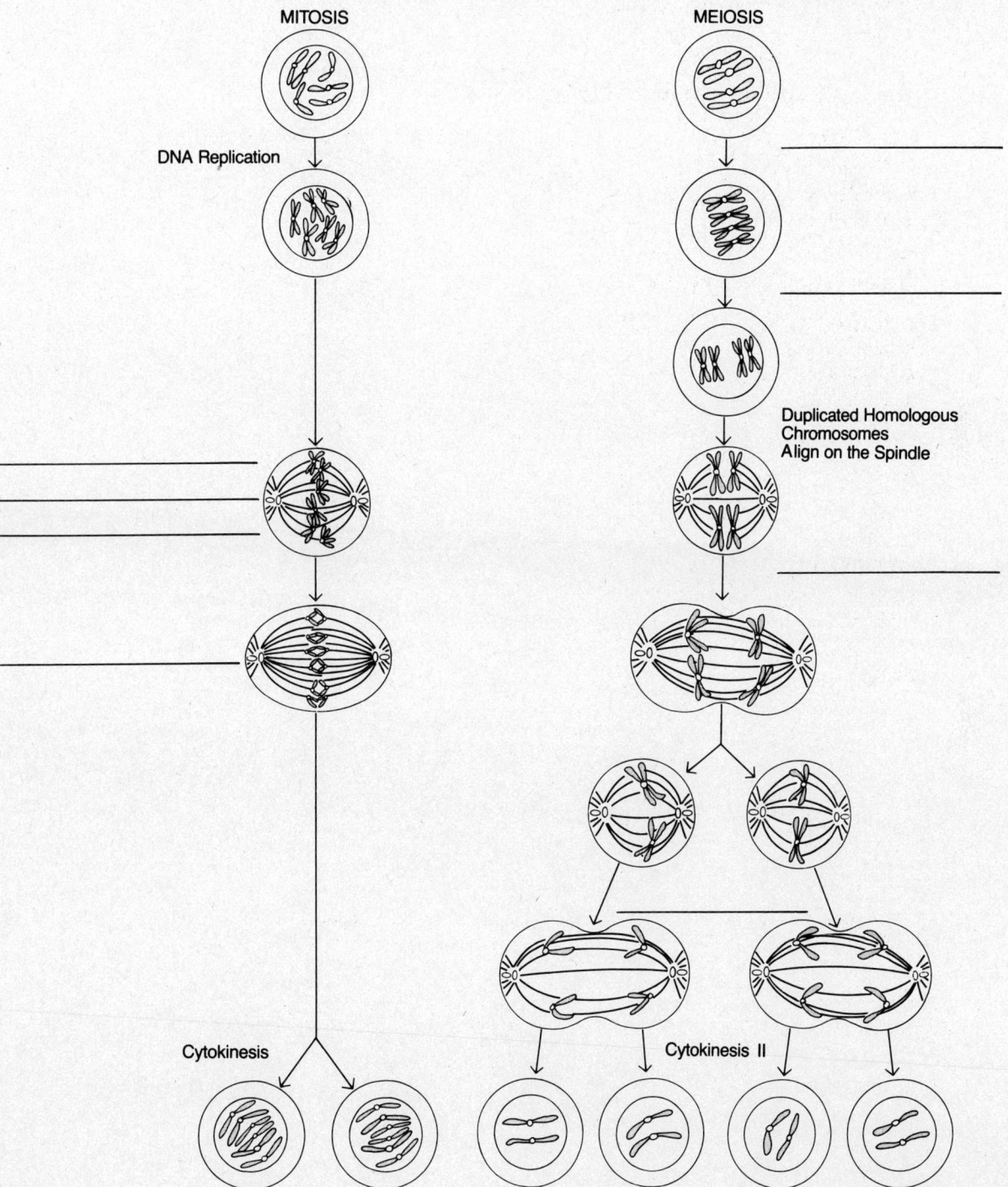

MITOSIS

MEIOSIS

DNA Replication

Duplicated Homologous
Chromosomes
Align on the Spindle

Cytokinesis

Cytokinesis II

# 10
# FOUNDATIONS OF GENETICS

---

## CHAPTER AT A GLANCE

---

## CHAPTER PREVIEW

Chapter 10, on the work and ideas of Gregor Mendel, traces the scientific explorations that laid the groundwork for modern genetics. Early concepts of heredity, including pangenesis, the germ plasm and blending theories, and vitalism (which has some adherents even today) are all discussed in the first section.

Next you will be introduced to Mendel, his life, and his particulate theory of inheritance. Mendel's discoveries are still fundamental today—and will be vitally important for an understanding of the genetics chapters that follow. In fact, Mendel's accomplishments—beginning with the identification of the inheritance "factors" that we know today as genes—were as significant as the famous modern description of the double-helix structure of DNA by James Watson and Francis Crick (Chapter 12). Be sure you understand how Mendel's classic experiments to uncover the mechanisms of inheritance yielded basic principles of genetics. Important ideas include the concepts of alleles, dominance, what test crossing accomplishes and how, and Mendel's laws of segregation and independent assortment.

The final main section of Chapter 10 describes the gains made in understanding genetic events that occurred almost as soon as Mendel's work was rediscovered at the beginning of the twentieth century. Beginning with the formulation of the chromosome theory of heredity by Sutton and Boveri, researchers were able to identify a number of phenomena, including sex-linked traits and the chromosomal defect known as nondisjunction. The explosion of research that followed this pioneering work—and the new knowledge it generated—will be the subjects of Chapters 11 and 12.

## LEARNING OBJECTIVES

When you have mastered the concepts of this chapter, you will be able to:

1. List and discuss at least three of the pre-Mendelian concepts of heredity.
2. Describe the life of Gregor Mendel, and point out why he was better able to study genetics because of his educational background.
3. Describe each of Mendel's laws, and outline how he arrived at the basis for each.
4. Explain incomplete dominance and why it was a problem for Mendel.
5. Tell why Mendel's ideas remained undiscovered for so long.
6. Outline the progress in genetics that has been made in the last century.

## CONCEPTS IN REVIEW

### Section I    Early Theories of Inheritance

Early ideas on inheritance included Hippocrates' theory of pangenesis and August Weismann's germ plasm theory. Based on experiments with mice, Weismann proposed that hereditary information in gametes transmitted traits to progeny (text p. 219). Both these early views incorporated the blending theory: they held that heritable traits of the two parents blend, so that the distinct characteristics of each are lost in offspring.

### Section II    Gregor Mendel and the Birth of Genetics

Gregor Mendel, an Augustinian monk in the monastery at Brunn, Austria, is known as the "father of genetics." Having been exposed to theories of the particulate nature of matter while a university student, and with a background in mathematics, Mendel carried out a series of carefully planned experiments that demonstrated the particulate nature of heredity. His revolutionary ideas were neither understood nor accepted until many years after Mendel died (text p. 221).

### Section III    Mendel's Classic Experiments

Mendel studied genetics through plant-breeding experiments using the garden pea, a plant species that is self-fertilizing and breeds true (each offspring is identical to the parent in the trait of interest). To test the blending theory, he focused his research on seven distinct characters (see Figure 10-4). For each of these characters, such as seed color and plant height, there are only two, clear-cut possibilities. He also recorded the type and number of all progeny produced from each pair of parent pea plants, and followed the results of each cross for two generations (text p. 223).

For each of the characters he studied, Mendel found that one trait was dominant while the other was recessive. In the second filial ($F_2$) generation, the ratio of dominant to recessive was 3:1. Mendel deduced that this result was possible only if each individual possesses only two hereditary units, one from each parent. The units Mendel hypothesized are today known as alleles, alternative forms of genes. Genes are the basic units of heredity (text p. 223). An organism that inherits identical alleles for a trait from each parent is said to be homozygous for that trait; if different alleles for a trait are inherited the organism is heterozygous for that trait (text p. 224).

When an organism is heterozygous for a trait, the resulting phenotype for that trait may be the expression of only one allele. The allele that is expressed is called dominant, and the one that remains hidden is called recessive. In this case, the organism's phenotype—its physical appearance and properties—will be different from its genotype, which includes both a dominant and a recessive allele. A pictorial representation of all the possible combinations of a genetic cross is known as a Punnett square (text p. 225).

The results of Mendel's experiments on dominant and recessive inheritance led to Mendel's first law: the law of segregation. This law states that, for a given trait, an organism inherits one allele from each parent. Together they form the allele pair. When gametes are formed during meiosis, the two alleles become separated (halving of chromosome number) (text p. 227). To gain evidence for his theory, Mendel performed test crosses, mating plants of unknown genotype to homozygous recessive plants for the trait of interest. The ratio of dominant phenotypes (if any) in the progeny makes clear whether the unknown genotype is heterozygous, homozygous dominant, or homozygous recessive (text p. 228).

Mendel also performed dihybrid crosses, which enabled him to consider in one cross how two traits are inherited (text p. 229). This work led to the law of independent assortment, which states that the alleles of genes governing different characters are inherited independently. (You will study exceptions to this rule in Chapter 11.) An apparent exception to Mendel's laws is incomplete dominance, a phenomenon in which offspring of a cross exhibit a phenotype that is intermediate between those of the parents. However, incomplete dominance reflects the fact that both alleles for the trait in question exert an effect on phenotype. The alleles themselves remain separate (text p. 230).

## Section IV    Mendel's Ideas in Limbo: A Theory Before its Time

Mendel presented his ideas in 1866 in a scientific paper published by the Brunn Society for Natural History. Unfortunately, the meaning of his research was not understood by other scientists of the day. His work was rediscovered in 1900 by Carl Correns and Hugo de Vries.

## Section V    Chromosomes and Mendelian Genetics

Soon after Mendel's work was rediscovered, Walter Sutton and Theodor Boveri independently proposed that the hereditary units might be located on chromosomes. Experiments to prove this hypothesis were carried out by Thomas Hunt Morgan and his students at Columbia University, in research on the sex chromosomes of fruit flies (text pp. 232–235). Morgan's studies were also the first exploration of sex-linked traits. They also led to the discovery in 1916 by Calvin Bridges of the phenomenon of nondisjunction, in which a chromosome pair fails to segregate during meiosis (text p. 235).

## KEY TERMS

allele    *text page 223*

blending theory of
    heredity    *220*

chromosome theory of
    heredity    *232*

dihybrid cross    *229*

dominant    *223*

$F_1$ (first filial) generation
    *223*

$F_2$ (second filial) generation
    *223*

gene    *223*

genotype    *224*

# SELF-QUIZ: TESTING WHAT YOU HAVE LEARNED

## Matching Key Terms

Match each term on the left with the most appropriate description on the right.

| | | |
|---|---|---|
| 1. dominant | a. | Hippocrates |
| 2. phenotype | b. | parental |
| 3. allele | c. | nondominant |
| 4. homozygous | d. | always expressed |
| 5. recessive | e. | alternative forms |
| 6. $P_1$ | f. | "grandchildren" |
| 7. dihybrid cross | g. | basic heredity units |
| 8. law of segregation | h. | different |
| 9. gene | i. | similar |
| 10. pangenesis | j. | Weismann |
| 11. $F_2$ | k. | total alleles |
| 12. sex-linked | l. | two characters |
| 13. heterozygote | m. | separate |
| 14. genotype | n. | X or Y chromosome |
| 15. germ plasm theory | o. | appearance |

## True or False?

1. _____ Vitalism is the notion that each body part produces a "seed."

2. _____ Breeding true means offspring are identical with parents in certain traits.

3. _____ Alleles are simply alternate forms of genes.

4. _____ Snapdragon color is an example of blending inheritance.

5. _____ An XO male may result from nondisjunction.

6. _____ Darwin fully rejected the idea of "pangenesis."

7. _____ Homozygous organisms produce only one type of gamete for a gene.

8. _____ Test crosses helped establish the law of segregation.

9. _____ Sutton and Boveri developed the chromosome theory of heredity.

10. _____ Mendel's second law states that characters are inherited dependently.

# Completion

1. Intermediate phenotypes are produced by the genetic phenomenon of _____.

2. _____ was the first to show that the units of heredity lie on the chromosomes.

3. Mendel developed his second law by performing a series of _____ in his pea plants.

4. Mendel's work was never recognized during his lifetime, but was later acknowledged by

   _____.

5. When dealing with a number of traits in genetic crosses, the simple checkerboardlike device

   of a _____ makes calculations easier.

6. Alleles are alternate forms of _____.

7. Mendel's peas were ideal for the selection of true breeding strains because of the fact that

   peas are _____.

8. The concept of _____, which was popular during Mendel's time, held that the natural laws governing living things were very different from the laws governing nonliving materials.

# Short Answer

1. What was the theory of pangenesis?

2. What are Mendel's two laws?

3. What is meant by dominant/recessive? Homozygous/heterozygous? Phenotype/genotype?

4. How can a test cross reveal heterozygosity?

5. What are two ways commonly used to calculate genotypes in dihybrid crosses?

6. How is incomplete dominance different from the "blending theory"?

## Multiple-Choice Review

In the following questions, fill in any blanks. Finish each sentence by circling the letter of the correct response.

1. The notion that each part of the body of an organism produces a "seed" which then travels to the reproductive organs is called _____, and it was first produced by:

    a. Hippocrates.
    b. Aristotle.
    c. Mendel.
    d. Darwin.
    e. Weismann.

2. The _____ theory of heredity held that each parent contributes half of a trait that would produce a perfect intermediate offspring. This idea was linked to a widespread belief in:

    a. blending.
    b. vitalism.
    c. pangenesis.
    d. Mendelism.
    e. Herodotus.

3. For some specific traits the phenotypes of offspring are intermediate from those of the parents' traits. This is a case of:

    a. blending.
    b. linkage.
    c. incomplete dominance.
    d. intermediate inheritance.
    e. none of the above

4. Sex chromosomes were first discovered in:

    a. peas.
    b. Manx cats.
    c. frizzle chickens.
    d. fruit flies.
    e. humans.

5. In humans, which of the following genotypes will produce a female?

    a. YY
    b. YO
    c. XY
    d. XO
    e. XX

## Exercise

Use this Punnett square to depict the F$_2$ results of a dihybrid cross. The parental characteristics are short (*SS*), round (*RR*), tall (*ss*), and oval (*rr*). One parent has only dominant alleles, the other only recessive ones. Check your work against Figure 10-9 in the text.

# 11
# MENDEL MODIFIED

---

## CHAPTER AT A GLANCE

---

## CHAPTER PREVIEW

Chapter 11 focuses on the concluding idea of Chapter 10: modification of Gregor Mendel's basic principles by twentieth-century geneticists. As you will see in the first section, the scientific detective work that clarified the mechanisms of heredity began in earnest with two bits of information: that genes are the units of heredity, and that genes are arranged on chromosomes in a linear manner. Drawing on these facts, researchers identified processes such as linkage and crossing over that alter Mendel's law of independent assortment. Understanding of the linear arrangement of genes also prompted development of the gene map—one of the geneticist's most useful tools.

  After describing the methods of gene mapping, the chapter surveys a variety of gene interactions. Here you will consider in detail several basic aspects of modern genetics: what is meant by "dominant" and "recessive" alleles; the significance of genes having more than two alleles; and the concepts of multiple allelic series, complementary genes, and epistatic genes. The next section describes two more phenomena, pleiotropy and the effects of lethal alleles. That "killer" genes can persist in populations is confusing to some students, but you should have no problem if you read carefully the discussion of the genetic basis of sickle cell anemia. Also, note the implications of the fact that a lethal allele may not take its toll until after an affected individual passes reproductive age.

The chapter ends with a brief survey of the role of mutation as a source of genetic variation, and some of the events that may produce mutations. You will encounter many of these ideas again in later chapters on the mechanisms of evolution.

# LEARNING OBJECTIVES

When you have mastered the concepts of this chapter, you will be able to:

1. Discuss the historical development of the idea that the genetic material lies on chromosomes.

2. Explain the processes of recombination and crossing over.

3. Describe how gene maps are constructed, and explain what information they relay and how accurate they are.

4. Explain the functioning of dominant and recessive alleles.

5. List the major ways genes act when more than two alleles are present.

6. Outline the various types of mutations, and describe the effects they have on genes and chromosomes.

# CONCEPTS IN REVIEW

## *Section I*    How Genes Are Arranged on Chromosomes

The analysis of exceptions to Mendel's laws by geneticists in the first half of the twentieth century yielded important insights into gene interactions, their arrangements on chromosomes, and the role of mutation as a source of genetic variation. One discovery was linkage: some genes that lie on the same chromosome move together during meiosis and do not separate (text p. 239). Genes may also be partially linked, in which case some but not all progeny inherit a recombinant genotype (text p. 240). In nature, nonlinkage, partial linkage, and complete linkage are all common.

Many genetic studies have been and continue to be carried out using fruit flies (*Drosophila*) as subjects. In the nomenclature of fruit fly genetics, the most common unmutated allele of any characteristic is termed the wild-type allele.

Although the genes for body color and eye color in *Drosophila* are linked, occasionally recombinant phenotypes arise. This occurs as a result of crossing over, the trading of pieces of chromosomes during meiosis, described in Chapter 9. A recombination frequency is the measure of how often crossovers occur between gene loci. Such frequencies are used in the technique of recombination analysis to determine how close two genes lie to each other on a chromosome (text p. 242).

Alfred Sturtevant was the first to use recombination frequencies to construct a gene map—a display that shows the relative distances between genes and where they lie along the chromosome. One rationale for Sturtevant's map was the idea that the frequency of crossing over between any two genes on a chromosome would be directly proportional to their distance from each other. Map units, the measure of distance between genes, are called centimorgans. Sturtevant's work was the first proof that genes on chromosomes have a linear relationship (text pp. 243–244).

Studies of giant *Drosophila* chromosomes, called polytene chromosomes, have yielded cytogenetic maps—maps on which the locations of particular genes can be pinpointed and labeled.

## Section II   How Genes Act and Interact

A gene may have many more than two alleles; a group of alleles determining many forms of the same trait is called a multiple-allelic series. *Drosophila* eye color and human blood type are determined by such series. In cases where two alleles are fully expressed when both are present, as with the human blood type alleles $I^A$ and $I^B$, the alleles are said to be codominant. The Rh trait in human blood is also an example of the effects of multiple alleles (text pp. 248–249).

One phenotype may be generated by the action of two or more complementary genes. These are genes whose products must act together to produce a given phenotype. In a third type of gene interaction, known as epistasis, the effects of one gene override or mask the effects of other, completely different genes (text pp. 249–250).

## Section III   Lethal Alleles and Pleiotropy

Mutated alleles that are capable of causing death are termed *lethal alleles*. Depending on the nature of the essential function that is missing, death may occur at any time in an organism's life (text p. 250). Lethal genes that persist in populations are usually recessive.

A situation in which an allele affects two or more phenotypes is termed *pleiotropy*. A well-known example is the allele that causes sickle cell anemia. The allele's fundamental effect is to cause the substitution of a single amino acid in hemoglobin. It is preserved in the human population because it also confers resistance to malaria and in heterozygotes does not produce severe sickling of red blood cells (text p. 253).

## Section IV   Mutation: One Source of Genetic Variation

A mutation is a change in the chemical structure of a gene or the physical structure of a chromosome. A point mutation alters the properties of a single gene and creates new alleles. Chromosomal mutations involve rearrangements of blocks of genes in the chromosome. Many mutations are hurtful to the organism, although a mutation may also confer some new advantage to its bearer. Mutations are ultimately the prime source of variation in genomes (text p. 254).

Mutation rate refers to the frequency with which mutations arise naturally in a population. Agents such as X-irradiation can significantly increase the mutation rate. Many chemical mutagens are now known as well. Researchers have also discovered a link between mutations and carcinogens—agents that can cause cancer (text pp. 254–255). A commonly used screening test for carcinogens is known as the Ames test.

## KEY TERMS

| | | |
|---|---|---|
| Ames test   *text page 255* | lethal allele   *250* | point mutation   *254* |
| carcinogen   *254* | linkage   *239* | polytene chromosome   *244* |
| chromosomal mutation   *254* | locus   *240* | recombinant genotype   *240* |
| codominant   *254* | multiple-allelic series   *246* | recombinant phenotype   *240* |
| complementary gene   *249* | mutagen   *254* | recombination analysis   *242* |
| cytogenetic map   *245* | mutation   *254* | recombination frequency   *242* |
| epistasis   *249* | mutation rate   *254* | wild-type allele   *241* |
| gene map   *243* | pleiotropy   *252* | |

# SELF-QUIZ: TESTING WHAT YOU HAVE LEARNED

## Matching Key Terms

Match each term on the left with the most appropriate description on the right.

| | |
|---|---|
| 1. carcinogen | a. unmutated allele |
| 2. pleiotropy | b. cause death |
| 3. epistasis | c. given location |
| 4. polytene chromosome | d. altered gene |
| 5. lethal allele | e. cause cancer |
| 6. Ames test | f. alters genes |
| 7. mutation | g. override |
| 8. linkage | h. single gene = several traits |
| 9. wild-type allele | i. identifies carcinogens |
| 10. mutagen | j. inherit together |
| 11. locus | k. giant |

## True or False?

1. _____ Wing length and body color are sex-linked in *Drosophila*.

2. _____ Gene maps show distances as relative.

3. _____ Gene map units are expressed as centmendels.

4. _____ Gene mapping had nothing to do with showing that genes are arranged in a linear fashion.

5. _____ Cytogenetic maps show actual genes.

6. _____ Wild-type alleles are usually dominant.

7. _____ Human blood types are an example of a multiple-allelic series.

8. _____ There is no way to detect or study genes that are lethal to an early embryo.

9. _____ A single pleiotropic gene can affect several traits.

10. _____ Inversion is not an example of chromosome mutation.

## Completion

1. The most common unmutated allele of any characteristic in *Drosophila* genetics is called the _____, and is designated by a _____.

2. The X chromosome gene maps developed by Sturtevant showed the _____ between various genes.

3. Both cytogenetic gene maps and recombination gene maps show that genes are distributed in _____ arrangement along the chromosomes.

4. The phenomenon of gene _____ on chromosomes explains why some traits are always inherited together.

5. The exact location of a gene on a chromosome is referred to as its _____.

6. _____ genotypes (or phenotypes) are those that occur in offspring but in neither parent.

7. _____ is the technique that allows recombination gene maps to be constructed.

8. Genes have both _____ and _____ roles in cells.

## Short Answer

1. What are the three possible types of linkage?

2. What is the relationship between crossing over and distance between genes on a chromosome?

3. What is the difference between recombination and cytogenic maps?

4. What is meant by the term *codominant?*

5. Compare and contrast epistasis and dominance.

6. What are lethal alleles? How do they vary? How can they be detected if they act on early embryos?

7. Use malaria as an example to explain how harmful recessive genes can be maintained in a population.

## Multiple-Choice Review

In the following questions, finish each sentence by circling the letter of the correct response.

1. The term *polytene* refers to:

    a. giant chromosomes.
    b. giant genes.
    c. many-threaded chromosomes.
    d. cytogenetic maps.
    e. none of the above

2. A human with type O blood has:

    a. both A and B antigens.
    b. only A antigens.
    c. O antigens.
    d. neither A nor B antigens.
    e. A and O antigens.

3. When a single phenotypic trait is generated by two or more separate genes, each with its own alleles, the trait is:

    a. additive.
    b. complementary.
    c. epistatic.
    d. pleiotrophic.
    e. none of the above

4. An individual gene that acts to determine several phenotypic traits at once is:

    a. pleiotropic.
    b. complementary.
    c. epistatic.
    d. polyphenic.
    e. none of the above

5. Mutations are easier to detect when they occur:

    a. in *Drosophila*.
    b. in humans.
    c. in the homozygous condition.
    d. on autosomes.
    e. on sex chromosomes.

# 12

# DISCOVERING THE CHEMICAL NATURE OF THE GENE

---

## CHAPTER AT A GLANCE

---

## CHAPTER PREVIEW

This chapter picks up the "detective story" begun in Chapter 11 that traces the history of the discovery of the chemical nature of the gene. As you read, you will see how the two fields of genetics and biochemistry each contributed pieces of the puzzle, and eventually exposed the molecular mechanisms of inheritance.

The first section introduces the early one-gene–one-enzyme hypothesis, and outlines its development and modification into the one-gene–one-polypeptide hypothesis. The next section describes the discovery of nucleic acids and the work by several teams of researchers that revealed that nucleic acid, and not protein, is the "stuff" of genes. Once the chemical composition of DNA was known, new studies by Erwin Chargaff yielded clear evidence that DNA had a structure complex enough to permit the coding of varied genetic information.

The search for the molecular structure of DNA, covered next, revolves around the fascinating story of the double helix—the revolutionary model of DNA structure discovered by James Watson and Francis Crick. Armed with their powerful model, Watson and Crick were able to describe how the two linked strands of the DNA double helix replicate. You'll find Figure 12-13 extremely useful in understanding how this remarkable process—the fundamental event of heredity—takes place.

# LEARNING OBJECTIVES

When you have mastered the concepts of this chapter, you will be able to:

1. Outline the historical development of modern definitions and concepts of the gene; name the workers who were involved, and briefly discuss their methods.

2. List Chargaff's rules, and describe how they apply to the components of DNA.

3. Outline the work of Watson and Crick, and explain how their theory took into account all the available evidence.

4. Outline the steps of DNA replication.

5. State the evidence for the semiconservative model of DNA replication.

6. Describe the chemical structure of the DNA molecule.

# CONCEPTS IN REVIEW

## Section I    Genes Code for Particular Proteins

The first scientist to investigate the question of how genes affect phenotype was Sir Archibald Garrod, whose studies of alkaptonuria implied a relationship between genes and enzymes. Thirty years later Beadle and Ephrussi showed a relationship between particular genes and the biosynthetic reactions responsible for eye color in fruit flies. Next, in a series of classic experiments on the effects of mutations in the bread mold *Neurospora crassa*, Beadle and Tatum explored the one-gene–one-enzyme hypothesis: the idea that each gene codes for a particular enzyme (text p. 260). Their work paved the way for other researchers to elucidate the precise ways in which enzymes affect complex metabolic pathways (text p. 261).

In 1949, in research on the role of hemoglobin in sickle cell anemia. Linus Pauling helped refine the one-gene–one-enzyme hypothesis into the one-gene–one-polypeptide hypothesis (text p. 263). Pauling's original results, which suggested that genes control proteins, were modified by Vernon Ingram to reflect the current view of gene activity: each gene brings about the formation of one polypeptide chain (text p. 264).

## Section II    The Search for the Chemistry and Molecular Structure of Nucleic Acids

Nucleic acid, originally isolated by Johann Meischer in 1871, was identified as a prime constituent of chromosomes through the use of the red-staining method developed by Feulgen in the early

1900s. Frederick Griffith's experiments with the R and S strains of pneumococci showed that an as yet unknown material from one set of bacteria could alter the physical traits of a second set. In the 1940s the team of Avery, MacLeod, and McCarty showed that this "unknown" material was DNA (text p. 267). At about the same time P. A. Levene discovered that DNA contained four nitrogenous bases, with each base attached to a sugar molecule and a phosphate group—a combination Levene termed a *nucleotide*.

Disagreement over whether DNA could carry complex genetic information was ended in the early 1950s by Martha Chase and Alfred Hershey, whose work with *E. coli* showed clearly that DNA, not protein, is the bearer of genetic information (text pp. 267–268).

Each DNA nucleotide contains a five-carbon sugar, deoxyribose, attached to one of four bases: adenine, guanine, cytosine, or thymine. Adenine and guanine molecules are double-ring structures called purines, while cytosine and thymine are single-ring structures called pyrimidines. The molecule made up of a base plus a sugar is termed a *nucleoside*. In each molecule of DNA a phosphate group links the five-carbon sugar of one nucleoside with the five-carbon sugar of the next nucleoside in the chain. This phosphate bonding creates a sugar–phosphate backbone (text p. 268).

Chargaff's rules describe the fact that (1) the amount of adenine is equal to the amount of thymine in DNA, and the amount of cytosine is equal to that of guanine; and (2) the ratios of A to T and C to G vary with different species.

## Section III    In Search of the Molecular Structure of DNA

In the late 1940s and early 1950s researchers looking for the structure of DNA drew upon Chargaff's insight, Levene's ideas on DNA components, and two other lines of evidence. One was the suggestion of Linus Pauling that DNA might have a helical structure held in place by hydrogen bonds, and the other was x-ray diffraction photos of DNA showing a helical structure with distance between the coils, taken by Franklin and Wilkins (text p. 271).

Based on this information Watson and Crick proposed the double helix model of DNA—a twisted ladderlike molecule with two outer sugar–phosphate chains, and rungs formed by nucleotide pairs. Paired nucleotides, which always occur as A–T or G–C, are linked by hydrogen bonds (text pp. 272–273). Watson and Crick also proposed that genetic information is encoded by the sequence of base pairs along the DNA molecule.

## Section IV    How DNA Replicates

In their model of DNA structure and function, Watson and Crick predicted that DNA replicates itself by "unzipping" along the hydrogen bonds joining A to T and C to G. This process would produce two opposite halves that could then serve as templates for the construction of new, complementary strands. This model of semiconservative replication—"conservative" because each new molecule has one-half of the former "parent" molecule—was later confirmed by the work of Meselson and Stahl (text p. 276).

In *E. coli* DNA replication begins with the formation of a bubblelike structure on the circular chromosome that is produced by replication forks. Studies of bacterial DNA replication have shown that a growing DNA chain lengthens only in the 5′ to 3′ direction (from the 5′ carbon of one sugar to the 3′ carbon of the next). The leading strand is synthesized continuously, while the lagging strand is synthesized in short stretches known as Okazaki fragments (text p. 279). The enzyme DNA polymerase links free nucleotides as they line up on the template formed by the original strand of the parent molecule.

In eukaryotes DNA replication follows the same general principles as in prokaryotes. On the long DNA molecules, replication proceeds (in two directions at once) from hundreds or thousands of points of origin (text p. 280).

## KEY TERMS

adenine   *text page 268*

base   267

Chargaff's rules   269

cytosine   268

DNA polymerase   278

double helix   272

5' to 3' direction   279

guanine   268

nucleoside   268

Okazaki fragment   279

one-gene–one-enzyme
   hypothesis   260

one-gene–one-polypeptide
   hypothesis   264

purine   268

pyrimidine   268

replication fork   279

semiconservative replication
   275

thymine   268

x-ray diffraction   271

# SELF-QUIZ: TESTING WHAT YOU HAVE LEARNED

## Matching Key Terms

Match each term on the left with the most appropriate description on the right.

1. replication fork
2. cytosine
3. Okazaki fragment
4. x-ray diffraction
5. nucleoside
6. semiconservative replication
7. 5' to 3' direction
8. Chargaff's rules
9. base
10. double helix

a. A = T
b. unwinding must occur
c. a DNA base
d. ring structure composed of carbon and nitrogen
e. shape proposed by Watson and Crick
f. a base plus a sugar
g. start of DNA replication
h. small pieces of DNA
i. DNA chain lengthens
j. photographic process

## True or False?

1. _____ Chase and Hershey showed that genes are composed of DNA.

2. _____ Complementation tests can be used to study metabolic pathways.

3. _____ Feulgen staining works on cell walls.

4. _____ Robert Feulgen discovered nucleic acid.

5. _____ Phages are made only of DNA or RNA and a protein coat.

6. _____ Dispersive replication requires breakage.

7. _____ Not all organisms exhibit semiconservative replication.

8. _____ Auxotrophs are mutant *Neurospora*.

9. _____ Virulent pneumonia bacteria secrete a capsule.

10. _____ Watson and Crick did not make use of x-ray diffraction data.

## Completion

1. Beadle and Tatum were the first to show a relationship between _____ and

   _____ .

2. Linus Pauling developed his one-gene–one-polypeptide hypothesis with the help of an electrical molecular separating process called _____ .

3. The fact that DNA is composed of four bases was revealed by the work of _____ .

4. A base and a sugar together make up a _____ .

5. The constant features of DNA were described in a set of rules proposed by

   _____ .

6. DNA can replicate itself because, when "unzipped," each strand can serve as a _____

   _____ .

7. The Watson and Crick model of DNA replication that requires unzipping of the double helix

   is the _____ model.

8. Okazaki fragments are joined by the enzyme _____ .

## Short Answer

1. Why are bacteriophages so useful in DNA research?

2. Why does DNA display a viscous-nonviscous behavior at different temperatures?

3. How can DNA carry such a wide variety of genetic information with only four bases?

4. What is the difference between semiconservative and conservative DNA replication?

5. What is meant by the term *density gradient*? How was this property/method used by Meselson and Stahl?

6. What is a replication fork?

## Multiple-Choice Review

Complete the following statements by circling the correct response.

1. The role that genes play in general cellular metabolism was determined by:

   a. Garrod.
   b. Watson.
   c. Tatum.
   d. Ingram.
   e. none of the above

2. Feulgen stain turns nucleic acids:

   a. deep red.
   b. pink.
   c. deep green.
   d. black.
   e. brown.

3. DNA chains lengthen in the:

   a. 3′ to 5′ direction.
   b. 5′ to 3′ direction.
   c. 3′ direction.
   d. 5′ direction.
   e. none of the above

4. The controversy over whether DNA or protein carries genetic information was resolved by:

   a. Meischer.
   b. Griffith.
   c. Watson and Crick.
   d. Chase and Hershey.
   e. Beadle and Tatum.

5. Disease-causing strains of microorganisms are:

   a. nonvirulent.
   b. lethal.
   c. virulent.
   d. disgenic.
   e. none of the above

# 13
# TRANSLATING THE CODE OF LIFE: GENES INTO PROTEINS

---

## CHAPTER AT A GLANCE

---

## CHAPTER PREVIEW

Chapter 13 presents the evidence that led to the statement of the "central theme of molecular biology": genetic information is stored in a linear message on nucleic acids and is expressed in a corresponding linear sequence of amino acids in proteins. As you study this evidence you will also learn the functional definition of a gene, and how the genetic information of organisms is organized in code.

The core of this chapter is its central section on DNA, RNA, and the steps of protein synthesis. This material is often difficult for many students—once again, you will find that the text illustrations are extremely useful in helping you visualize complex events. You will also find it helpful to make a chart of the different kinds of RNA, and the characteristics and function of each. A simple outline will help you gain command of the four steps of protein synthesis.

# LEARNING OBJECTIVES

When you have mastered the concepts of this chapter, you will be able to:

1. List the evidence for a colinear genetic code.

2. Explain how DNA can, with only four bases, code for the vast diversity of life on Earth.

3. List the three basic features of the genetic code.

4. Explain where and how proteins are synthesized in a cell.

5. Outline the functions of the three types of RNA.

6. Outline the four basic steps in protein synthesis, and explain what happens in each.

7. Give an up-to-date definition of a gene.

# CONCEPTS IN REVIEW

## Section I    Genetic Information Must Occur in Code

Studies of the structure and functioning of DNA suggested strongly that genetic information is stored in a linear message—the sequence of nucleic acids—and is expressed in a corresponding linear sequence of amino acids in proteins. In the 1960s Charles Yanofsky and his colleagues confirmed that the sequence of gene codons—sequences of nucleotides coding for particular amino acids—is colinear with the amino acids in the corresponding polypeptides (text pp. 285–286).

The genetic code, then, is colinear, sequential, and based on the sequence of codons. Prime features of the code are: (1) it does not overlap; (2) it is deciphered by reading frames; (3) it is degenerate (different codons can code for the same amino acid); and (4) each codon consists of three nucleotides (text pp. 287–289). There are sixty-four possible codons to code for the twenty amino acids in organisms. A frameshift mutation is one that causes a reading frame in a DNA molecule to shift by one codon, thereby altering all subsequent reading frames and leading to the production of radically different polypeptides (text p. 288).

## Section II    DNA, RNA, and Protein Synthesis

Elucidation of the structure of DNA showed, at the molecular level, how genes are translated into biological molecules. Because proteins are made only in the cytoplasm of a cell, the DNA message must be carried to the cytoplasm by RNA (Chapter 3). The genetic information encoded in DNA is read during the process of transcription into several types of RNA. Messenger RNA (mRNA) carries information from DNA in the nucleus to the sites of protein synthesis in the cytoplasm. mRNAs are single-stranded chains of nucleotides that are synthesized on single strands of the DNA molecule. Nucleotides are linked to the growing RNA strand by RNA polymerases. DNA sequences known as promoters serve as the binding site for RNA polymerase near each gene, and transcription proceeds from that site. The result in eukaryotes is a primary transcript containing the coded instructions of the DNA molecule (text p. 290). In prokaryotes mRNA is transcribed directly from DNA and is used directly in translation, the process of protein synthesis.

The second type of RNA, transfer RNA (tRNA), transports individual amino acids to the sites of protein synthesis in the cytoplasm. Different tRNAs bond only to specific amino acids. Amino acids attach to tRNAs at aminoacyl attachment sites, which consist of the base sequence CCA. An anti-codon on the tRNA—specific for each amino acid—binds to a specific, complementary three-base codon on an mRNA molecule. Thus, a tRNA can transport a particular amino acid to an appropriate site on mRNA. The amino acids are then held in the correct position to be added to a growing polypeptide (text pp. 290–291).

Proteins and rRNA molecules assemble in the nucleolus to form the two major subunits of ribosomes, the organelles responsible for ensuring proper assembly of amino acids—coded in mRNA and transported by tRNA—into proteins in the cell cytoplasm (text pp. 291–292).

## Section III    Protein Synthesis: Translating the Genetic Code

The four general steps of protein synthesis are (1) amino-acid activation; (2) initiation; (3) elongation; and (4) termination.

During amino-acid activation, amino acids are attached to appropriate tRNAs by high-energy bonds. The attachment is catalyzed by enzymes known as aminoacyl–tRNA synthetases (text p. 293). Polypeptide synthesis begins at initiation with the attachment of a small ribosomal subunit to a binding site on mRNA. An AUG codon at this site functions as a start signal for protein synthesis. A large ribosomal subunit then binds to the smaller one, forming a complete ribosome attached to the mRNA. The ribosomal P (peptidyl) site is bound to an initiator tRNA, and when the A (aminoacyl) site becomes bound to an appropriate amino-acid–bearing tRNA, synthesis begins (text p. 294). Overall, the small subunit carries out the task of ordering amino acids according to the mRNA codon sequence, while the large subunit links amino acids to the polypeptide chain.

Elongation of the polypeptide chain is catalyzed by the enzyme peptidyl transferase. As the enzyme catalyzes successive peptide bonds between amino acids, the mRNA moves along the ribosome until its entire sequence of codons has been "read" (text p. 295). Elongation continues until a termination codon is reached on the mRNA molecule. Enzyme reactions then release the completed polypeptide (text p. 295).

## Section IV    Cracking the Genetic Code

Research in the 1960s enabled biologists to match particular codons with particular amino acids. An initial breakthrough came from Nirenberg and Matthaei, who inadvertently discovered the codon for polyphenylalanine. Subsequent work by Khorana and others yielded indentification of the meanings of all sixty-four codons (text pp. 297–298).

The modern definition of a gene is a unit of DNA that codes for a polypeptide or a structural RNA molecule (text p. 300).

---

## KEY TERMS

amino-acid activation
    *text page*  293

aminoacyl attachment site
    290

aminoacyl–tRNA synthetases
    293

anticodon  290

A site  294

codon  285

elongation  293

frameshift mutation  288

genetic code  286

initiation  293

messenger RNA (mRNA)
    290

peptidyl transferase  294

# SELF-QUIZ: TESTING WHAT YOU HAVE LEARNED

## Matching Key Terms

Match each term on the left with the most appropriate description on the right.

| | | | |
|---|---|---|---|
| 1. activation | a. | molecular emissary |
| 2. uracil | b. | protein synthesis |
| 3. mRNA | c. | amino acids join to tRNA |
| 4. termination | d. | three nucleotides |
| 5. anticodon | e. | amino acids form a polypeptide chain |
| 6. codon | f. | specific for each amino acid |
| 7. frameshift mutation | g. | found in ribosomes |
| 8. P site | h. | DNA read out |
| 9. tRNA | i. | central theme, or molecular grammar |
| 10. translation | j. | specific amino acid links to specific tRNA |
| 11. elongation | k. | stop signal on mRNA |
| 12. genetic code | l. | found in RNA, but not DNA |
| 13. rRNA | m. | peptidyl |
| 14. initiation | n. | amino-acid "tugboat" |
| 15. transcription | o. | incorrect reading of triplet |

## True or False?

1. _____ Single-nucleotide mutations alter just one amino acid.

2. _____ A degenerate code (in biological terms) means only one codon for one amino acid.

3. _____ Only one kind of protein is made in the nucleus.

4. _____ Prokaryotic cells transcribe mRNA directly from DNA.

5. _____ Peptidyl transferase is a peptide-bond catalyst.

6. _____ Protein synthesis is, or can be, a very slow process.

7. _____ There are sixty-four possible codons.

8. _____ A gene can be considered as "a unit of mutation."

9. _____ UGA is a termination codon.

10. _____ Elongation is energetically inexpensive.

11. _____ Simultaneous transcription and translation occurs in eukaryotes.

## Completion

1. The first principle of the genetic code is that it is _____ .

2. In a _____ code each amino acid will be specified by a single unique codon.

3. DNA information is "read out" during the process of _____ .

4. Molecules of tRNA have a very distinctive _____ shape.

5. Polypeptide _____ takes place as amino acids are joined, one by one, to form a lengthening chain.

6. The joining of amino acids with peptide bonds is catalyzed by the enzyme _____ .

7. The most up-to-date definition of a gene would state that genes are units of DNA that code

   for a _____ or a _____ RNA molecule.

8. Proteins are made only in the _____ of a cell.

## Short Answer

1. List three definitions of the term *gene*.

2. Why is it important that the genetic code be degenerate rather than nondegenerate?

3. Why is the rate of protein synthesis important to cells and organisms?

4. What happens in "termination"?

5. What are the basic differences in protein synthesis in prokaryotic and eukaryotic cells?

6. List three types of RNA and their basic functions.

## Multiple-Choice Review

In the following sentences, fill in any blanks. Complete each statement by circling the correct response.

1. The DNA of a gene codes for a specific RNA, which in turn codes for a:

   a. protein.
   b. polypeptide.
   c. enzyme.
   d. amino acid.
   e. RNA polymerase.

2. If there are four nucleotide bases, and it takes a set of four to code for each amino acid, the number of possible amino acid correspondences is:

   a. 4.
   b. 20.
   c. 22.
   d. 256.
   e. 500.

3. If the genetic code were nondegenerate, we would have _____ nonsense codons.

   a. twenty
   b. twenty-two
   c. forty-four
   d. sixteen
   e. none of the above

4. The smallest RNA molecules are those of:

   a. tRNA.
   b. rRNA.
   c. mRNA.
   d. prokaryotic RNA.
   e. the nucleus.

5. Simultaneous transcription and translation takes place in:

   a. prokaryotic cells.
   b. the cytoplasm.
   c. the nucleus.
   d. eukaryotic cells.
   e. none of the above

## Exercise

Supply the missing labels in the figure below. Color tRNAs red, amino acids yellow, and elongating polypeptide chains green.

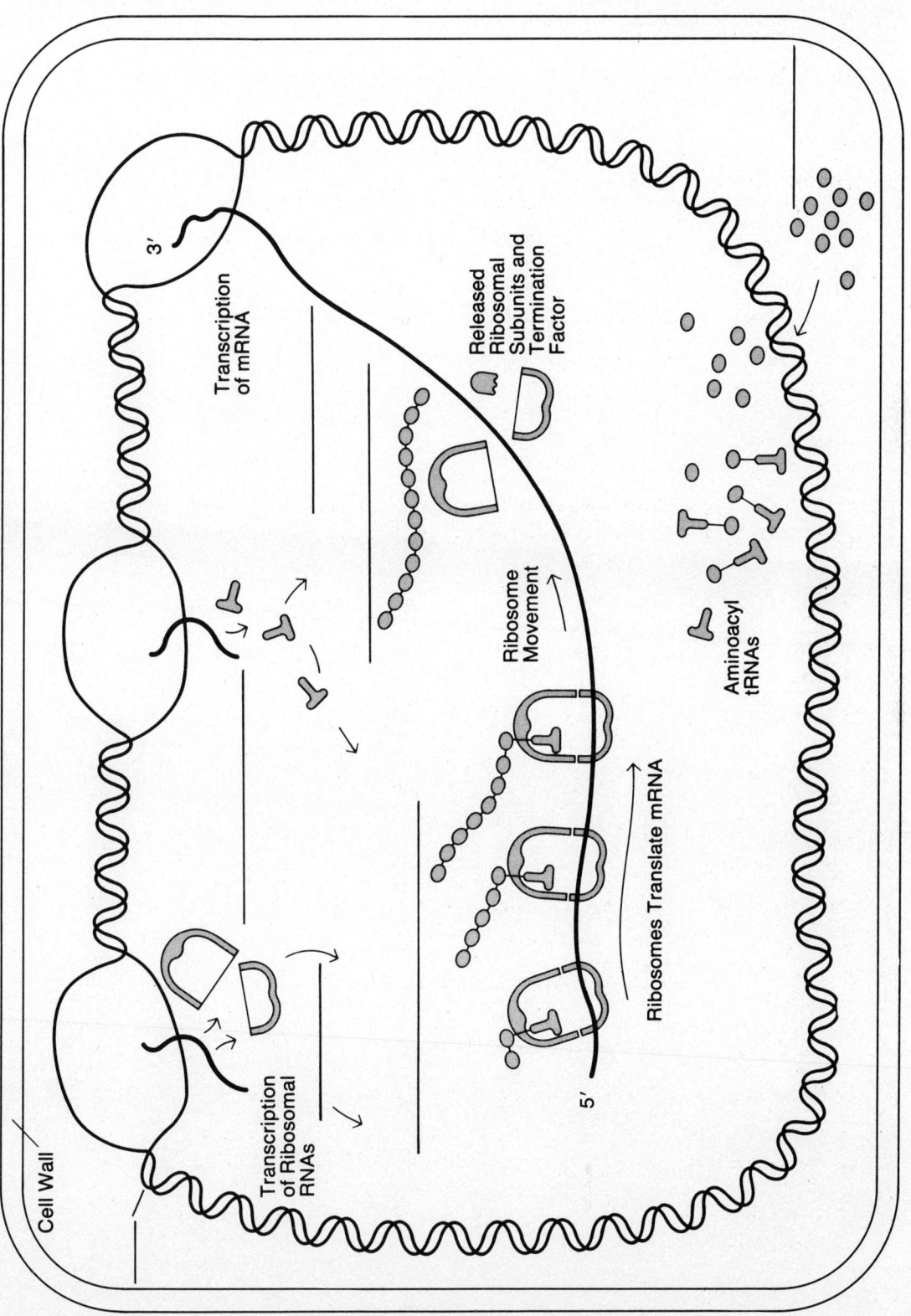

# 14
# BACTERIAL GENETICS, GENE CONTROL, AND GENETIC ENGINEERING

## CHAPTER AT A GLANCE

# CHAPTER PREVIEW

Chapter 14 is remarkable—the techniques and advances in genetics it describes would have been impossible to find in a general biology text only a few years ago. The first section outlines some of the methods by which biologists distinguish between different types of bacteria according to their nutritional requirements, ability to absorb dyes and stains, and so forth. Then the chapter surveys the transmission of genes in bacteria, and the direct and indirect processes that accomplish it: conjugation, transformation, and transduction. A subsequent short section describes how biologists may use DNA transfer processes to create a map of the genome of that favorite prokaryote, *E. coli*.

Chapter 14 then turns to the mechanisms that regulate gene expression in bacteria. You will want to focus especially on the operon model of gene regulation, which sometimes gives students trouble. The questions and problems at the end of the text chapter should help you learn this material. An interesting section on the differences between the genomes of prokaryotes and eukaryotes follows. As you'll see, different species show big—and sometimes surprising—differences in the amount of DNA they possess. The chapter's final section outlines the tools and methods of genetic engineering.

# LEARNING OBJECTIVES

When you have mastered the concepts of this chapter, you will be able to:

1.  Explain bacterial reproduction, and outline the three sexual processes found in bacteria.
2.  Describe how DNA transfer processes in bacteria have been exploited to produce gene maps.
3.  Outline the operon model of gene regulation.
4.  List the differences in gene regulation between eukaryotes and prokaryotes, and explain why eukaryotes have so much more DNA.
5.  Explain why eukaryotes have fragmented genes, and be able to name and list the functions of the various fragments.
6.  Define recombinant DNA, and outline how the process works in a simple gene transfer.
7.  Discuss the future of DNA transfer and genetic engineering, and list at least four areas in which society will feel its impact.

# CONCEPTS IN REVIEW

## Section I    Bacteria Exchange Genes Sexually and Asexually

The genetic discoveries described in Chapters 12 and 13 were made possible, in part, by research on the genetics of bacteria. A bacterium has a single circular chromosome, lacks a nuclear envelope, and does not undergo mitosis before it divides in two (fission). The single chromosome is duplicated (review Chapter 13) before the bacterial cell divides. Bacteria also have a means of exchanging genes sexually.

To study bacterial phenotypes and gene variation, geneticists focus on mutant bacteria having certain biochemical characteristics. These characteristics include resistance or sensitivity to certain antibiotics, such as streptomycin; abnormal nutritional requirements; and the inability to use certain compounds for growth (text pp. 305–306). Auxotrophs are mutants that are unable to make all the necessary growth factors from a minimal medium. Mutants that can grow on the sugar glucose but not on lactose played a major role in studies on gene control. A technique known as replica plating is used to separate normal from mutant cells.

Once mutants for different biochemical phenotypes were isolated, it was possible for researchers to study how different traits are inherited in bacteria. Pioneering work by Lederberg and Tatum has shown that bacteria exchange genetic information in several ways, including conjugation, transformation, and transduction (text p. 306).

Conjugation, discovered by Lederberg and Tatum, involves the transfer of DNA from one cell to another by direct contact. DNA is transferred from the "male" donor cell to a "female" recipient through a cytoplasmic bridge known as a sex pilus. Male cells contain F factor, a segment of DNA that codes for the pilus, and are thus designated $F^+$; female cells are $F^-$. The male F factor resides on a plasmid, a small circle of double-stranded DNA that is separate from the bacterium's chromosome (text p. 307).

During conjugation, the one strand of the F factor is transferred to the $F^-$ cell, where a complementary strand is synthesized to form a double-stranded F factor plasmid. The $F^-$ cell then becomes an $F^+$ cell. The F factor plasmid is an element that can exist freely in the cytoplasm or be integrated into the bacterial chromosome and replicate with it. Such elements are called episomes (text p. 307). The integration of the plasmid into a bacterial chromosome converts the $F^+$ cell into an Hfr cell—one that exhibits a high frequency of recombination when mixed with genetically different $F^-$ cells. The mixing of Hfr and $F^-$ cells allows geneticists to map the genes on the bacterial chromosome in an interrupted mating experiment (text p. 309).

In the process of transformation, DNA that has been released from one cell into the surrounding medium is taken up by another cell. In transduction, DNA is carried from one bacterium to another by a virus. Viruses that infect bacteria are termed *bacteriophages* (text p. 310). A final type of gene exchange in bacteria involves genetic elements called transposons, which contain insertion sequences of DNA that can insert a gene into a variety of chromosomal sites (text p. 310).

## *Section II*   Use of DNA Transfer Processes to Map Prokaryotic Genes

Knowledge of conjugation, transformation, and transduction has enabled researchers to map the genome of *E. coli*. This map has revealed that genes for enzymes involved in related tasks are frequently grouped close together. Such clustering allows precise control of gene expression.

## *Section III*   Bacterial Genes: Subject to Precise Regulatory Control

In prokaryotes, gene expression is controlled by regulatory molecules that exert their influence over batteries of genes. Studies of the utilization of lactose by *E. coli*, carried out by Jacob and Monod, revealed the presence of a lactose operon—a cluster of genes that function together to generate the enzymes for breaking down lactose. The operon consists of an operator locus, a regulator gene, and the protein-coding genes they control. According to the operon model of gene regulation, four elements interact: (1) a regulator gene that codes for a repressor molecule that can act as an "off" signal; (2) an operator gene that receives the "off" signal; (3) a set of protein-coding genes, all of which are transcribed onto the same mRNA molecule; and (4) a promoter site that partially overlaps the operator (text p. 313). Regulation takes place at the level of mRNA synthesis, and is an example of transcriptional-level regulation.

Catabolite repression refers to a genetic mechanism whereby a bacterial cell preferentially uses glucose in lieu of other sugars to fuel its metabolism (text pp. 315–316). It is an example of positive control, whereas transcriptional regulation, wherein a repressor acts by binding an operator, is an example of negative control (text p. 316).

Enzymes, and the gene activity that generates them, can be induced; lactose is an example of such an inducible enzyme. Other enzymes are repressible: their synthesis is repressed when their pathway end product is present. An example of a repressible system is the *trp* operon. As shown in Figure 14-12, when the amino acid tryptophan is present, transcription of the *trp* operon sets in motion a series of steps that inhibit the synthesis of tryptophan by the bacterium (text p. 316).

## Section IV    Gene Regulation in Eukaryotes

Several features distinguish eukaryotic from prokaryotic DNA. Eukaryotes have repetitive DNA sequences; also, multiple gene families are present in eukaryotic DNA. The families are genes related in both structure and function. Lastly, most eukaryotic genes consist of protein-coding regions called exons interspersed with introns—regions that do not pair with mRNA nucleotides and thus do not code for proteins (text pp. 318–320).

Both introns and exons are transcribed into a primary RNA transcript; introns are then cleaved enzymatically, and exon-encoded RNA sequences are spliced together to form an RNA molecule, which, after further processing, passes to the cytoplasm (text p. 321). This extensive processing occurs only in eukaryotes. Unlike the much simpler situation in prokaryotes, eukaryotic gene expression can thus be regulated at the levels of processing of gene messages and transport of information via mRNA to the cytoplasm.

## Section V    Recombinant DNA Technology

Recombinant DNA technology encompasses laboratory techniques for moving genes from one chromosome to another and for creating new genomes and genetically altered organisms. Genes are removed from their locations on chromosomes by restriction endonucleases. DNA ligases can rejoin the complementary cohesive ends of DNA fragments (text p. 322). Molecular vectors, such as plasmids, are pieces of DNA that can carry a foreign gene into a host cell, where such genes may be replicated along with the cell's own DNA. DNA that results from the enzymatic splicing of a bacterial plasmid and host DNA is termed a recombinant DNA molecule (text p. 323). In the technique of DNA–RNA hybridization, unzipped DNA strands are mixed with radioactively labeled RNA for a particular gene. Complementary sequences of the DNA and RNA then pair, effectively isolating the gene (DNA) of interest (text p. 324).

## Section VI    Genetic Engineering: Promises and Prospects

Recombinant DNA technology can be used to benefit human society in several ways. Microbes may be genetically engineered to produce large amounts of desirable proteins, such as insulin. Genes for desirable traits, such as drought or disease tolerance, may be inserted into species such as food plants. Finally, gene therapy may allow cures of genetic diseases (text pp. 326–328).

## KEY TERMS

auxotroph    *text page 305*

catabolite repression    *314*

clone    *323*

conjugation    *307*

DNA ligase    *322*

DNA–RNA hybridization    *324*

episome    *307*

exon    *319*

F factor plasmid    *307*

gene regulation    *312*

genetic engineering    *304*

genome    *308*

Hfr cell    *307*

inducible enzyme    *312*

intron    *319*

operator    *313*

operon    *319*

plasmid    *304*

promoter site    *313*

R factor plasmid    *310*

recombinant DNA molecule    *323*

recombinant DNA technology    *304*

repressible enzyme    *316*

repressor    *313*

restriction endonuclease    *322*

transduction    *310*

transformation    *309*

# SELF-QUIZ: TESTING WHAT YOU HAVE LEARNED

## Matching Key Terms

Match each term on the left with the most appropriate description on the right.

| | | | |
|---|---|---|---|
| 1. episome | | a. | sexual bacteria |
| 2. repressor | | b. | DNA carried by virus |
| 3. Hfr cell | | c. | an operator gene and its associated structural genes |
| 4. plasmid | | d. | mutant |
| 5. transformation | | e. | prevents transcription |
| 6. operator | | f. | small circle of DNA |
| 7. genome | | g. | protein-coding parts of genes |
| 8. clone | | h. | parts of genes that are not expressed |
| 9. operon | | i. | DNA picked up from medium |
| 10. F factor plasmid | | j. | receives the "turn off" signal |
| 11. exon | | k. | identical |
| 12. transduction | | l. | shows a high frequency of recombination |
| 13. intron | | m. | replicates independently of chromosome |
| 14. conjugation | | n. | full genetic complement |
| 15. auxotroph | | o. | fertility |

## True or False?

1. _____ Mutant bacterial cells are separated by replica plating.

2. _____ Plasmids contain a bacterial chromosome.

3. _____ Transformation is usually a result of cell death.

4. _____ The operon model of gene regulation resulted from the work of Jacob and Monod.

5. _____ Multiple gene families exist only in eukaryotes.

6. _____ Introns are the protein-coding parts of eukaryotic genes.

7. _____ DNA ligases are the molecular scissors of the genetic engineer.

8. _____ "Probes" are molecules that recognize certain genes.

9. _____ Biochemical warfare is a possible use of DNA technology.

10. _____ Gene changes at the germ line are currently illegal.

11. _____ Introns are gene segments that are always expressed.

## Completion

1. The passage of genes from one generation to the next may be loosely termed _____.

2. Mutant bacteria that lack the ability to use a minimal medium are called _____.

3. The process of _____ is a type of bacterial gene exchange discovered by Lederberg and Tatum.

4. Hfr cells are simply cells that exhibit a high frequency of _____.

5. Short segments of DNA that can insert a gene into a number of different sites on a chromosome are called _____.

6. Jacob and Monod developed the _____ model of gene regulation.

7. Eukaryotic genes that are related in both structure and function comprise _____ gene _____.

8. The segments of eukaryotic genes that code for proteins are the _____.

## Short Answer

1. What is recombinant DNA technology?

2. List three characteristics that can be used to define a bacterial phenotype.

3. Why are bacteria so useful in DNA technology?

4. List and describe the three kinds of bacterial genetic ("sexual") transfer.

5. Why do eukaryotic organisms have extra DNA?

6. List the basic steps in a gene-transfer experiment.

## Multiple-Choice Review

In the following sentences, fill in any blanks. Complete each statement by circling the correct response.

1. Molecules that are used to specifically recognize desired genes are called:

   a. recombinant DNA molecules.
   b. DNA ligases.
   c. probes.
   d. plasmids.
   e. transposons.

2. In addition to bacteria, other organisms may prove useful hosts for genetic engineering plasmids. These may include:

   a. agricultural crops.
   b. yeast cells.
   c. giant mice.
   d. fruit flies.
   e. *Paramecium*.

3. In the production of lactose, the _____ gene codes for a repressor substance.

   a. i
   b. z
   c. y
   d. o
   e. p

4. The F factor is found:

   a. in the cytoplasm.
   b. in the nucleus.
   c. on the cell wall.
   d. on the plasmid.
   e. none of the above

5. The process of transferring an entire strand of a bacterial chromosome from an Hfr cell to an F⁻ cell takes:

   a. about 30 minutes.
   b. an hour.
   c. about 90 minutes.
   d. several hours.
   e. as long as six months.

## Exercise

From top to bottom in the figure below, name the four steps that yield mature messenger RNA from an ovalbumin gene.

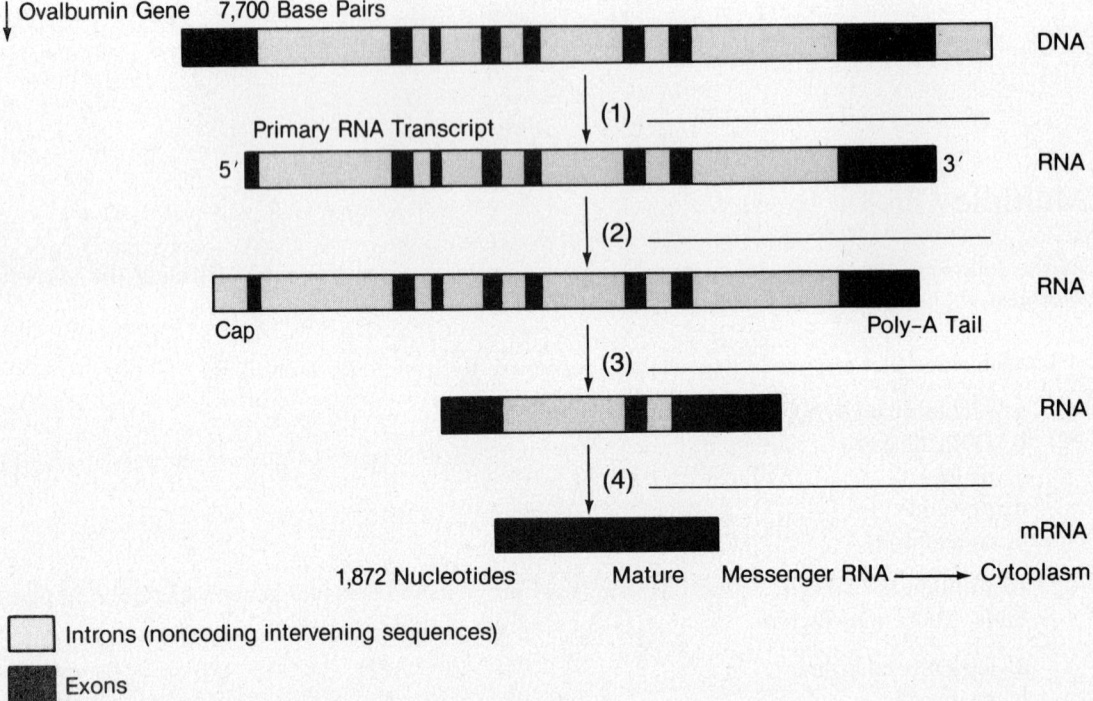

Introns (noncoding intervening sequences)

Exons

# 15
# HUMAN GENETICS

---

## CHAPTER AT A GLANCE

---

## CHAPTER PREVIEW

Julian Huxley, one of the world's most noted biologists, once remarked that "humans are the most variable species in the world." Although we now know that variability in humans is actually about average, it is still fascinating to consider ourselves as genetic examples. Chapter 15 begins by discussing the reasons why humans are difficult genetic subjects (reasons such as ethical problems in performing test crosses), and then delves into the various methods that researchers do use to explore the human genome. These include family trees, twin studies, karyotypes, biochemistry, and gene mapping.

Next the chapter surveys chromosomal abnormalities and the human genetic traits that have been the subjects of classic studies. Included are sex-linked traits such as hemophilia and color blindness, recessive traits on autosomes such as albinism, and dominant traits such as pattern baldness. The final two sections describe the approaches of genetic counseling and—where they exist—treatments for genetic diseases.

## LEARNING OBJECTIVES

When you have mastered the concepts of this chapter, you will be able to:

1. List the three main reasons why humans generally make poor genetic subjects.
2. List the methods of studying human genetics, and briefly outline the following: pedigree analysis, karyotyping, biochemical analysis, human gene mapping, and human twin studies.
3. List and define the six categories of human chromosomal abnormalities.
4. Explain what is meant by a sex-linked trait, and give examples of several human sex-linked diseases.
5. Discuss some of the modern methods of treating genetic diseases.

## CONCEPTS IN REVIEW

### Section I    Humans as Genetic Subjects

Humans are challenging subjects for genetic research, for a variety of reasons. It is not possible to do test crosses to help determine genotypes, and human matings generally do not produce a true $F_2$ generation—a cross between brother and sister. Also, a human family rarely produces enough children to represent a statistically significant sample (text p. 334).

### Section II    What Are the Methods of Human Genetics?

Human genetics is studied through several indirect methods. One is pedigree analysis, the preparation of a formal representation of a set of traits for all members of a family lineage. Another, karyotyping, involves examining the chromosomes themselves for evidence of genetic abnormality. The chromosomes are first stained during mitotic metaphase, then photographed, and then the images are arranged on the basis of size and shape. Karyotype analysis reveals the third chromosome 21 characteristic of Down's syndrome. The extra chromosome results from nondisjunction, the failure of homologous chromosomes to separate properly during meiosis (text p. 337).

The most direct tool for determining human genotypes is the analysis of specific metabolic pathways. This method is most effective in studies involving blood factors and other phenotypes in which there is a direct link between a single gene and a single gene product (text p. 337).

Methods of somatic cell genetics are used to create hybrids of body cells other than germ cells and to permit geneticists to map the human genome. This technique can be used in combination with others to detect conditions such as genetic birth defects. Somatic cells from a fetus or its parents are grown in culture and analyzed for the presence of a gene or DNA sequence that serves as a marker, denoting the site on a chromosome of another, defective gene (text p. 339). Such markers can also be located with restriction endonucleases (Chapter 14). Another powerful gene-mapping technique, called in situ hybridization, yields extremely precise maps of genes coding for many proteins.

In reverse genetics, molecular biologists identify specific genes on chromosomes, then the proteins coded by such genes, and ultimately the biological role of the protein (text p. 340).

A final method of study in human genetics is the twin study. Such work is aimed at minimizing the developmental unknowns arising from the nature–nurture problem. Another hurdle for researchers is the fact that many human traits are polygenic (controlled by several genes).

## Section III   Chromosomal Abnormalities: A Major Source of Genetic Disease

Successful embryonic development requires the normal complement of human chromosomes. The loss or gain of autosomes (non–sex chromosomes) results in the general condition known as aneuploidy. Down's syndrome, caused by an extra chromosome 21 that results from nondisjunction during meiosis, is a form of aneuploidy (text p. 342).

Other abnormalities arise from the loss or gain of sex chromosomes. These include Klinefelter's syndrome (XXY males); Turner's syndrome (XO females); and X chromosome inactivation, with its resulting mosaicism (text p. 344).

Partial changes in chromosomes result from the processes of chromosome translocation, deletion, and duplication. In translocation, part of a chromosome is moved to the end of one arm of another chromosome (text p. 344). Parts of chromosomes may also be deleted entirely; if the loss is too significant, death is the usual result. Other deletions are thought to be responsible for certain cancers. Duplications in human chromosomes occur, but have thus far been difficult to study.

## Section IV   Human Genetic Traits: A Survey of the "Classics"

A number of human illnesses and abnormal characteristics are linked to the sex chromosomes. One sex-linked trait, associated with the maternal X chromosome, is color blindness. Another, also associated with the maternal X, is hemophilia (text pp. 346–347).

Most abnormal human traits are specified by recessive alleles on autosomal chromosomes. A number of defects, including phenylketonuria (PKU), arise from a mutant genotype that results in abnormal metabolism of the amino acid phenylalanine (text p. 348). Other mutations result in various anemias, including sickle cell anemia and the group of diseases classified under the heading of thalassemia.

Certain mutations in humans are dominant rather than recessive. Pattern baldness occurs in both homozygous and heterozygous males. Because it does not cause baldness in heterozygous females, the condition is an example of sex-influenced inheritance. The dominant mutant that causes extra digits (polydactyly) shows variable expressivity—that is, there are differences in the way the gene is expressed among individuals with identical genes (text p. 351). It also shows incomplete penetrance: the dominant allele is present but not expressed at all in certain individuals. Huntington's disease is caused by a lethal mutant allele that does not die out because its effects are manifested after reproductive age.

## Sections V and VI   Treating Genetic Diseases and Genetic Counseling

Efforts to treat genetic diseases currently focus on altering the phenotype of affected individuals. Such alterations may include eyeglasses, diet therapies, and surgical procedures, among others. However, mutant alleles are still present in afflicted individuals and may be passed on to future generations. Genetic counseling generally involves screening for adverse genotypes and working with prospective parents of children that might be born with the gene defect. Many defects can be detected prenatally by way of amniocentesis and chorionic villi sampling (text p. 354).

## KEY TERMS

albinism    *text page 349*

amniocentesis    *354*

aneuploidy    *342*

Barr body    *343*

chorionic villi sampling    *354*

Down's syndrome    *336*

genetic counseling    *354*

hemophilia    *346*

incomplete penetrance    *351*

karyotyping    *336*

Klinefelter's syndrome    *343*

mosaicism    *344*

nondisjunction    *337*

pedigree    *335*

phenylketonuria (PKU)    *348*

polygenic    *340*

restriction fragment polymorphisms    *339*

sex-influenced inheritance    *350*

somatic cell genetics    *338*

Tay-Sachs disease    *337*

thalassemia    *350*

Turner's syndrome    *343*

variable expressivity    *351*

# SELF-QUIZ: TESTING WHAT YOU HAVE LEARNED

## Matching Key Terms

Match each term on the left with the most appropriate description on the right.

1. Barr body
2. nondisjunction
3. Turner's syndrome
4. incomplete penetrance
5. hemophilia
6. variable expressivity
7. pedigree
8. Tay-Sachs disease
9. albinism
10. mosaicism
11. Down's syndrome
12. aneuploidy
13. thalassemia
14. polygenic
15. amniocentesis

a. absence of pigment
b. hexosaminidase
c. dominant allele not expressed
d. XO
e. hemoglobin defect
f. forty-five or forty-seven chromosomes in humans
g. extra chromosome 21
h. multiple genes for a single trait
i. different expression of same gene
j. failure to separate
k. test of fetal amniotic fluid
l. royal disease
m. inactive X chromosome
n. alternate cellular phenotypes
o. family tree

## True or False?

1. _____ Trisomy 21 results from nondisjunction.

2. _____ Tay-Sachs disease in humans is sex-linked.

3. _____ Low IQ shows strong correlation to very low birth weight.

4. _____ Human triploids have sixty-nine chromosomes.

5. _____ The probability of nondisjunction decreases with the mother's age.

6. _____ Male somatic cells possess one Barr body.

7. _____ XXY sex chromosomes produce Klinefelter's syndrome.

8. _____ Pattern baldness is sex-linked.

9. _____ Chorionic villi sampling may be performed early in pregnancy.

10. _____ The majority of human traits are polygenic.

11. _____ Sex-linked traits are easier to identify in females.

## Completion

1. _____ are simply formal representations of sets of genetic traits in all members of a family lineage.

2. The process of karyotyping involves making a pictorial display of stained chromosomes that are best seen in cells during the mitotic phase of _____.

3. Down's syndrome is also called _____.

4. Somatic cell genetics makes use of fused cells that have two diploid nuclei. These are called _____.

5. The human traits of height, weight, and IQ are each caused by interactions of many genes and are therefore _____.

6. After fertilization, approximately _____ percent of human embryos are spontaneously aborted.

7. XXY embryos will develop into individuals of the _____ sex.

8. Inactivated X chromosomes are known as _____.

## Short Answer

1. What are the major types of chromosomal abnormalities?

2. List the three reasons why genetic studies of humans are difficult.

3. What are three classic human sex-linked traits?

4. What are three classic recessive traits carried on human autosomes?

5. What is meant by the term *incomplete penetrance?* Give an example.

6. Compare the process of amniocentesis with chorionic villi sampling.

# Multiple-Choice Review

Complete the following statements by circling the correct response.

1. Errors in the metabolism of sugars can lead to:

   a. porphyria.
   b. galactosemia.
   c. Tay-Sachs disease.
   d. cystinosis.
   e. premature heart disease.

2. Down's syndrome is a classic example of:

   a. mosaic sex chromosomes.
   b. polygenic effects.
   c. delection.
   d. nondisjunction.
   e. XXY syndrome.

3. Pattern baldness in human males is a classic example of:

   a. sex-linked traits.
   b. variable expressivity.
   c. sex-influenced inheritance.
   d. polygenic effects.
   e. none of the above

4. The genetic origin of thalassemia anemias is probably:

    a. deletion.
    b. nondisjunction.
    c. translocation.
    d. point mutation.
    e. a rare disease.

5. Barr bodies in females are a result of:

    a. chromosome inactivation.
    b. translocation.
    c. deletion.
    d. duplication.
    e. none of the above

## Exercise

Supply the missing labels in the figure below. Note that this genome consists of two pairs of homologous chromosomes, one "large" pair and one "small" pair. Each of the five unlabeled arrangements depicts a chromosome abnormality.

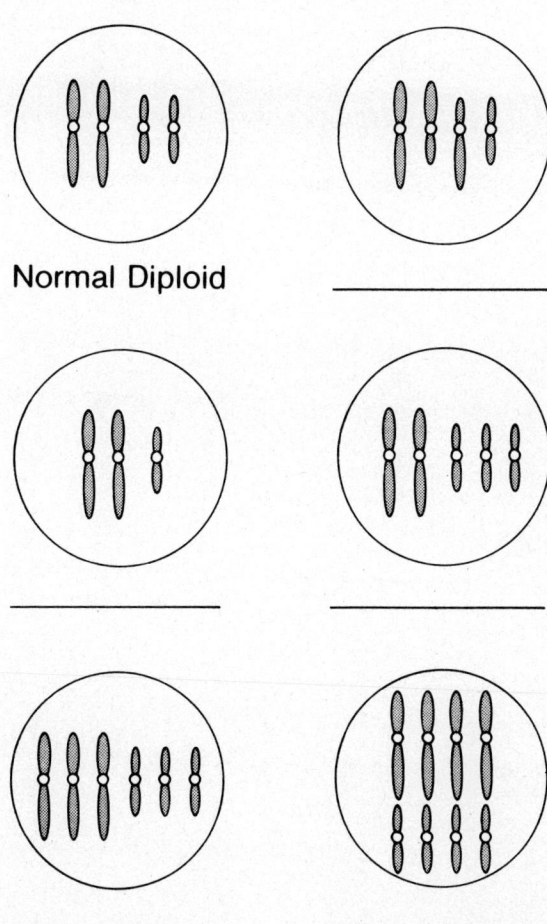

Normal Diploid

_____

# 16
# ANIMAL DEVELOPMENT

## CHAPTER AT A GLANCE

## CHAPTER PREVIEW

In this chapter there are many new anatomical terms, as well as concepts that traditionally give some biology students problems. However, once you have learned this material you will be able to draw upon your knowledge in Chapter 18 (Reproduction) and other chapters to come.

The first section presents a detailed look at the processes that generate sperm and eggs. These are concepts that you will meet again in Chapter 18. The next section describes the phenomenon of fertilization—a process that is, most likely, much more complex than you have imagined. The chapter then explores the developmental events that take place after an egg is fertilized, beginning with cleavage. Here, you will want to note carefully the various cleavage patterns and their relationships to the amount of yolk present in the egg. This material and the following section on gastrulation introduce the subject matter of classical embryology.

In studying organogenesis you will trace the remarkable events of differentiation—the process that involves the functional maturation of cells. Then, after considering the types and roles of the different coverings that surround vertebrate embryos, the chapter closes with the ultimate outcome of development: aging and death. Development is truly a lifelong process.

# LEARNING OBJECTIVES

When you have mastered the concepts of this chapter, you will be able to:

1. Trace the development of sperm from gonial cells to functioning germ cells.
2. Trace the development of an egg from an oogonium to a mature germ cell.
3. Describe what takes place during fertilization, including the sequence of changes that occurs in the egg surface to prevent multiple fertilizations.
4. Describe the patterns of cleavage, and explain how each is related to the amount of yolk present.
5. Outline the sequence of events in gastrulation.
6. Outline the main features of organogenesis from the three tissue layers of the embryo.
7. Name the various embryonic coverings and membranes, and give the function of each.
8. Characterize growth at the embryonic and adult stages of an organism's life.
9. Discuss the process of biological aging, and list the four factors thought to contribute to cell aging and death.

# CONCEPTS IN REVIEW

## Section I    Production of Sperm and Eggs

In sexually reproducing organisms, males and females produce sex cells, known as gametes. These are swimming sperm in males and ova (eggs) in females.

The process of sperm production, spermatogenesis, takes place in testes. The sperm originate in gonial cells (spermatogonia) in the walls of seminiferous tubules. Spermatocytes produced by mitosis in spermatogonia divide meiotically to generate haploid spermatids. The mature sperm has a tail, a nucleus containing haploid chromosomes, and a front end with an acrosome, the storage site for enzymes that will aid fertilization (text p. 361).

Ova, produced during oogenesis, are generated in gonial cells (oogonia) of the female's ovaries. Oocytes enter a stage of arrest in early meiosis. At a species-specific later time, there is a final ripening, ovulation, and the first meiotic division. A second meiotic division, followed by development of the embryo, takes place if the egg is fertilized.

Eggs vary greatly in size from species to species and have complex structures. Virtually all developing animal ova are surrounded by helper cells, either follicle cells or nurse cells (text pp. 363–364). Depending on the species, eggs also store varying amounts of yolk, a reservoir of nutrients produced by digestive-gland cells in the mother's body. Finally, follicle cells or cells of the maternal oviduct provide protective coatings for the egg, including albumen (egg white) and various types of outer membranes and shells (text p. 364).

Frog oocytes serve as model systems for studies of oocyte development. While maturing, they produce huge numbers of ribosomes during the process of gene amplification (text p. 365). Much mRNA may also be made and stored.

## Section II    Fertilization: Initiating Development

Fertilization unites male and female gametes and initiates development. In some species fertilization is external; in others, including most terrestrial animals, it takes place internally. The first contact of the sperm head with the egg's jelly coat triggers the acrosome reaction, in which enzymes are released to digest a hole through the egg's protective layers, and the plasma membrane of the sperm is brought into position to bind to the ovum's surface (text p. 366). After fusion of the egg and sperm plasma membranes, the haploid male nucleus with its chromosomes moves into the egg cytoplasm. Fusion also triggers the egg's final meiotic reduction divisions. When sperm and egg nuclei unite, the two sets of chromosomes mingle to create a diploid set. The fertilized egg is now a zygote.

The egg's cortical reaction serves as a barrier to the entry of more than one sperm. Initially, there is a temporary change in the egg's electrical state, and the egg cell is activated. The final stage of the reaction, the rapid elevation of the fertilization membrane, prevents further sperm penetration (text pp. 368–369).

In some species fertilization is not necessary. Instead, parthenogenesis takes place: the egg is spontaneously activated and proceeds to normal embryonic development. Normal chromosome number in parthenogenetic organisms is restored through specializations of meiosis (text p. 369).

## Section III    Cleavage: An Increase in Cell Number

Cleavage, the major developmental event immediately following fertilization, is a special form of cell division (mitosis). Cleavage produces a blastula, a sheet of cells rounded into a sphere that in most species surrounds a cavity. In the process, the single-celled zygote is divided into many small cells, and yolk, mRNA, ribosomes, and other materials are distributed to each cell in precise ways (text p. 369). The cells of the blastula, called blastomeres, also each receive a full diploid set of chromosomes.

There are different patterns of cleavage in different species, as shown in text Figure 16-12. The amount of yolk present in the egg is a major factor in determining the pattern: in species having little yolk, such as mammals, the zygote cleaves completely through, forming cells that are roughly equivalent in size. In frogs, in which the egg has somewhat more yolk, cleavage proceeds more rapidly in regions of the embryo that have less yolk. In bird eggs, the yolk is so massive that cleavage divisions are restricted to a tiny area of cytoplasm (text pp. 369–370).

In many species, the precise distribution to blastomeres of molecular determinants in the cytoplasm is crucial to proper development of different cell types in the embryo. Maternally derived developmental information is allocated to some blastomeres but not to others. In mammal and bird species, the fate of a cell is determined by its position late in the cleavage process (text p. 372).

## Section IV    Gastrulation: Rearrangement of Cells

The rearrangement of the blastula into a three-dimensional organism with inner, middle, and outer layers occurs during gastrulation. The resulting gastrula consists of an outer ectoderm, an inner endoderm, and a mesoderm layer positioned between them (text p. 373). Review text Figure 16-15.

Each layer gives rise to specific tissues during embryonic development. A variation in gastrulation, involving the movement of cells into endodermal and mesodermal positions through the thickened primitive streak, arose in reptiles and can still be seen in bird and mammalian embryos (text p. 374).

## Section V    Organogenesis: Formation of Functional Tissues and Organs

The organs and tissues of the embryo arise during organogenesis, as cells inside the embryo and on its surface become specialized. Organogenesis actually includes two closely linked processes: morphogenesis and differentiation. During morphogenesis, cells and cell populations change shape; an example is neurulation in vertebrate embryos, in which the edges of the flat neural plate fold upward and fuse, forming the beginnings of the hollow brain and spinal cord. During differentiation, cells mature so they may perform separate functions. This maturation may include taking on a function-related shape, such as the long, spindly shape of skeletal muscle cells (text pp. 374–375). Cell differentiation also results in responsiveness, the ability of a cell to be regulated within the organism through the action of hormones, neurons, and other signals (text p. 376).

## Section VI    Embryonic Coverings and Membranes

The embryos of land vertebrates are enclosed within four extraembryonic membranes that afford protection while still permitting the exchange of gases, nutrients, and other materials. These membranes, which you will study again in Chapter 18, include the yolk sac, the chorion, the amnion, and the allantois (text p. 377).

## Section VII    Growth: Increase in Size

Growth in embryos is largely due to an increase in the number of cells rather than to an increase in the size of individual cells. In many species the extent of embryonic growth is limited by the availability of food (yolk). In animals that develop entirely free of the maternal body, such as frogs and insects, the embryo does not grow; instead it gives rise to a larval stage that can feed itself (text p. 377) and later undergo metamorphosis to the adult stage. In many species the most spectacular growth phase takes place during the juvenile and adolescent phases of the life cycle (text p. 378). Actual growth generally stops once the organism reaches its typical adult size, although replacement of dead cells may continue.

A special type of growth, regeneration of lost body parts, can take place in adults of some species. Prior to regenerating, cells in stump tissue undergo dedifferentiation. They lose their functional phenotype, divide rapidly, and generate a population of cells that will regenerate the lost part (text pp. 378–379). Compensatory hypertrophy is a different, temporary growth response in which residual tissue increases in mass and cell number.

## Section VIII    Aging and Death: Both Developmental Processes

Aging is an ongoing, time-dependent developmental process in which body parts deteriorate. Theories as to its causes include the degeneration of collagen, the fibrous proteins of the connective tissues, and limits on the number of times cells can divide. Other theories are based on a decline of the immune system and on the accumulation of lipofuscins, or aging pigments (text p. 380).

## KEY TERMS

acrosome reaction  *text page 365*

allantois  377

amnion  377

blastomere  369

blastula  369

chorion  377

cleavage  369

compensatory hypertrophy  379

cortical reaction  368

dedifferentiation  378

differentiation  374

ectoderm  373

endoderm  373

fertilization  365

gastrula  372

gastrulation  372

gene amplification  365

gonial cell  361

mesoderm  373

metamorphosis  377

morphogenesis  374

neurulation  374

oogenesis  361

organogenesis  374

ovum  360

ovary  363

oviduct  364

parthenogenesis  369

primitive streak  374

regeneration  378

sperm  360

spermatogenesis  361

stem cell  378

testis  361

yolk  364

yolk sac  377

zygote  365

# SELF-QUIZ: TESTING WHAT YOU HAVE LEARNED

## Matching Key Terms

Match each term on the left with the most appropriate description on the right.

| | | | |
|---|---|---|---|
| 1. | yolk | a. | gamete, or egg |
| 2. | amnion | b. | spermatogonia |
| 3. | regeneration | c. | homologous with ovaries |
| 4. | parthenogenesis | d. | egg tube |
| 5. | cleavage | e. | prevents multiple fertilizations |
| 6. | zygote | f. | "virgin" birth |
| 7. | chorion | g. | individual blastula cells |
| 8. | testis | h. | divides a single-celled zygote into many small cells |
| 9. | cortical reaction | i. | food |
| 10. | ova | j. | fertilized egg |
| 11. | primitive streak | k. | associated with large yolks |
| 12. | blastomere | l. | cushions embryo |
| 13. | allantois | m. | trash dump |
| 14. | oviduct | n. | fuses with allantois |
| 15. | gonial cell | o. | replacement of lost parts |

## True or False?

1. _____ Sertoli cells are helper cells in sperm production.

2. _____ The sperm head contains a small number of mitochondria.

3. _____ Oogenesis may be arrested for years in some species.

4. _____ An ostrich egg can be considered a single cell.

5. _____ Chromosomes in the "lamp brush" phase produce little mRNA.

6. _____ The acrosome reaction is triggered by the egg's jelly coat.

7. _____ Prior to fertilization, the egg's electrical charge is positive.

8. _____ Parthenogenesis produces only females.

9. _____ The so-called vegetable pole contains the most yolk.

10. _____ The gut cavity is derived from an archenteron.

11. _____ *Amphioxus* eggs show a prominent primitive streak.

## Completion

1. The heads of spermatids are embedded in _____ cells.

2. The sperm acrosome is derived from the _____.

3. Another name for "egg white" is _____.

4. The outermost protective layer of the unfertilized egg is the _____.

5. The last step of the cortical reaction is the elevation of the _____.

6. After fertilization the mammalian zygote cleaves to form the _____, a simple ball of cells.

7. The change in shape of cells and cell populations is _____.

8. The neural tube forms during _____.

## Short Answer

1. List the four extraembryonic membranes and give the function of each.

2. What is compensatory hypertrophy? Give an example.

3. What are the three theoretical causes of aging?

4. Trace the development of a sperm from a diploid gonial cell to a motile sperm.

5. List the three germ layers of a gastrula.

6. What happens during the acrosome reaction?

## Multiple-Choice Review

In the following questions, finish each sentence by circling the letter of the correct response.

1. Sperm are produced and developed in the:

   a. acrosome.
   b. Sertoli cells.
   c. seminiferous tubules.
   d. vas deferens.
   e. follicle.

2. Energy to power the sperm flagellum is generated by:

   a. mitochondria.
   b. the acrosome.
   c. glucose.
   d. glycogen.
   e. none of the above

3. The largest eggs are produced by the:

   a. whale.
   b. elephant.
   c. ostrich.
   d. mouse.
   e. fruit fly polytene strains.

4. The most important factor determining whether eggs and sperm meet is:

   a. the pH of the vagina.
   b. mobility of the sperm.
   c. sperm numbers.
   d. mating behavior.
   e. the egg's time of release.

5. The zygote progresses to a three-layered stage during:

    a. amplification.
    b. gastrulation.
    c. morulation.
    d. blastulation.
    e. cleavage.

# Exercises

## Exercise 1

In the chart below, list three structures that arise from each germ layer.

| Endoderm | 1. _____ |
|          | 2. _____ |
|          | 3. _____ |
| Mesoderm | 1. _____ |
|          | 2. _____ |
|          | 3. _____ |
| Ectoderm | 1. _____ |
|          | 2. _____ |
|          | 3. _____ |

## Exercise 2

This figure shows the developmental events leading to the blastula stage of a monkey embryo. Fill in the missing labels.

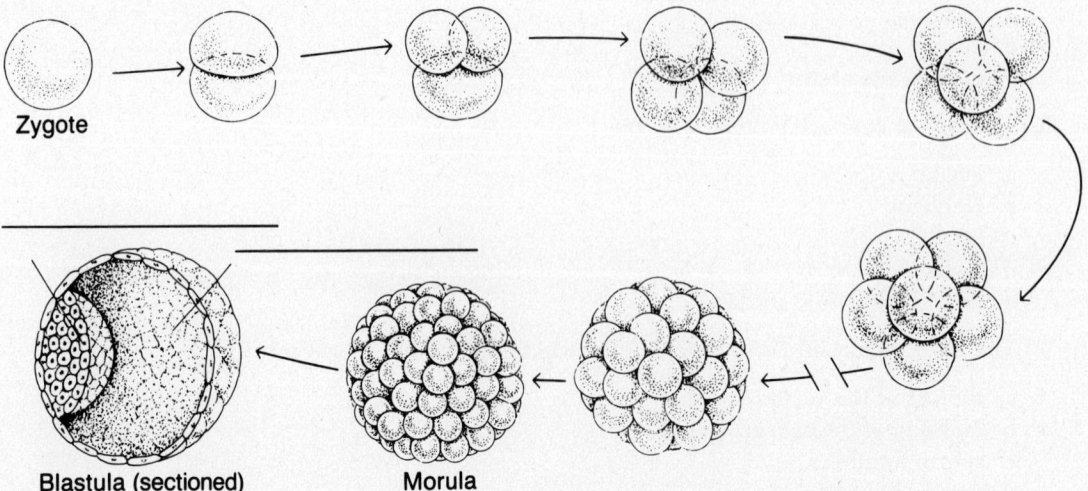

Zygote

Blastula (sectioned)　　　Morula

What is the origin of each part?

What is the future fate of each?

# 17

# DEVELOPMENTAL MECHANISMS AND DIFFERENTIATION

---

## CHAPTER AT A GLANCE

# CHAPTER PREVIEW

Chapter 17 confronts one of the most significant problems of biology: What governs the development of organisms from the zygote to maturity? Although what you will learn in this chapter is fascinating, you should realize that it is only a bare introduction to this provocative and rapidly changing subfield of biology.

The chapter begins with the concept of determination—that is, commitment to a specific type of differentiation. As you will learn, commitment is a "once and for all" process—otherwise, you might have arms, legs, or other body parts that continually change shape and function! To present the concept of larger morphological change, the chapter then traces the development of the vertebrate limb along its three axes. Here, you will come to understand why your fingers and thumbs are positioned the way they are, and not in reverse order. The next two sections introduce the role of regulatory genes—genes that control blocks of other genes that, in turn, guide development and pattern formation. The chapter ends by examining cancer, an example of what can happen when the regulators of cell growth malfunction.

# LEARNING OBJECTIVES

When you have mastered the concepts of this chapter, you will be able to:

1. Define determination, state its developmental characteristics, and explain the processes by which it may be modified.

2. Discuss the stability of determination, and explain how nuclear transplantation experiments affected our understanding of determination stability.

3. List the five classes of causes in determination, and the four characteristics of differentiated cells.

4. List the six basic mechanisms of morphogenesis, and provide examples of each.

5. Describe pattern formation, using the vertebrate limb as your example.

6. Discuss the relationship between structural and regulatory genes, and explain how regulatory genes work in pattern formation.

7. Explain how gene regulation is important in DNA processing and RNA synthesis.

8. Explain how protein synthesis is related to developmental processes.

9. Tell how cancer is thought to develop, and give the evidence for its suspected causes.

# CONCEPTS IN REVIEW

## *Section I*    Determination: Commitment to a Type of Differentiation

Before differentiation and morphogenesis take place (Chapter 16), most cell lines in the embryos of higher animals become determined for one particular type of differentiation. Determination is the final selection in cells of a single developmental pathway from among several alternatives, such as the "decision" to form muscle or liver cells. Once a subpopulation of cells is determined, the condition is usually permanent. This state is generally passed from a cell to its progeny during mitosis—a fact which shows that the determined condition is an inherited state of gene control (text pp. 384–386).

In some cases determination can be modified. Studies using cells from the imaginal disks of fruit flies showed that such cells are transdetermined, and under proper conditions can shift from one determined state to another. Using the nuclear transplantation approach, researchers have demonstrated that the genes required for many determined states and modes of differentiation are present in the nuclei of fully differentiated cells. Thus determination of a cell appears to depend on factors in the cytoplasm (text p. 387).

## Section II    Differentiation: Building Cell Phenotype

As you learned in Chapter 16, differentiation is a cellular maturation process. As the work of Spemann and Mangold showed, tissue interactions may induce differentiation (text p. 387). They also ensure proper spatial relationships among developing organs. Protein hormones such as prolactin and erythropoeitin can induce differentiation in specific tissue cells.

A differentiated cell has four specific characteristics that set it apart from undifferentiated cells. First, it makes and uses a specific set of proteins that enable it to carry out its functions. Examples include the hemoglobin of red blood cells and the actin, myosin, and troponin of muscle cells. Second, a differentiated cell is metabolically active. Third, as it matures, a differentiating cell assumes a characteristic shape that enables it to function effectively in the tissue to which it belongs. And fourth, a differentiated cell often ceases to undergo further cell division (text pp. 389–390).

## Section III    Morphogenesis: The Organization of Cells into Functional Units

The many types of morphogenesis in embryos are produced by only a few mechanisms. One of these is single-cell movements, in which individual cells migrate from one site to another and set up populations that will later form organs. Cell adhesion (Chapter 6) is a key factor in governing such movements. A second mechanism involves the interaction between cells and extracellular substances that serve as substrates to which cells can stick, or on which cells can move. A third mechanism is the movement of entire cell populations, such as the folding of epithelial sheets. A fourth is localized relative growth, in which the mitotic rates of local areas of cells increase or decrease. A fifth mechanism is localized cell death—the process that, in humans, eliminates the embryonic webbing between fingers and toes. A sixth and final mechanism of morphogenesis is the deposition of extracellular matrix, which contributes to the shapes of bones, cartilage, corneas, and other structures and regions (text pp. 391–394).

## Section IV    Pattern Formation: The Vertebrate Limb

Pattern formation is the gradual process through which an organism's overall body plan emerges. It clearly shows the importance of control factors, including the positional information resulting from cleavage divisions that was described in Chapter 16.

The development of the body's three primary axes—anterior–posterior, dorsal–ventral, and right–left—is another feature of pattern formation. British biologist Louis Wolpert has suggested that the parts of a body limb arise from a region called the progress zone—an area of specialized mesodermal cells located just beneath the tip of the elongating limb bud. As the zone is moved outward during limb growth, successive groups of cells behind it are assigned positional values such as upper arm, forearm, and so on. According to this model, the zone of polarizing activity (ZPA) establishes the anterior/posterior axis (thumb/little finger) (text pp. 395–397).

## Section V    Regulatory Genes at Work During Pattern Formation

Pattern formation is based on the activity of regulatory genes, which in turn control structural genes. Studies of regulatory genes in *Drosophila* have shown that a single mutation can change the order or number of appendages, and have other effects. Regulatory genes exert their influence on body pattern and form early in development (text pp. 397–398).

## Section VI    Gene Regulation in Development

Biologists have begun to see the outlines of the mechanisms through which genes are controlled during early development. Two such mechanisms—gene amplification and gene rearrangement—operate at the level of DNA processing. Just as ribosomes can be manufactured in large numbers in oocytes (Chapter 16), recent studies show that in at least some cases structural genes can be amplified. These "extra" DNA templates enable a larger amount of mRNA to be manufactured, and hence more protein generated. Gene rearrangement occurs in the embryonic immune system, and permits a huge number of antibody proteins to be manufactured from a limited set of genes (text p. 399).

A second level of regulation occurs during RNA synthesis and processing. Differential gene activity at this level—the functioning of some genes and not others—is thought to account for most types of cell differentiation. Such activity can be observed in the puffing patterns of polytene chromosomes (Chapter 10). Thus, because some genes in a cell are active while others are not, a specific set of mRNAs is manufactured to support a given type of differentiation. Nonhistone chromosomal (NHC) proteins may also participate in regulating specific genes during determination and differentiation (text pp. 400–401).

Gene regulation may also take place via the processing of RNA transcripts, as sequences coded by introns are deleted (Chapter 14), and in the storage and activation of mRNAs (text p. 401).

Finally, gene expression may be regulated at the level of protein synthesis. For example, a protein may be manufactured only if a cofactor or necessary precursor is present in the cytoplasm; newly manufactured proteins may be modified, and degradation of a specific protein may be speeded or slowed (text p. 402).

## Section VII    Cancer: Normal Cells Running Amok

Cancer is a large group of diseases characterized by a breakdown in the regulatory mechanism that prevents excess mitotic activity. Uncontrolled mitosis and long-lived cells result in enlarged tumors, leukemia, or other malignant conditions. Cancerous cells may also metastasize, spreading to distant sites in the body.

There are many types of carcinogens, including radiation and numerous chemicals. Cancer-causing genes are known as oncogenes; proto-oncogenes are existing genes that cause cancers when altered in some way by mutation or chromosome translocation. Some families may have genetic tendencies toward cancer; oncogenes may also be carried by viruses. In addition, if viral genetic material becomes inserted near a cell's proto-oncogene, the result may be transformation of the cell into a cancerous state (text pp. 403–404). Recent research has also shown that cancer-causing genes in viruses can originate from proto-oncogenes in animal cells.

Other recent work suggests that the uncontrolled growth of a cancer-transformed cell arises when the products of an oncogene act as an enzyme to catalyze the addition of phosphate to cell proteins (text p. 405).

## KEY TERMS

cell death    *text page 393*        differential gene                    differentiation    *387*
                                     activity    *399*
determination    *384*                                                    extracellular matrix    *394*

gene rearrangement   399          nuclear transplantation   387          tissue interaction   387

imaginal disk   386               oncogene   403                        transdetermination   387

leukemia   403                    pattern formation   395               tumor   403

lymphoma   403                    progress zone   395                   tyrosine phosphokinase   405

metastasis   403                  protein degradation   402             zone of polarizing activity
                                                                          395
nonhistone chromosomal            proto-oncogene   403
   (NHC) protein   400

---

# SELF-QUIZ: TESTING WHAT YOU HAVE LEARNED

## Matching Key Terms

Match each term on the left with the most appropriate description on the right.

| | | | |
|---|---|---|---|
| 1. | tumor | a. | found in fruit fly larvae |
| 2. | ZPA | b. | a gradual emergence |
| 3. | transdetermination | c. | determines anterior–posterior axis |
| 4. | differentiation | d. | operates in developing immune system |
| 5. | protein degradation | e. | chance destruction |
| 6. | gene rearrangement | f. | circulating cancer |
| 7. | metastasize | g. | solid mass |
| 8. | proto-oncogene | h. | cancer-causing gene |
| 9. | pattern formation | i. | determines proximal–distal axis |
| 10. | imaginal disk | j. | protein implicated in gene regulation |
| 11. | determination | k. | may cause cancers if mutated |
| 12. | progress zone | l. | spread of cancer |
| 13. | lymphoma | m. | maturation |
| 14. | oncogene | n. | change of determination state |
| 15. | NHC | o. | selection of single developmental pathway |

## True or False?

1. _____ Stability of determination depends on the nucleus.

2. _____ Differentiated cells undergo rapid division.

3. _____ Division rate is closely linked to a cell's life span.

4. _____ Posterior may be simply defined as near the ZPA.

5. _____ Structural genes control sets of regulatory genes.

6. _____ In most cases, regulatory genes act fairly late in development.

7. _____ Differential gene activity is observable in polytene chromosomes.

8. _____ Sarcomas are connective tissue cancers.

9. _____ The term *metastasis* refers to a cancer confined to the middle of the body.

10. _____ Carcinogens may include certain genes.

11. _____ Carcinogens are less critical cancer types than melanomas.

## Completion

1. Hans Spemann and Hilde Mangold discovered the mechanism of induction, which we now call _____.

2. All differentiated cells make and use specific sets of _____.

3. The life span of cells is linked to their _____ rate.

4. One of the most common materials used in extracellular matrix is _____.

5. _____ genes control structural genes by switching them on and off.

6. The word *cancer* refers to the large, red _____ that develop on large tumors.

7. Circulating cancers are called _____ or _____.

8. The determined state of a cell can be changed by exposure of the determined cell's nucleus to the _____ of a different cell type.

## Short Answer

1. What are the four characteristics that define a differentiated cell?

2. List four processes of morphogenesis and give a specific example of each.

3. What is meant by differential gene activity? Give an example.

4. What are the two basic types of cancerous tumors, and what tissue types do they invade?

5. List and define three ways by which protein synthesis may be altered.

6. What is meant by gene rearrangement?

## Multiple-Choice Review

Complete the following statements by circling the correct response.

1. Gene rearrangement operates:

    a. during synthesis of RNA.
    b. only with mRNA.
    c. in the nucleus.
    d. in the immune system.
    e. at the level of the gonial cell.

2. Briggs and King studied determination in/through:

    a. imaginal disks.
    b. the immune system.
    c. tissue interaction.
    d. nuclear transplantation.
    e. none of the above

3. The gradual emergence of a body plan is termed:

    a. morphogenesis.
    b. traction morphogenesis.
    c. pattern formation.
    d. progress.
    e. axial patterning.

4. The anterior–posterior axis is controlled by:

    a. the ZPA.
    b. the progress zone.
    c. the ectoderm.
    d. the first forelimb axis.
    e. none of the above

5. The shift from one determined state to another is called:

    a. transplantation.
    b. transdetermination.
    c. interaction.
    d. differentiation.
    e. pattern formation.

## Exercise

In the chart below, fill in the name of the embryonic tissue layer that gives rise to each structure listed at the left.

| Structure | Origin |
| --- | --- |
| Liver | |
| Adrenal cortex | |
| Lungs | |
| Bones | |
| Blood | |
| Sweat glands | |
| Pharynx | |
| Thyroid gland | |
| Cartilage | |
| Spinal cord | |
| Brain | |
| Heart | |
| Skin (epidermis) | |
| Urinary bladder | |
| Gonads | |
| Gut | |

# 18
# HUMAN REPRODUCTION AND DEVELOPMENT

---

## CHAPTER AT A GLANCE

---

## CHAPTER PREVIEW

Although you may think you already know quite a bit about the subject matter of Chapter 18, most students discover that much of the biology of human reproduction is new to them. The chapter begins with a review of the anatomy and reproductive strategies of nonhuman species, including reptiles and early mammals. As you read, note the differences in the role of estrus between humans and their primate relatives.

The next two sections detail the male and female reproductive systems, including basic anatomy, gamete formation, the role of hormones, and the stages of sexual response. These topics are followed by a discussion of the origins of sex differences and the roles of sex chromosomes and hormones during development. The text then focuses on human embryology as it traces the events of pregnancy, prenatal development, and birth. A boxed essay presents the all-important topic of birth control. Finally, you will consider the phenomenon of multiple births and the typical mammalian trait of milk production.

# LEARNING OBJECTIVES

When you have mastered the concepts of this chapter, you will be able to:

1. Outline the vertebrate reproductive strategy, from reptiles to mammals.

2. Describe the role of estrus among the primates, and explain how humans depart from the typical primate strategy.

3. Describe the reproductive anatomy of the human male, and detail sperm production and transport.

4. List the Masters and Johnson sexual response phases, and describe what takes place in each.

5. Describe the reproductive anatomy of the human female, and detail production and transport of the egg.

6. List the hormones important for sexual development in males and females, and discuss the functioning of each.

7. Describe the secondary sex characteristics of males and females, and explain the mechanisms that ultimately determine sex in humans.

8. Explain the main events of each trimester of human pregnancy.

9. Describe how the placenta supports the fetus, and explain its role as a hormone producer.

10. List the sequence of events during birth.

11. Discuss the various methods of birth control, including which methods are most effective and why.

12. Explain the causes of multiple births.

13. Explain the relationship between milk production and hormonal control.

# CONCEPTS IN REVIEW

## Section I    Reproduction in the Vertebrates: Anatomy and Strategy

Much of the anatomy and physiology that ensures successful human reproduction originated and was shaped in our reptilian and early mammalian ancestors. Basic features include the four membranes that surround and protect the developing embryo, the ability to retain the developing embryo within the female's body for long periods, and the nursing of young from mammary glands. Vertebrate evolution also gave rise to estrus, a period of sexual receptivity in females that is seen in most mammalian species, though not as distinctly in modern humans (text p. 409).

## Section II   The Male Reproductive System

For background on this section, you may want to reread the first section of Chapter 16, on the development of sperm. The male reproductive system consists of several internal and external structures that are needed to produce sperm and the fluids necessary to carry and protect them, and to discharge the sperm within the female reproductive tract. Sperm are produced in the seminiferous tubules of the testes. Mature sperm are moved from the seminiferous tubules into the tubular epididymis, where they are stored before being released into the vas deferens. Sperm become motile in the vas deferens, where secretions generated in the prostate gland and seminal vesicles mix with the sperm to form an alkaline semen. During sexual activity sperm and semen are ejaculated via the urethra (text pp. 412–413).

Male and female humans experience a four-stage sexual response, including the excitement phase, plateau phase, orgasmic phase, and resolution. In males, erection occurs during the excitement phase, and ejaculation takes place at the end of the orgasmic phase (text p. 413). Proper sexual functioning relies on hormones (gonadotropins) secreted in the brain and pituitary gland. In a series of hormonal events, the release of hormones produced in the hypothalamus triggers the release of luteinizing hormone (LH) and follicle-stimulating hormone (FSH). LH and FSH govern sexual maturation, sperm production, and much male sexual behavior.

Testosterone, produced in the testes in response to LH, is one of the male sex hormones known as androgens. It causes body changes including the development of male secondary sex characteristics. In embryos and newborns it also exerts brief, organizational effects on the development of male sex characteristics; by contrast, its longer term activational effects—maintenance of secondary sex characteristics and masculine sexual behaviors—are exerted during puberty and adulthood (text pp. 413–414).

## Section III   The Female Reproductive System

As with the previous section, to refresh your memory of background details, you should reread the material on the development of eggs in Chapter 16.

Human eggs are produced in the ovaries. Egg release occurs at ovulation in response to a burst of LH; after the egg is expelled, remnants of the follicle collapse and form a corpus luteum. The ovulated egg moves through the Fallopian tube (oviduct) into the muscular uterus. If the egg is not fertilized, the uterine lining, the highly vascularized endometrium, is shed during menstruation (text pp. 414–415).

The female external genitals are known collectively as the vulva. They consist of the vagina, the clitoris, and the labia minor and majora. The upper end of the vagina, which is also the base of the uterus, is the cervix.

Females have the same sexual response phases as males. As in males, heightened physiological responses include increases in heart rate, breathing rate, and blood pressure during the excitement phase. At orgasm, the release of oxytocin triggers contractions of the uterus that help propel sperm toward the Fallopian tubes, where they may encounter eggs (text p. 416).

Reproductive cycles and sexual functioning in females are coordinated by a variety of hormones. The menstrual cycle, monthly preparation of the uterus to receive an embryo, begins at puberty with menarche and ends years later at menopause, when the ovaries cease to function. During a normal cycle, the gonadotropin FSH stimulates egg maturation, and the female sex hormone (gynogen) estrogen causes thickening of the endometrium. Roughly halfway through the cycle a burst of LH from the pituitary triggers ovulation. The corpus luteum then secretes the gynogens estrogen and progesterone, which inhibits maturation of additional follicles in the ovary and stimulates increased development of blood vessels in the uterus. If the ovulated egg is not fertilized, it dies within a few days and the corpus luteum regresses. Progesterone levels drop, and the endometrium is shed. The decline in progesterone results in renewed secretion of FSH and LH, and the cycle begins once more (text p. 416).

## Section IV    Origins of Sex Differences: Femaleness and Maleness

Male and female embryos have identical sexual structures during early development. In males this indifferent stage ends at about five weeks' gestation, when genes on the Y chromosome cause the primitive gonads to develop into testes. Male embryonic Wolffian ducts develop into the epididymis and vas deferens, while the Mullerian ducts degenerate. In females the early gonads develop into ovaries, the Wolffian ducts degenerate, and the Mullerian ducts develop into Fallopian tubes, uterus, and vagina. You can get a clear view of these events in Figure 18-7 (text p. 418).

Sexual development in embryos is controlled by hormones. Specifically, the "neutral" female body develops in the absence of hormones, while maleness depends on the presence, at critical times, of testosterone, dihydrotestosterone, and Mullerian inhibiting substance (MIS) (text p. 418). Errors in hormonal control result in intersexes, such as hermaphrodites and pseudohermaphrodites (text p. 419).

## Section V    Pregnancy and Prenatal Development

By about the fourth day after fertilization the mammalian embryo consists of an outer trophoblast and an inner cell mass. At implantation, at about eleven to twelve days, the embryo becomes secured to the wall of the uterus. The trophoblast cells then secrete human chorionic gonadotropin (HCG), which stimulates the corpus luteum to continue secreting progesterone and estrogen, which in turn inhibit menstruation and loss of the embryo (text p. 420). During the first trimester of pregnancy embryonic organs form; after that time the embryo is referred to as a fetus.

The fetus grows spectacularly during the second and third trimesters. This growth is supported by the feeding efficiency of the placenta, through which nutrients and gases are exchanged and fetal wastes are excreted. The placenta also serves as a barrier between mother and fetus, and secretes hormones necessary to the well-being of the fetus (text pp. 422–423).

The placenta is connected to the abdomen of the fetus by the umbilical cord, which is all that human embryos retain of the fourth embryonic membrane, the allantois described in Chapter 16. Other embryonic membranes are the amnion, a protective, fluid-filled sac that surrounds the embryo; the chorion, which grows thousands of chorionic villi that provide a huge surface area for gas, nutrient, and waste exchange; and the yolk sac, an early site of development of red blood cells and primitive germ cells (text p. 422).

## Sections VI, VII, and VIII    Birth, Milk Production, and Lactation

Birth occurs about 266 days after fertilization. Along with other prebirth changes in both mother and infant, the hormone relaxin, secreted by ovaries and placenta, acts to loosen the mother's pelvic bones so that the baby can safely pass through. Changing hormone levels are also thought to initiate the birth process. In particular, adrenocorticotropic hormone (ACTH) produced by the fetal pituitary stimulates the secretion of steroids, which in turn trigger labor in the mother (text p. 424). After the newborn emerges, uterine contractions continue and expel the placenta and attached membranes, collectively called the afterbirth.

Mammalian females lactate, and they nurse their young via milk-producing mammary glands. Before the pregnancy comes to term, the glands begin to secrete colostrum, a fluid rich in maternal antibodies that may be the first food of the newborn. A few days after birth, true milk is synthesized in response to the infant's suckling and the pituitary hormone, prolactin. Suckling also triggers the release of oxytocin, which causes contractions in the uterine muscles just as during labor and delivery. The human breast will continue to produce milk as long as nursing continues (text pp. 427–428).

## KEY TERMS

activational effect   *text page* 414

adrenocorticotropic hormone (ACTH)   424

afterbirth   424

androgen   413

cervix   415

chorionic villus   422

colostrum   427

corpus luteum   414

endometrium   415

estrogen   416

estrus   409

excitement phase   413

Fallopian tube   415

fetus   419

follicle-stimulating hormone (FSH)   413

gynogen   416

hermaphrodite   419

human chorionic gonadotropin (HCG)   419

implantation   419

intersex   419

lactation   426

luteinizing hormone (LH)   413

mammary gland   426

menarche   416

menopause   417

menstrual cycle   416

menstruation   415

Mullerian duct   418

Mullerian inhibiting substance (MIS)   418

organizational effect   414

orgasmic phase   413

ovarian cycle   416

ovulation   414

oxytocin   415

penis   413

placenta   420

plateau phase   413

progesterone   416

prolactin   427

prostaglandin   413

pseudohermaphrodite   419

relaxin   424

resolution   413

secondary sex characteristics 413

semen   413

testosterone   413

trimester   419

trophoblast   419

umbilical cord   423

uterus   415

vagina   415

vulva   416

Wolffian duct   418

# SELF-QUIZ: TESTING WHAT YOU HAVE LEARNED

## Matching Key Terms

Match each term on the left with the most appropriate description on the right.

| | | |
|---|---|---|
| 1. estrus | a. | ninety days |
| 2. relaxin | b. | male erection |
| 3. excitement phase | c. | milk production |
| 4. plateau phase | d. | upper end of vagina |
| 5. trophoblast | e. | mosaic male/female characteristics |
| 6. intersex | f. | cessation of ovulation |
| 7. lactation | g. | regresses Mullerian system in males |
| 8. trimester | h. | stimulates synthesis of milk |
| 9. MIS | i. | attachment of blastocyst |
| 10. cervix | j. | major androgen |
| 11. menopause | k. | period of sexual receptivity |
| 12. prolactin | l. | loosens pelvic bones |
| 13. testosterone | m. | increase in heart rate and respiration |
| 14. implantation | n. | first food |
| 15. colostrum | o. | outer layer of blastocyst |

## True or False?

1. _____ Sealed eggs originated with the reptiles.
2. _____ Most primate females mate only in estrus.
3. _____ A vasectomy removes the epididymis.
4. _____ The resolution phase is similar in males and females.
5. _____ Fallopian tubes are connected directly to the ovaries.
6. _____ In ectopic pregnancy, an egg develops outside the uterus.
7. _____ The female hymen can be ruptured only by sexual intercourse.
8. _____ The Wolffian duct gives rise to the vas deferens.
9. _____ The chorion is commonly called the "bag of waters."
10. _____ Maternal and fetal blood mix in the placenta.
11. _____ Milk production stops with the secretion of oxytocin.

## Completion

1. The bulk of the penis is composed of two types of spongelike masses of tissue: the corpora _____ and the corpus _____.
2. After the egg is released, the remnants of the follicle form the _____.
3. The _____ glands are found near the vaginal opening.
4. In early development, or the _____ stage, the gonads and external genitals of all embryos are identical.
5. When ovaries develop from the early gonad, the _____ develops into Fallopian tubes, uterus, and vagina.
6. When testes develop from the early gonad, the _____ form the epididymis and vas deferens.
7. By definition, pregnancy starts with _____.
8. The placenta and its attached membranes are expelled in the _____.

## Short Answer

1. What are the three most effective means of birth control? The three least effective?

2. What are the two mechanisms that produce multiple births?

3. What is the role of colostrum in transfer of antibodies in humans?

4. What event actually triggers the uterine contractions of birth?

5. What are the three stages of human pregnancy, and what fetal changes take place in each?

6. What is the difference between a "true" and a "pseudo-" hermaphrodite? How is each produced?

## Multiple-Choice Review

In the following sentences, fill in any blanks. Complete each statement by circling the correct response.

1. The outer layer of cells in a mammalian blastocyst is the:

    a. trophoblast.
    b. mesoderm.
    c. ectoderm.
    d. endoderm.
    e. blastoderm.

2. The female sexual response parallels that of males except for:

    a. the excitement phase.
    b. the plateau phase.
    c. the orgasm phase.
    d. the resolution phase.
    e. no differences

3. Normal fertilization in humans takes place in:

    a. the upper Fallopian tube.
    b. the uterus.
    c. the lower vagina.
    d. the upper vagina.
    e. the cervix.

4. Sperm traveling in the female reproductive system are helped by:

    a. viscous cervical mucus.
    b. a noncontractile uterus.
    c. Fallopian cilia.
    d. an IUD.
    e. cervical ATP.

5. During the sterilization procedure of _____, the surgeon severs the:

    a. glans.
    b. scrotum.
    c. vas deferens.
    d. seminiferous tubules.
    e. prostate gland.

# Exercise

In the figure below, use colored pencils to color the maternal blood supply blue, the chorion yellow, and the amnion brown. You may use text Figure 18-10 as a reference.

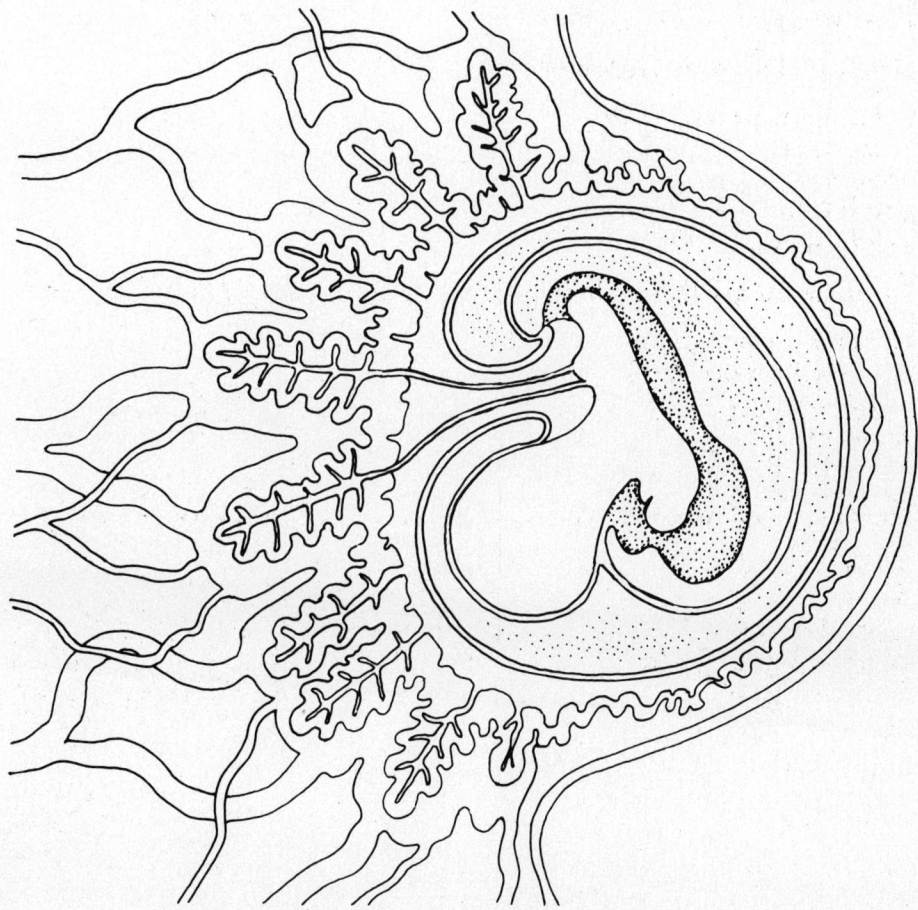

# 19

# THE ORIGIN AND DIVERSITY OF LIFE

---

## CHAPTER AT A GLANCE

---

## CHAPTER PREVIEW

This chapter begins on a cosmic note, as it considers life on Earth in the context of the Big Bang—the hypothesized "explosion" that gave rise to the entire universe. Although you will encounter a fair amount of chemistry here, it is almost all conceptual—there are no formulas to memorize.

After discussing the Big Bang, the chapter moves on to describe the close relationship between physical and biological evolution, and the emergence of organic and biological molecules. In the section on early cells you will see clearly how early life and the physical characteristics of ancient Earth were intertwined. This section also includes a look at the origin of eukaryotic cells and their organelles, and a discussion of the powerful explanatory theory of endosymbiosis. Another important concept is continental drift, and in the "Changing Face of Planet Earth" section you will learn how shifting continents have affected both evolving life forms and continental weather patterns.

The final sections of Chapter 19 present the basic concepts of taxonomy—the methods biologists use to classify living organisms. The chapter closes with a summary of the five-kingdom classification system devised by R. H. Whittaker.

# LEARNING OBJECTIVES

When you have mastered the concepts of this chapter, you will be able to:

1. Describe the basic physical traits of the Earth that may explain the appearance of life.
2. Give three possible explanations for the origin of life, and explain which theories are testable with the scientific method.
3. List the four basic steps of the emergence of organic and biological molecules, and detail what must happen in each.
4. Explain how the physical conditions of the early Earth are related to the emergence of biological molecules, and eventually cells.
5. List the metabolic options of the first living cells, and describe the sequence of emergence for heterotrophs and autotrophs.
6. Describe the interrelated effects that life forms and the physical planet have had on each other.
7. Explain the endosymbiont theory of eukaryote origins, and cite the evidence for it.
8. Outline the early history of taxonomy, and explain how the theory of evolution modified taxonomic methods.
9. List the five kingdoms of organisms, and discuss the justification for each group.

# CONCEPTS IN REVIEW

## Section I   A Home for Life: Formation of the Solar System and Planet Earth

The story of life's origins begins with the formation of the Earth. The sequence of events that gave rise to our planet began, in turn, with the cosmic explosion physicists call the Big Bang. The sun at the center of our solar system condensed from a cloud of primordial matter roughly 5 billion years ago; the planets, including Earth, condensed about 4.6 billion years ago (text p. 435). The Earth is composed of a number of layers: a solid crust, a semisolid mantle, and a largely molten (liquid) core that has a solid center (text p. 436). Basic physical features of Earth that may have made the emergence of life possible include the planet's size, temperature, composition, and distance from the sun (p. 437).

## Section II   The Question of How Life Began on Earth

The major current hypothesis holds that life arose spontaneously on early Earth by means of chemical evolution from nonliving substances (text p. 437). Other ideas on the origin of life, including creation myths and religious concepts, are not testable. Thus they lie outside the realm of science.

## Section III   The Emergence of Organic and Biological Molecules

Evidence for prelife stages of chemical organization comes from laboratory experiments that try to duplicate the physical environment and chemical resources of early Earth. These experiments, including the pioneering work of Miller and Urey, have successfully produced organic monomers such as amino acids, simple sugars, and nucleic acid bases (text p. 438). The probable next step toward life was the spontaneous linking of such monomers into polymers such as proteinoids and nucleic acids. Current research suggests that likely sites for this polymerization were clay or rock surfaces (text p. 440).

Researchers have found that, when energy is available to a system, they can generate three kinds of organic molecular aggregates. The Russian Aleksandr Oparin obtained polymer-rich droplets, called coacervates, from solutions of polymers. Sidney Fox generated proteinoid microspheres from mixtures of amino acids and water. A third laboratory structure is the liposome, a spherical lipid bilayer that forms from phospholipids. A structure similar to one or more of these aggregates may have been the precursor of true cells (text p. 440).

Further steps in the appearance of cells on Earth included the development of RNA and DNA as biological information molecules. Evidence suggests that RNA, which can form spontaneously under conditions mimicking those of the early Earth, was the first informational molecule. The discovery of RNA ribozymes—RNA that can act as an enzymelike catalyst—suggests that such catalytic RNA also could have assembled new RNAs from early nucleotides. Certain catalytic RNAs can also carry out sexlike exchanges of pieces of RNA (text p. 442).

Following the development of a lipid-protein surface layer and replicating RNA and DNA informational molecules, the events leading to the emergence of living cells would have included the origin of the genetic code; the sequestering of RNA or DNA into cell-like structures, and the development of metabolic pathways (text pp. 443–444).

## Section IV   The Earliest Cells

The oldest fossils that may represent living cells are found in rocks that are about 3.4 to 3.5 billion years old. The first cells were probably anaerobic heterotrophs, with autotrophs arising much later. The first autotrophs produced their own nutrients and released $O_2$—a metabolic by-product that had a crucial impact on later life forms. The resulting ozone layer in the Earth's atmosphere reduced the penetration of ultraviolet light. As a result, cells could survive in shallow water and on the land surface (text p. 445). The increasing quantity of atmospheric oxygen also permitted cellular respiration to originate, which in turn signaled the beginning of the global carbon cycle (text p. 446).

The earliest cells were all prokaryotes, but, by about 1.5 billion years ago, eukaryotes appeared. Eukaryotes probably arose through the hypothetical process of endosymbiosis, described in more detail in Chapter 20 (text p. 446).

## Section V   The Changing Face of Planet Earth

Changes in land masses, the seas, and climate have greatly affected the evolution of life on Earth. The basic parts of the planet include a light, solid crust over a hot, semisolid mantle and an inner, partially molten core. Massive segments or plates of the crust move over the mantle in the process of continental drift. By way of this movement the early land masses of Laurasia and Gondwana fused to form Pangaea. The breakup of Pangaea during the past 500 million years has resulted in the present continents and their distributions. Climatic changes that greatly affected living organisms accompanied these plate movements; the period was marked by massive extinctions of living creatures. Organisms were also affected by periods of glaciation that followed variations in Earth's orbit and the output of energy by the sun (text p. 447).

## Section VI    Taxonomy: Categorizing the Variety of Living Things

Biologists use the binomial system of nomenclature to categorize the varieties of life on Earth. The system assigns each type of organism to a genus and species. Organisms are then further classified into higher taxonomic categories—family, order, class, division (plants), phylum (animals), and kingdom. Evidence from many subfields of biology, such as biochemistry and comparative anatomy, helps define species and higher taxa. And whereas species were originally defined in terms of morphological traits, today biologists generally use the criterion of a reproductively isolated population (text pp. 448–450). You will study this concept in detail in the chapters on evolution.

Taxonomy reveals a great deal about the evolutionary relationships among organisms. A clade is a taxonomic unit whose members are derived from a common ancestor (text p. 451).

## Section VII    The Five Kingdoms

A phylogenetic tree is a graphic representation of evolutionary relationships. Your text uses a common five-kingdom arrangement: organisms are grouped into the kingdoms Monera, Protista, Fungi, Plantae, and Animalia. Although this system is a convenient organizational tool, the kingdoms are probably not true clades. As you will see in the next few chapters, certain taxonomic groups are extremely difficult to place (text pp. 451–452).

## KEY TERMS

| | | |
|---|---|---|
| Animalia *text page* 451 | division 448 | Pangaea 447 |
| Big Bang 435 | endosymbiosis 446 | phylum 448 |
| binomial system of nomenclature 448 | family 448 | Plantae 451 |
| | Fungi 451 | plate tectonics 447 |
| chemical evolution 435 | genus (genera) 448 | proteinoid 440 |
| clade 451 | kingdom 448 | proteinoid microsphere 440 |
| cladistics 451 | liposome 448 | Protista 451 |
| class 448 | mantle 446 | species 448 |
| coacervate 440 | Monera 451 | stromatolite 445 |
| continental drift 447 | monophyletic 451 | taxon 448 |
| core 436 | order 448 | taxonomy 448 |
| crust 436 | ozone layer 445 | |

## SELF-QUIZ: TESTING WHAT YOU HAVE LEARNED

### Matching Key Terms

Match each term on the left with the most appropriate description on the right.

1. stromatolite
2. coacervate

a.  made of iron and nickel
b.  middle layer of Earth

3. clade
4. ozone layer
5. Pangaea
6. crust
7. genus
8. core
9. plate tectonics
10. cladistics
11. Big Bang
12. monophyletic
13. binomial system
14. mantle
15. endosymbiosis

c.  fossil marine algae
d.  origin of eukaryotic organelles
e.  drifting of crust
f.  first binomial name
g.  monophyletic group
h.  branching of closely related groups
i.  share a common ancestor
j.  Linnaeus
k.  single land mass
l.  filters out ultraviolet light
m.  polymer-rich spheres
n.  origin of universe
o.  outer rock skin

# True or False?

1. _____ The Earth and sun were formed shortly after the Big Bang.

2. _____ The Earth's core is composed of molten iron and nickel.

3. _____ The "greenhouse effect" is caused by $H_2O$ in the atmosphere.

4. _____ Special creation is an untestable hypothesis.

5. _____ Lipid bilayers cannot be formed in the laboratory.

6. _____ The first living cells were probably aerobic heterotrophs.

7. _____ The first cells could not have appeared without the protection of the ozone layer.

8. _____ Endosymbiosis probably occurred at least 3 billion years ago.

9. _____ Mountains are believed to be produced by the movement of Earth's crustal plates.

10. _____ The mantle is semisolid.

11. _____ Molecular taxonomy analyzes proteins and DNA to generate taxonomic trees.

# Completion

1. Free, replicating biological molecules are sometimes referred to as _____.
2. The most widely accepted theory to explain the origin of eukaryotic cell organelles is

_____.

3. The three geological eras are the _____, _____, and

_____.

4. Pangaea was a single supercontinent formed from two other land masses, _____

and _____.

5. The binomial system of Linnaeus requires a two-part name for each organism: the first name

is the _____, and the second the _____.

6. The largest, most inclusive taxonomic category is the _____.

7. The biological definition of a species is based on the criterion of _____ isolation.

8. Groups of organisms that are derived from only one common ancestral source are said to be

_____, or in other words, of a single lineage.

## Short Answer

1.  What is molecular taxonomy?

2.  What are the five kingdoms of R. H. Whittaker, and what kinds of organisms are in each?

3.  Why do scientists think that clay surfaces might have been ideal sites for the formation of the first life?

4.  Why is it almost certain that the first cells had to be anaerobic heterotrophs?

5.  What factors make the planet Earth an ideal site for the appearance of life?

6.  How has the concept of evolution changed the field of taxonomy?

# Multiple-Choice Review

In the following sentences, fill in any blanks. Complete each statement by circling the correct response.

1. Plate tectonics, as a mechanism, is driven by activity in:

   a. the core.
   b. the crust.
   c. the mantle.
   d. the oceans.
   e. the atmosphere.

2. The study of how closely related groups branched and separated is:

   a. generative biology.
   b. classification.
   c. taxonomics.
   d. cladistics.
   e. bioanalysis.

3. Based on what you have learned in this chapter, _____ probably arose on the Earth:

   a. about 5,000 years ago.
   b. between 3 and 4 billion years ago.
   c. between 2 and 3 billion years ago.
   d. about 500 billion years ago.
   e. about 10 billion years ago.

4. The energy to drive organic polymerization in _____ may have come from:

   a. sunlight.
   b. radioactive minerals.
   c. coal.
   d. friction.
   e. electrovoltaic cells.

5. Liposomes are critical for the formation of:

   a. a lipid bilayer.
   b. lipid storage.
   c. microspheres.
   d. monomers.
   e. none of the above

# 20
# MONERA AND VIRUSES: THE INVISIBLE KINGDOM

---

## CHAPTER AT A GLANCE

# CHAPTER PREVIEW

Chapter 20 deals with organisms that are extremely small. One result of this microscopic size is that it is difficult to discern physical differences among them that might help in their classification. Thus, as you will see, biologists have used a variety of special criteria to study bacteria—the ability to take up cell wall stains, appearance of the colony, and so forth.

All bacteria are prokaryotes, and the chapter's first section outlines general features of their structure, metabolism, and reproduction. The next section covers the bacterial taxonomic groupings, which are neatly summarized in Table 20-1. The chapter then surveys the various ways that bacteria and humans interact, and, in Table 20-2, sums up a range of bacterially caused diseases.

As the chapter shifts its focus to the viruses, you may want to review the seven characteristics of life noted in Chapter 1. This section presents simple concepts about viral structure, reproduction, diseases, and the role of viral agents in cancer. The chapter closes with a renewed discussion of a most intriguing idea in modern biology: the endosymbiont theory.

# LEARNING OBJECTIVES

When you have mastered the concepts of this chapter, you will be able to:

1. Give the basic criteria for classifying organisms as members of kingdom Monera, and describe the size ranges of each group.
2. Describe the basic anatomy of prokaryotic bacteria.
3. Describe the four basic types of bacterial movement, and explain how each works.
4. Outline asexual and sexual reproductive strategies seen in bacteria.
5. List the major groups of bacteria, and describe them in terms of movement, reproduction, and relationship with humans.
6. Describe the cyanobacteria, and explain their role in determining the composition of Earth's early atmosphere.
7. List the main characteristics of the methanobacteria, and discuss the evidence that they are an independent lineage from other bacteria.
8. Describe the ways in which bacteria are important to humans, including food production and disease.
9. Discuss the "anatomy" and composition of a virus, and the question of whether a virus meets the criteria for life.
10. Discuss reproduction in the viruses, and their role in human disease.
11. Outline the evolution of the monerans.

# CONCEPTS IN REVIEW

## Section I   Monera: Tiny but Complex Cells

All monerans are prokaryotes. They lack a nuclear envelope, other membrane-bound organelles, and a cytoskeleton. And although monerans are also extremely diverse, all grow and reproduce extremely rapidly.

Bacteria exemplify the general features of monerans. Each bacterial cell is bounded by a cell wall composed of large peptidoglycan polymers. Some bacteria, including *E. coli,* have a second, outer lipid bilayer. This difference is exploited in Gram staining: bacteria having only the cell wall are Gram-positive and stain purple, while those that have the additional lipid bilayer stain red and are termed Gram-negative (text p. 459). In some species a protective capsule surrounds the cell wall.

Within the dense cytoplasmic area called the nucleoid, bacteria have a single circular chromosome (introduced in Chapter 14). A membranous mesosome projects into the cell from the plasma membrane, which lies just beneath the cell wall (text p. 460). Many bacteria move about propelled by flagella, which rotate clockwise or counterclockwise. The bacterial flagellum is a unique, hollow structure consisting of a filament, a midpiece, and a rod that extends through the cell wall and is anchored in the plasma membrane by protein rings. Two other means of bacterial movement are the axial filament locomotion of spirochetes, and the gliding movements of myxobacteria (text p. 462).

Bacteria reproduce nonsexually. The basic reproductive process is binary fission (text p. 462); in addition, some bacteria exchange genetic information in a pseudosexual interaction that was described in Chapter 14. Under adverse conditions many bacterial species can form tough endospores, a resting stage that contains only genetic material and other cellular components necessary to start active metabolism when conditions improve.

Bacteria may be heterotrophs (decomposers or parasites) or autotrophs. Photoautotrophs, including green sulphur bacteria and two species of purple bacteria, derive energy from a form of anaerobic photosynthesis using modified chlorophylls. Chemoautotrophs oxidize inorganic compounds (text p. 466). All bacteria have extremely high metabolic rates, although differences exist in the amounts of oxygen various species can tolerate. Some are obligate anaerobes, and generate ATP only through fermentation. Others are facultative anaerobes able to grow with or without oxygen, while obligate aerobes must have oxygen for metabolic processes (text p. 466). This varied menu of metabolic options is one reason for the great evolutionary success of prokaryotes.

## *Section II*   Microorganism Diversity: The Variety of Monerans

The moneran subkingdom Schizomycete includes a number of distinct taxonomic groups. The eubacteria or "true" bacteria are the most abundant and diverse, and are traditionally classified by shape. Subgroups include the rod-shaped bacillus, spherical coccus, and spiral-shaped spirillum. A second major group, the spirochetes, are bacteria with a spiral or corkscrew shape. Myxobacteria are small, unflagellated, rod-shaped cells that sometimes swarm together to form fruiting bodies and generally live in soil. Actinomycetes comprise a variety of types of filamentous microorganisms, including members of the genus *Streptomyces* from which several antibiotics are derived (text p. 468). Among the tiniest prokaryotes are rickettsiae, which live as parasites in alternating hosts (arthropods and mammals or birds). They cause several human diseases. Finally, the smallest living cells known are the mycoplasmas—organisms that lack rigid cell walls, grow in colonies, and are resistant to antibacterial agents (text p. 469).

A second subkingdom of monerans, photosynthesizers known as cyanobacteria, includes 200 species once known as blue-green algae. They exist as single rod-shaped or spherical cells in clusters or long filamentous chains. Cyanobacteria lack flagella and move by gliding or spinning; they contain a variety of pigments, including chlorophyll *a*. In addition to carrying out photosynthesis, many species are capable of fixing nitrogen. Species that have adapted to extremes of temperature, salinity, or pH may represent the kinds of cells that inhabited the early Earth (text p. 470).

A third, unique group of monerans are the chloroxybacteria, which contain several types of photosynthetic pigments. Some biologists have proposed that chloroxybacteria are similar to ancient microorganisms that may have given rise to chloroplasts.

The fourth major group of monerans are the archaebacteria—a group so distinct that some biologists classify them as an independent kingdom. Included are anaerobes that produce methane (methanobacteria), others that are sulfur-dependent, and a third group that lives in environments high in salt (text p. 471). To some researchers, the presence in archaebacteria of several unique enzymes, as well as other molecules and structures similar to some found in eukaryotes, suggests that ancestral cell populations gave rise to three, rather than two, main lineages: prokaryotes, eukaryotes, and archaebacteria.

## Section III    Bacteria and Humans

Bacteria are important to human societies in a number of ways. Production of many foods—cheeses, pickles, yogurt, and others—depends on bacterial action. Bacteria are also important pathogens, producing disease states via destruction of tissues, irritation by bacterial wastes, and action of toxins released by live bacteria (exotoxins) or by the bursting of dead bacterial cells (endotoxins) (text p. 473). Many bacteria tend to develop resistance to modern antibiotics, posing an increasing problem for human medicine (text p. 474).

## Section IV    Viruses: Noncellular Molecular Parasites

Viruses are not cells but particles of genetic material and protein that can invade living cells, take over their metabolic machinery, and reproduce. Each metabolically inert particle—termed a *virion*—consists of a core of nucleic acid (DNA or RNA) surrounded by a protein coat or capsid. Some animal, plant, and bacterial viruses are also enclosed within a membranelike envelope (text p. 475).

A virus may invade a host cell by injecting its genetic core into the host, or by being engulfed by the host. In some cases, lysogeny takes place: the virus is nonvirulent when it first enters the host cell, then its DNA is inserted into the host's chromosome, and the viral DNA is replicated along with the host's chromosome each time the host cell passes through a growth and division cycle (text p. 475). Later the lysogenic cycle may break down and be replaced by a lytic pathway—that is, the viral DNA reproduces independently and viral particles within the cell assemble, leading to the subsequent bursting (death) of the host cell (text pp. 476–477).

Various virus-caused diseases are based on the destruction of host cells by the lytic pathway. Other viruses produce toxins that inhibit cell metabolism. Viral infection is often tissue-specific; it may be that specific cell types bear receptors that bind virus particles. These "virus receptors" probably serve other functions for the cells that possess them (text p. 478).

A promising approach to combating viral diseases is the use of interferons, naturally occurring mammalian proteins liberated after cells have been infected with certain types of viruses. Interferons bind to infected cells, where they stimulate the production of proteins that prevent viral reproduction. As described in Chapter 17, viruses are implicated in some cancers (text p. 479).

## Section V    Evolution of Monerans, Viruses, and Eukaryotic Cells

The endosymbiont theory attempts to explain the evolutionary relationships among prokaryotes, viruses, and eukaryotes. It proposes that one or more simple progenitor cell types led to three early forms: the monerans, the archaebacteria, and a preeukaryote cell type that lacked organelles but possessed multiple chromosomes and other biochemical attributes of eukaryotes. This organism's evolution included symbiotic associations with one or more types of bacteria, including the ancestor of the mitochondrion (text p. 480). Other steps in the process may have included the evolution of cilia and flagella from spirochetes or bacteria attached to the preeukaryote.

In the lineage leading to plants, a photosynthetic, aerobic prokaryote may have been engulfed and then may have evolved into the chloroplast (text p. 481). The evolution of viruses is especially problematic. One hypothesis is that viruses may be "escaped" fragments of chromosomes; another is that they are highly evolved parasites (text p. 481). The earliest viruses may have resembled modern viroids, particles of DNA with no protein coat.

# KEY TERMS

actinomycete   *text page 468*

archaebacteria   *471*

binary fission   *462*

chemoautotroph   *465*

chloroxybacteria   *410*

cyanobacteria   *469*

endospore   *463*

endosymbiont theory   *480*

endotoxin   *473*

eubacteria   *467*

exotoxin   *473*

facultative anaerobe   *466*

Gram-negative bacteria   *459*

Gram-positive bacteria   *459*

interferon   *478*

lysogeny   *475*

lytic pathway   *476*

methanobacteria   *471*

mycoplasma   *469*

myxobacteria   *468*

obligate aerobe   *466*

obligate anaerobe   *466*

parasite   *464*

pathogen   *472*

peptidoglycan polymer   *459*

photoautotroph   *464*

prokaryote   *458*

rickettsia   *469*

schizomycete   *466*

spirochete   *467*

virion   *475*

viroid   *481*

virus   *474*

# SELF-QUIZ: TESTING WHAT YOU HAVE LEARNED

## Matching Key Terms

Match each term on the left with the most appropriate description on the right.

|   | | | |
|---|---|---|---|
| 1. interferon | a. | smallest monerans |
| 2. facultative anaerobe | b. | produces identical daughter cells |
| 3. endotoxin | c. | true bacteria |
| 4. pathogen | d. | a tough, resting cell |
| 5. cyanobacteria | e. | with or without oxygen |
| 6. endospore | f. | corkscrew |
| 7. virus | g. | blue-green photosynthesizer |
| 8. lysogeny | h. | Rocky Mountain spotted fever |
| 9. binary fission | i. | causes a specific disease |
| 10. exotoxin | j. | poison from living bacteria |
| 11. rickettsia | k. | composed of DNA and protein coat |
| 12. eubacteria | l. | viral DNA replicates with hosts |
| 13. mycoplasma | m. | extracellular virus particle |
| 14. virion | n. | virus-inhibiting protein |
| 15. spirochete | o. | poison from lysed bacteria |

## True or False?

1. _____ All monerans are prokaryotic.

2. _____ Most monerans possess a cell wall.

3. _____ Gram-positive bacteria stain dark purple.

4. _____ Bacterial and eukaryotic flagella are identical.

5. _____ Most bacteria reproduce by binary fission.

6. _____ The rickettsiae are motile as adults.

7. _____ Spirochetes move with a twisting motion.

8. _____ Leprosy is caused by an actinomycete.

9. _____ Q fever is transmitted by a bite.

10. _____ Endotoxins are released when bacterial cells die.

11. _____ Antibiotics are the best defense against viral infection.

## Completion

1. The smallest living cells are called _____. They have no outer cell _____ and contain only about half as much _____ as larger bacteria.

2. Many bacteria have a protective structure surrounding the cell wall. This is the _____, which prevents attack from a host's immune system.

3. Although there is no true sexual reproduction among bacteria, genetic information is exchanged by the processes of _____, _____, and _____.

4. _____ are bacteria that can live in harsh environments and oxidize inorganic molecules.

5. _____, such as *Treponema pallidum*, the causative agent of syphilis, have a spiral shape and move with twisting motions.

6. Nonvirulent viruses can lie dormant in the host, and replicate each time the host's cells undergo division in a process of _____.

7. _____ are actual genes that appear programmed to produce cancers.

8. The most common method of bacterial reproduction is _____.

## Short Answer

1. List three ways bacteria differ in, and therefore may be classified by, their tolerance to oxygen.

2. Compare and contrast the two main modes of viral reproduction.

3. List and briefly describe four ways that bacteria reproduce.

4. What is an endospore and what is its value to bacteria?

5. What are Gram-negative and Gram-positive bacteria? What actual physical differences allow bacteria to be classified by this method?

6. List four major bacterial pathogens.

## Multiple-Choice Review

Complete the following statements by circling the correct response.

1. The methanobacteria are sometimes placed in a separate kingdom, the:

   a. Cyanobacteria.
   b. Archaebacteria.
   c. Chloroxybacteria.
   d. Schizomycetes.
   e. none of the above

2. The outer protein coat of a virion is called the:

   a. plasma shell.
   b. cell wall.
   c. capsule.
   d. capsid.
   e. sheath.

3. The germ theory of disease was proposed by:

   a. Pasteur.
   b. Koch.
   c. Ames.
   d. Woese.
   e. van Leeuwenhoek.

4. One of the major *functional* differences between eukaryotic and bacterial flagella is that the flagella of bacteria can:

    a. lash back and forth.       d. move in a spiral.
    b. undulate.               e. retract.
    c. rotate.

5. Bacteria that must have oxygen in order to survive are:

    a. facultative anaerobes.    d. obligate aerobes.
    b. obligate anaerobes.     e. methanobacteria.
    c. facultative aerobes.

# Exercise

Fill in the chart below, giving the phylum or subkingdom and ecological role of each organism listed.

| Organism | Phylum | Ecological Role |
| --- | --- | --- |
| *Vibrio cholerae* | | |
| *Anabaena* | | |
| Rous sarcoma virus | | |
| *Herpes simplex* | | |
| *Streptococcus* | | |
| Mycoplasma | | |
| Methanobacteria | | |

Why might the term *organism* not strictly apply to some of these?

# 21
# PROTISTA: THE KINGDOM OF COMPLEX CELLS

---

## CHAPTER AT A GLANCE

Protist Classification: By Life Styles
The Producers: Plantlike Protists
*Euglenophyta*
*Pyrrophyta*
  Glowing waves and red tides
*Chrysophyta*
  Golden-brown algae
  Diatoms
The Decomposers: Funguslike Protists
*Myxomycetes: True slime molds*
*Acrasiomycetes: Cellular slime molds*
The Consumers: Animal-like Protists
*Mastigophora*
*Sarcodina*
*Sporozoa*
*Ciliophora*
Protistan Evolution

---

## CHAPTER PREVIEW

When it comes to classification, protists are true biological puzzles—almost nothing is known about their evolutionary relationships. As a result, members of the kingdom Protista are grouped by life style rather than by following the cladistic methods you studied in Chapter 19. As it turns out, however, this method is a good organizing tool: all of the organisms you will learn about in Chapter 21 fit into one of three basic categories: the producers, the decomposers, and the consumers. The features of each of these categories are summarized in Tables 21-1 and 21-2.

The plantlike producers introduced first are the most important of the three groups; along with green plants, they are the basis of Earth's ecosystems. Decomposers, the funguslike protists, play vital roles in the cycling of nutrients, as do the fungi you will encounter in Chapter 22. Protistan consumers are tiny hunters that have many characteristics we associate with animals. As you proceed through the chapter you may find it useful, as an aid to learning—and to mastering the Key Terms— to create your own chart of taxonomic groups and characteristics.

# LEARNING OBJECTIVES

When you have mastered the concepts of this chapter, you will be able to:

1. List the three main categories of protists, and explain why protists are classified by life style.
2. List the three basic groups of plantlike protists, and give an example of each type.
3. List the two basic groups of funguslike protists, and give an example of each type.
4. List the four groups (phyla) of the animal-like protists, and give an example of each type.
5. Describe the anatomy of a typical zooflagellate, and explain how the reproductive strategies of several species make them serious human parasites.
6. Describe locomotion in the Sarcodina, and explain how the various phyla members are important to humans.
7. Explain the life cycle of the malaria-causing sporozoan *Plasmodium vivax*.
8. Describe the anatomy of a ciliate, and outline the unique process of ciliate reproduction.
9. Outline the evidence for the evolutionary relationships among members of the protista.

# CONCEPTS IN REVIEW

## Section I    Protist Classification: By Life Styles

The kingdom Protista is wonderfully diverse, but has proven extremely difficult to classify. Using the criterion of life style, Chapter 21 groups them broadly into (1) plantlike autotrophs; (2) funguslike heterotrophs; and (3) animal-like heterotrophs. Despite their differences, these main subgroups all share the basic eukaryotic cell structure.

## Section II    The Producers: Plantlike Protists

The photosynthetic phytoplankton represent plantlike protists. Members of one phylum, Euglenophyta, have both animal-like and plantlike features. Nearly all are photosynthesizers, but animal features include the absence of a rigid cellulose cell wall. Euglenoids also have a characteristic eyespot and one or two flagella for locomotion, and they reproduce asexually by mitosis (text p. 488).

A second phytoplankton division is Pyrrophyta, the dinoflagellates. Each is a single cell endowed with a set of flagella that causes the organism to spin while it swims. Like euglenoids, most species have contractile vacuoles and chloroplasts. Some nonphotosynthetic species produce dartlike trichocysts that can be used to kill prey or for protection. Dinoflagellates are the sources of "red tides" and "glowing waves." They usually reproduce asexually (text p. 489).

The division Chrysophyta includes diatoms and golden-brown algae. Both are extremely abundant; species may be unicellular or colonial, and while most reproduce asexually, some species produce flagellated gametes that carry out sexual production. Golden-brown algae get their golden color from the carotenoid pigment fucoxanthin (text p. 490).

The great numbers of diatoms in the sea may be the most important source of atmospheric oxygen. Diatoms are also colored by fucoxanthin, and have intricately sculpted cell walls containing both pectin and silica. They reproduce via mitosis, with the progeny of successive generations becoming smaller and smaller. When a lower size limit is reached, sexual reproduction is triggered (text p. 491). The whitish sediments containing innumerable diatom skeletons make up diatomaceous earth.

## Section III    The Decomposers: Funguslike Protists

To represent this group of decomposer protists, Chapter 21 focuses on the Gymnomycota, the slime molds. The true slime molds (class Myxomycetes) and the cellular slime molds (class Acrasiomycetes) can both exist as a single, motile cells or as large sluglike masses. However, while myxomycetes reproduce sexually, generating one huge cell with many nuclei (a pseudoplasmodium), acrasiomycetes are asexual. The mass they produce is an aggregate of individual cells called a plasmodium (text p. 491).

## Section IV    The Consumers: Animal-like Protists

Protozoans are protists that have primarily animal features. The four commonsense groupings presented in this chapter are mastigophorans, or zooflagellates (class Mastigophora), sarcodines (class Sarcodina), sporozoans (class Sporozoa), and a true phylum, Ciliophora, the ciliophores. The four locomote in different ways, but all have food vacuoles that contain digestive enzymes, and contractile vacuoles that discharge excess water (text p. 494).

Many zooflagellates move about by means of whiplike flagella; some species move by way of pseudopods—extensions of the cell surface that can be used for movement and for ingesting food. A representative species is *Trypanosoma gambiense*, the protozoan that causes one form of African sleeping sickness (see Figure 21-9, text p. 495). Sarcodines range from amoebas to hard-shelled, diatomlike creatures including foraminiferans and radiolarians. The latter are marine protozoans and members of the zooplankton. Sarcodines such as *Amoeba* use pseudopods for locomotion and feeding (text p. 496).

Members of Sporozoa all have a sporelike stage in their life cycle, and all are parasites. Adult forms are nonmotile, while immature cells can move via flagella or pseudopods. *Plasmodium falciparum* is one sporozoan species responsible for malaria. The most complex single-celled organisms on Earth are members of the phylum Ciliophora, generally called ciliates. Most are free-living, aquatic species, and move by way of rows or bands of cilia. A typical ciliate such as *Paramecium* also has specialized organelles, including micro- and macronuclei (text pp. 499–500).

A final protistan phylum, Caryoblastea, has only a single species. This primitive organism has amoebalike features but lacks mitochondria. Also, instead of undergoing mitosis, it reproduces through a process that pinches the nucleus in two after chromosomes have duplicated. It is a creature at the borderline between prokaryotes and eukaryotes (text p. 501).

## KEY TERMS

| | | |
|---|---|---|
| Acrasiomycete   *text page 491* | diatom   *490* | Gymnomycota   *491* |
| Caryoblastea   *500* | diatomaceous earth   *490* | Mastigophora   *494* |
| cellular slime mold   *492* | dinoflagellate   *488* | myxamoeba   *492* |
| Chrysophyta   *489* | Euglenophyta   *487* | Myxomycetes   *491* |
| ciliate   *499* | foraminiferan   *496* | *Pelomyxa*   *501* |
| Ciliophora   *499* | golden-brown alga   *489* | phytoplankton   *487* |

# SELF-QUIZ: TESTING WHAT YOU HAVE LEARNED

## Matching Key Terms

Match each term on the left with the most appropriate description on the right.

1. diatom
2. Sporozoa
3. plasmodium
4. phytoplankton
5. golden-brown alga
6. zooplankton
7. ciliate
8. Euglenophyta
9. pseudopod
10. Pyrrophyta
11. diatomaceous earth
12. acrasiomycete
13. Mastigophora
14. true slime mold
15. radiolarian

a. floating autotrophs
b. a slimy layer
c. pseudoplasmodium
d. false foot
e. move by flagella
f. all nonmotile as adults
g. fossil remains of silica shells
h. pseudopods and silica shells
i. most complex single-celled organisms
j. myxomycete
k. colored by fucoxanthin
l. form diatomaceous earth
m. a taxonomic puzzle
n. "fire plants"
o. floating heterotrophs

## True or False?

1. _____ Protist taxonomy is based on modern evolutionary principles, as outlined in Chapter 19.

2. _____ Although euglenoids are photosynthetic, they lack cellulose.

3. _____ The stigma of higher plants originates in *Euglena*.

4. _____ Dinoflagellates and euglenoids share identical chlorophylls.

5. _____ Although small, diatoms produce most of Earth's atmospheric oxygen.

6. _____ Myxamoebae are diploid.

7. _____ Cellular slime molds reproduce sexually.

8. _____ The white cliffs of Dover were formed of dinoflagellate shells.

9. _____ Ciliates are the most complex single-celled animals.

10. _____ All sporozoans are parasitic.

11. _____ Cytoplasmic inheritance is best developed among the members of Sarcodina.

# Completion

1. Euglenoids have a plasma membrane that is strengthened by a _____, or series of underlying protein bands.
2. Autotrophic euglenoids store carbohydrates in a unique substance called _____.
3. Dinoflagellates can function as heterotrophs by the use of tiny poison dart–like structures called _____.
4. The color of golden-brown algae comes from the specific carotenoid pigment _____.
5. The slime molds are composed of two taxonomic classes: the _____, or true slime molds, and the _____, or cellular slime molds.
6. The secondary host, or vector, of *Trypanosoma gambiense* is the _____.
7. _____ and _____ are shelled sarcodines that make up a portion of the zooplankton of marine waters.
8. The best known cases of how cytoplasmic inheritance influences cellular structures are found among the _____.

# Short Answer

1. List the three protistan life styles. Why do we base protistan classification on life styles?

2. What protistan phyla make up the phytoplankton? Of what importance are phytoplankton in our planet's ecology?

3. What is meant by the names "Pyrrophyta" and "dinoflagellate"? What do these terms tell us about the organisms?

4. What behavioral, structural, and reproductive differences are there between true and cellular slime molds?

5. What methods of locomotion are used to distinguish the four protozoan phyla?

6. What is the life history of the organism that causes malaria?

## Multiple-Choice Review

In the following sentences, fill in any blanks. Complete each statement by circling the correct response.

1. The first microorganism discovered, and pictured by van Leeuwenhoek, was a:

   a. sporozoan.
   b. ciliate.
   c. sarcodine.
   d. radiolarian.
   e. diatom.

2. Euglenoids are considered _____ because they lack:

   a. chlorophyll.
   b. chloroplasts.
   c. a cell wall.
   d. pigments.
   e. a capsid.

3. Euglenoids counteract their lack of a _____ and the effects of a hypertonic cytoplasm with the aid of:

   a. a stigma.
   b. chloroplasts.
   c. flagella.
   d. a contractile vacuole.
   e. none of the above

4. Toxic "red tides" are produced by periodic "blooms" of:

   a. dinoflagellates
   b. radiolarians.
   c. golden-brown algae.
   d. diatoms.
   e. free carotenoid pigments.

5. Glasslike cell walls of silica are found among:

   a. the dinoflagellates.
   b. the diatoms.
   c. members of Euglenophyta.
   d. members of Sporozoa.
   e. none of the above

## Exercise

For each protist listed in the chart below, give its mode of nutrition and several distinguishing characteristics.

| Organism | Mode of Nutrition | Characteristics |
|---|---|---|
| Euglena | | |
| Trypanosoma | | |
| Plasmodium | | |
| golden-brown alga | | |
| Pelomyxa | | |
| Dictyostelium | | |
| Amoeba | | |
| Protogonyaulax | | |

# 22

# FUNGI: THE GREAT DECOMPOSERS

---

## CHAPTER AT A GLANCE

---

## CHAPTER PREVIEW

Like many biology students, you may not be very familiar with fungi—aside from the ones you see sliced on top of pizza. Much of the anatomy of a fungus is hidden underground, or inside dead and decaying organisms or tissues. Nevertheless, fungi are a vitally important group of organisms because of their abundant numbers and their crucial role as decomposers. Without them, the recycling of nutrients and materials in ecosystems (Chapter 44) might be seriously disrupted.

Beginning in the first section of Chapter 22 you will see that reproductive differences are the prime features that distinguish the various fungal groups. Also, note the two basic types of fungal spores. Much of this section describes the six classes of fungi and their reproductive life styles; Table 22-1 clearly summarizes these features. The next section surveys the peculiar combinations of fungi and algae known as lichens. Aside from being composite organisms, lichens are of interest because of their ecological importance. As you will learn in Chapter 44, they are often the first colonizers of new landscapes.

The chapter's final section, on fungal evolution, is brief, but it contains a great deal of useful information about characters and classification that will help you organize and better understand the various fungal groups.

# LEARNING OBJECTIVES

When you have mastered the concepts of this chapter, you will be able to:

1. List the characteristics of the various groups of fungi, and discuss the basic life style of each group.
2. Discuss the basic fungal growth pattern, and explain the unique character of fungal mitosis.
3. Outline the spore-based survival strategy seen in fungi, and characterize the two basic spore types.
4. Describe the kinds of organisms belonging to the class Chytridiomycetes, and explain why this group is thought to represent a link between protists and the fungi.
5. Describe the kinds of organisms belonging to the class Oomycetes, and outline the reproductive strategy of this group.
6. Describe the kinds of organisms belonging to the class Zygomycetes, and outline their reproductive strategy.
7. Explain what is meant by the term *mycorrhizal associations*, and why such arrangements are important.
8. Describe the kinds of organisms belonging to the class Ascomycetes, and outline their reproductive strategy and economic importance.
9. Describe the kinds of organisms belonging to the class Basidiomycetes, and outline their reproductive strategy and economic importance.
10. Describe the kinds of organisms belonging to the class Deuteromycetes, and explain why they are commercially important.
11. Discuss the relationships between algae and fungi that give rise to lichens.

# CONCEPTS IN REVIEW

## Section I   Characteristics of Fungi

The approximately 175,000 species of fungi make up some of the simplest multicellular organisms. Fungi have a variety of life styles. They may be saprobes, which decompose dead organic matter; parasites, which obtain nutrients from living hosts; or symbionts, which live in symbiotic relationships with algae or with the roots of higher plants. In spite of these variations, however, all fungi carry out extracellular digestion—they secrete enzymes that digest organic matter and then absorb the resulting nutrients (text p. 506).

Most fungi have the same basic body structure: a main body or thallus that is composed of filaments called hyphae. In most species the walls of hyphal cells contain chitin. Hyphae in certain species may become specialized to form rhizoids, which serve as rootlike anchors, or may become the feeding structures known as haustoria. Finally, hyphae may or may not be septate, that is, have cross walls that segregate independent cells, each with at least one nucleus. Lower fungi are coenocytic; that is, they are one mass of cytoplasm containing multiple nuclei (text p. 507).

Hyphae grow and branch to form a filamentous network called a mycelium. Food is digested and absorbed at the tip of each hypha; more hyphae are generated as this process continues. As a result, fungi may grow very rapidly. Growth depends on mitosis and the rapid manufacture of cytoplasm;

fungal mitosis is unique in that it occurs within the nucleus (text p. 508). Hyphae from genetically distinct individuals may fuse to form a heterokaryon—a single cytoplasm with dissimilar nuclei.

As nonmotile heterotrophs, fungi must eventually be able to find new sources of nutrients. This function is fulfilled by spores, the fungal reproductive bodies. Spores may be born on aerial hyphae, which discharge spores into the air. Depending on the species, they may be haploid or diploid. The two main categories of spores are (1) dispersal spores, which are usually short-lived and are produced in large numbers during active fungal growth, and (2) survival spores, which are usually produced in lesser numbers and at a time in the life cycle when the fungus is under some kind of environmental stress (text p. 509).

## Section II    Classification of Fungi

As with some of the groups you have studied in previous chapters, it is difficult to group fungi according to actual evolutionary relationships. In general, they are classified according to morphology, methods of reproduction, and modes of spore production. Relying on these features, the single division of the kingdom Fungi, Mycota, is divided into six principal classes (text p. 510; see Table 22-1).

The lower fungi comprises the chytridiomycetes, oomycetes, and zygomycetes. All lack septate hyphae and are commonly coenocytic; spores are formed by asexual means. Of all six fungal classes, only the oomycetes usually have a diploid vegetative state. Oomycetes and chytrids, sometimes called water molds, produce motile, flagellated spores in sporangia. They also produce gametes in gametangia; oomycetes are distinguished by their large, immobile egg cells. In fact, both these groups have such distinctive features that some biologists prefer to classify them as protists rather than fungi (text pp. 511–513). Zygomycetes resemble the other two classes in this group, but have nonmotile spores. They are also completely terrestrial, and some form mycorrhizal associations with plant roots (text p. 513).

The higher fungi include members of Ascomycetes, the largest class of fungi. Most are either saprobes or parasites. Asexual reproduction produces spores called conidia, which develop on the tips of specialized aerial hyphae. In the ascomycete sexual cycle, hyphae of different mating strains fuse, giving rise to ascospores, which form in a small, saclike ascus. Groups of asci form fruiting bodies. Ascomycetes of interest to humans include truffles, yeasts, and *Penicillium* species (text pp. 516–518).

Most basidiomycetes—the second group of higher fungi—form visible fruiting bodies. A prime difference from ascomycetes is the dense mass of dikaryotic hyphae called the basidiocarp—the "mushroom" seen on damp lawns and the forest floor. Club-shaped basidia, each bearing four haploid basidiospores, line the surfaces of the gills on the underside of the mushroom cap. Members of this group have both sexual and asexual reproductive processes at different times in the life cycle, and as a result of varying environmental influences (text pp. 521–522).

The class Deuteromycetes, or Fungi Imperfecti, includes a variety of fungi that lacks modes of sexual reproduction. Most are known to reproduce asexually by means of conidia. Deuteromycetes important to humans include those used to ferment soybeans to make soy sauce and sake, and the fungi that produce citric acid and the highly dangerous aflatoxin (text p. 523).

## Section III    Lichens: The Ultimate Symbionts

Lichens are composite organisms in which about 90 percent of the lichen mass consists of one species of fungus, while the remaining 10 percent is made up of one or two species of algae. The algal portion of the lichen provides the fungal portion with essential nutrients; the presence of the fungal component may enable the alga to exploit an otherwise unavailable ecological niche. Lichen fungi are usually ascomycetes, although the other two higher fungi are sometimes found. Lichens have a number of remarkable features, including the ability to become almost completely dessicated and to absorb inorganic nutrients. Reproduction in lichens is not well understood (text p. 524).

## Section IV   Fungal Evolution

The various fungi may have arisen independently from prokaryotes; among the evidence for this hypothesis is some that suggests that asco- and basidiomycetes did not evolve from known lower forms. However, all fungi do show the same dependence on nutrients produced by plants, animals, or algae (text p. 524).

## KEY TERMS

aerial hypha   *text page 509*

Ascomycetes   *515*

ascus   *517*

basidiocarp   *520*

Basidiomycetes   *519*

basidium   *521*

Chytridiomycetes   *511*

coenocytic   *507*

conidium   *515*

Deuteromycetes   *522*

dikaryon   *511*

ergot   *518*

extracellular digestion   *506*

fruiting body   *517*

Fungi Imperfecti   *522*

gametangium   *511*

haustorium   *507*

heterokaryon   *508*

heterokaryosis   *508*

hypha   *507*

lichen   *523*

mycelium   *507*

mycorrhiza   *513*

Mycota   *510*

Oomycetes   *512*

rhizoid   *507*

saprobe   *506*

septate   *507*

sporangium   *511*

Zygomycetes   *513*

## SELF-QUIZ: TESTING WHAT YOU HAVE LEARNED

### Matching Key Terms

Match each term on the left with the most appropriate description on the right.

1. rhizoid
2. ergot
3. oomycete
4. mycorrhiza
5. aerial hypha
6. conidium
7. sporangium
8. basidiomycetes
9. lichen
10. haustorium
11. dikaryon
12. mycelium
13. septate
14. ascus
15. hypha

a. having cross walls
b. cellular fungal filaments
c. a fungal feeding structure
d. "water molds"
e. a hyphal cell with two nuclei
f. "fungus roots"
g. spore sac
h. analogous to plant roots
i. spore case
j. network of hyphae
k. dustlike
l. contains edible mushrooms
m. composite organisms
n. source of drug LSD
o. tall spore-producing structures

## True or False?

1. _____ Fungal mitosis, unlike mitosis in all other organisms, occurs within the nucleus.

2. _____ Survival spores must be produced in large numbers.

3. _____ Oomycetes are usually diploid.

4. _____ Zygomycetes are usually diploid.

5. _____ The dikaryons include the worst fungal pests of humans.

6. _____ Oomycetes contain cellulose.

7. _____ Mycorrhizal associations occur among 80 percent of land plants.

8. _____ Most basidiomycetes undergo asexual reproduction.

9. _____ Crustose lichens look like tiny leaves.

10. _____ Lichens do poorly in dry climates.

11. _____ The ratio of fungus to algae in lichens is about 9:1.

## Completion

1. Fungal chromosomes differ from those of other eukaryotes by having extremely small amounts of _____, the basic proteins involved in DNA coiling.

2. *Plasmodara viticola*, which causes downy mildew on grapes, and almost wiped out the French wine industry in 1880, is a member of the class _____.

3. The associations of fungi and higher plant roots are called _____.

4. About _____ percent of all land plants develop these fungal associations.

5. Single-celled yeasts usually reproduce by _____.

6. The common mushroom shape of "toadstools" and mushrooms of the class Basidiomycetes is actually a reproductive structure called a _____.

7. Lichens have a remarkable ability to filter out airborne substances such as sulfur and phosphorus, and hence are sometimes used as _____ indicators.

8. Fungal cell walls contain _____, and in some species, _____.

## Short Answer

1. What are heterokaryons? How are they formed?

2. Compare and contrast dispersal and survival spores.

3. What are lichens? Are they really true "species"? How do they "reproduce"?

4. List two fungal diseases of humans, and four fungal agricultural pests.

5. What structures are found in the Basidiomycetes? Outline a typical basidiomycete reproductive cycle.

6. What are the advantages of a mycorrhizal association to both partners?

## Multiple-Choice Review

In the following sentences, fill in any blanks. Complete each statement by circling the correct response.

1. Genetic _____ in fungi arises most effectively in/by:

   a. aerial hyphae.
   b. crossing over.
   c. heterokaryosis.
   d. recombination.
   e. rapid manufacture of cytoplasm.

2. Dispersal spores are usually:

   a. slow to germinate.
   b. short-lived.
   c. produced in low numbers.
   d. made only in the spring.
   e. endowed with thick cell walls.

3. Fungi provide all of the following for human societies except:

   a. food.
   b. medicines.
   c. fermenting agents.
   d. industrial filaments.
   e. commercial enzymes.

4. The biomass of a _____ is approximately _____ percent fungal tissue.

   a. 10
   b. 30
   c. 40
   d. 80
   e. 90

5. Fungal haustoria are:

   a. feeding structures.
   b. reproductive organs.
   c. used for support.
   d. the edible parts of mushrooms.
   e. produced in cycles.

# 23

# THE PLANT KINGDOM: ALGAE AND LOWER LAND PLANTS

---

## CHAPTER AT A GLANCE

---

## CHAPTER PREVIEW

This short but important chapter introduces the kingdom Plantae, with emphasis on the algae and lower land plants. The first section lays out the basic divisions of the plant kingdom. Most of the important information is summarized in Table 23-1. As you proceed through the chapter, you'll find it helpful to build a chart that shows each plant group and the reasons (taxonomic characters) why biologists place it in its particular taxonomic category. For example, different types of algae are placed in different divisions based on the combinations of photosynthetic pigments and energy storage products they contain. For the land plants you will see that there is a hierarchy of characters—the presence or absence of vascular tissue; among vascular plants, the presence or absence of seeds; among seed plants, the presence of naked seeds versus flowers; and so on. The characteristics of seed plants are the main subject of Chapter 24.

Next the chapter takes up a crucial concept: the unique feature of the plant life cycle called alternation of generations. You will find the text figures especially helpful in understanding this remarkable phenomenon. The last two sections survey the basic features of algae and the primitive land plants. As you read about the algae, note especially that they are among the primary producers in aquatic systems, and take care to understand algal life cycles. In the final section, you might try listing the advantages and disadvantages to plants of colonizing land, and the kinds of structures that were needed to successfully make the transition away from water.

# LEARNING OBJECTIVES

When you have mastered the concepts of this chapter, you will be able to:

1. List the fifteen divisions in the plant kingdom, and explain how they are further divided on the basis of habitat, and the presence or absence of vascular tissues, seeds, and flowers.
2. Compare and contrast the life cycle of plants with that of animals.
3. List the various groups of algae and their adaptations for gathering sunlight at various water depths.
4. Explain which groups of algae are likely ancestors of the land plants.
5. Identify the period in which plants first colonized land, and describe the terrestrial environment of that time.
6. List and discuss the advantages and challenges of terrestrial life.
7. List the anatomical adaptations necessary in order for a plant to make the transition from an aquatic to a dry environment.
8. List the nonvascular land plants, and briefly summarize their reproductive strategies.
9. Explain why nonvascular land plants are restricted to small size and moist habitats.
10. List the four groups of seedless vascular plants, and outline their reproductive strategies.

# CONCEPTS IN REVIEW

## Section I    The Plant Kingdom and Its Major Divisions

In order to be classified as a plant, an organism must usually display four characteristics: (1) it must be multicellular (some species of algae are single-celled); (2) photosynthetic pigments such as chlorophyll must be present; (3) the organism must be nonmotile; and (4) it must show alternation of generations. Kingdom Plantae comprises fifteen plant divisions (containing some 300,000 species) that can be broken down as follows: algae (three divisions) and the land plants (twelve divisions). The land plants are split into nonvascular and vascular plants; the vascular plants are then subdivided into seedless and seed plants; and finally, the seed plants are divided into gymnosperms and angiosperms (text pp. 528–529). This is an evolutionary classification, as described in Chapter 19.

## Section II    The Basic Plant Life Cycle

Some animals and animal-like species share many of the four characteristics of plants (including chlorophyll), but the *alternation of generations* is wholly unique to members of the kingdom Plantae. In animals, haploid gametes fuse at fertilization to form a diploid zygote, which in turn develops into a diploid adult. The gametes (eggs and sperm) are almost always very small, even microscopic in many species. In plants, however, it is possible to have very large adults that may be either haploid or diploid. The haploid organism produces gametes and so is called the gameto-phyte; the diploid organism produces spores and is called the sporophyte (text p. 529). Plants are also different from most animal species in that they may propagate by asexual means, and may go into extended periods of dormancy.

## Section III    The Algae: Diverse Aquatic Producers

The algae comprise an extremely large group of aquatic (both marine and fresh water) plants that range from single-celled species to huge multicellular organisms. From an evolutionary perspective, they are the oldest of Earth's plants. Unicellular algae are extremely important because they are primary producers in virtually every aquatic ecosystem (text p. 532).

Green, red, and brown algae derive their names from the colors of their photosynthetic pigments; the pigment colors reflect adaptations to the different water depths at which the plants live. The various lineages reveal clear evolutionary trends in their structures, modes of reproduction, and the relative development of their haploid and diploid generations.

Green algae (Chlorophyta) live near the water surface, have the pigments chlorophyll *a* and *b*, and store food as starch. They may occur as siphonous, volvocine, or filamentous species (text pp. 532–535). *Chlamydomonas*, a well-studied volvocine alga, shows the reproductive complexity typical of this group. The unicells may reproduce asexually, producing motile zoospores that mature into haploid adults. Under stressful conditions, however, sexual reproduction takes place (text p. 533). *Chlamydomonas* and other volvocine algae can also aggregate into colonies, which may exhibit oogamy—sexual reproduction by means of large immobile eggs and smaller, swimming male gametes. By contrast, most algae show the more primitive isogamy—that is, there is no true egg, and male and female gametes are the same size (text p. 534).

Filamentous algae show specializations in body form, such as the holdfast cell, and in alternation of generations. In *Ulva*, both the gametophyte and sporophyte are multicellular, as they are in land plants (text p. 535).

The red algae (Rhodophyta) occur at a variety of depths, store food as floridean starch, and have chlorophylls *a* and *d* as well as the blue and red pigments known as phycobilins. Coral reefs are built from the calcium carbonate skeletons of coralline algae. Most red algae have a life cycle that includes alternation of generations.

The brown algae (Phaeophyta) can occur at very deep environments, contain chlorophylls *a* and *c,* and store their food as laminarin. Some brown algae are among the largest of all plants—giant kelps can reach lengths of over 100 meters. As a result of parallel evolution, many of the brown algae possess anatomical structures that are analogous to structures in land plants. These include large fronds, or areas of tissue for the collection of sunlight at different depths (analogous to leaves), stemlike stripes for frond support, and a number of "holdfast" structures that provide anchorage (analogous to the roots of higher land plants). Some cells of brown algae may even be specialized for the transport of materials, similar to the true vascular tissue of land plants (text p. 537).

In most brown algae, both alternating generations are multicellular. However, *Fucus* and its relatives show a life cycle that is unique in the plant kingdom—they lack alternation of generations (text p. 538).

## *Section IV*    Plants That Colonized Land

Biologists believe that land plants arose from species of geen algae some 400 million years ago. These first colonizers faced several problems: the need for support of the plant body (formerly supplied by water); the need for waterproofing (to halt dessication); the need for a mechanism of gas exchange in air and for nutrient uptake from soil; and problems associated with reproduction away from water (text pp. 539–540). Land plants evolved differing solutions to these challenges, so that the various groups, like the algae, follow a pattern of progressive structural specialization.

The nonvascular land plants lack specialized tissue for conducting water and nutrients. They are also relatively small in size, and they occupy moist environments. The most conspicuous generation is the gametophyte, which has structures analogous to roots, stems, and leaves. The three modern divisions are Hepatophyta (liverworts), Anthocerotophyta (hornworts), and Bryophyta (mosses). In all three groups reproduction requires water, which is necessary in order for the motile sperm to reach the egg. The most numerous and familiar nonvascular land plants are the mosses (text pp. 541–542).

The primitive vascular land plants, including the divisions Psilophyta (whisk ferns), Lycophyta (club mosses), Sphenophyta (horsetails), and Pteridophyta (ferns), all possess two kinds of vascular tissue made up of tracheid cells (xylem and phloem). They also have a superficial layer of cutin to inhibit water loss, and multicellular embryos that are protected within the archegonium on the diploid sporophyte. All lack seeds, however, and require water for reproduction. At one time members of this group comprised a major portion of Earth's land flora.

The whisk ferns lack roots and leaves, although these structures, along with stems, appear in the club mosses. The horsetail rushes (also called joint grass, or snake grass) contain a large amount of silica to add strength to their stems. Ferns are probably the most "normal" in appearance among the primitive vascular plants, largely because of their leaflike fronds (text pp. 542–545).

## KEY TERMS

| | | |
|---|---|---|
| alga  *text page* 528 | holdfast  534 | Pteridophyta  544 |
| alternation of generations  529 | homosporous  544 | Rhodophyta  535 |
| anisogamete  534 | isogamete  533 | siphonous alga  532 |
| antheridium  538 | isogamy  534 | sorus  545 |
| Anthocerotophyta  541 | Lycophyta  543 | Sphenophyta  544 |
| Bryophyta  540 | megaspore  544 | sporophyll  543 |
| Chlorophyta  532 | microspore  544 | sporophyte  529 |
| coralline alga  536 | nonvascular plant  540 | stipe  537 |
| filamentous alga  534 | oocyte  534 | strobilus  544 |
| frond  537 | oogamy  534 | tracheid  542 |
| gametophyte  529 | oogonium  538 | vascular plant  542 |
| gemma cup  541 | Phaeophyta  537 | volvocine alga  534 |
| Hepatophyta  540 | phycobilin  535 | zoospore  533 |
| heterosporous  544 | Psilophyta  543 | zygospore  533 |

# SELF-QUIZ: TESTING WHAT YOU HAVE LEARNED

## Matching Key Terms

Match each term on the left with the most appropriate description on the right.

| | |
|---|---|
| 1. sorus | a. dominant haploid or diploid stages |
| 2. gametophyte | b. plant that produces diploid spores |
| 3. holdfast | c. analogous to leaves |
| 4. alternation of generations | d. attaches algae to a substrate |
| 5. isogamy | e. stemlike |
| 6. oocyte | f. male spore |
| 7. megaspore | g. lycophyte spore-producing organ |
| 8. stipe | h. plant that produces haploid cells |
| 9. Psilophyta | i. resemble the first land plants |
| 10. gemma cup | j. groups of fern sporangia |
| 11. microspore | k. female spore |
| 12. sporophyte | l. water transport |
| 13. strobilus | m. liverwort asexual reproduction |
| 14. fronds | n. male and female gametes the same size |
| 15. tracheid | o. a large, immobile egg cell |

## True or False?

1. _____ The gametophyte generation of plants is always small and inconspicuous.

2. _____ Among algae, different pigments usually correspond to different water depths.

3. _____ Green algae are believed related to land plants because of their chlorophyll types and modes of energy storage.

4. _____ Zygospores are always diploid.

5. _____ Kelp vessel cells serve the same function as the phloem of higher plants.

6. _____ Plants colonized the land before the breakup of Pangaea.

7. _____ Liverwort guard cells are crescent-shaped.

8. _____ Gemma cups can be considered clones.

9. _____ Leptoids transport water and other fluids.

10. _____ *Psilotum* looks much like the early land plants, but it lacks roots.

11. _____ The adult fern is diploid.

## Completion

1. The basic plant life cycle is summed up as _____ of _____.

2. In the plant life cycle, the gamete-producing generation is called the _____.

3. The oldest plant group is the _____.

4. Algae are _____ in almost all aquatic food chains.

5. The most likely ancestors of the first land plants are found in the division _____.

6. Red algae photosynthetic pigments that absorb light well at all wavelengths are collectively called _____.

7. Plants first colonized dry land over 400 million years ago, during the _____ period.

8. Mosses belong to the division _____.


## Short Answer

1. How is the basic plant life cycle different from that of animals?

2. What are the adaptations that allow various species of algae to live at various depths?

3. What adaptations of large, multicellular algal species are analogous to anatomical parts of higher land plants?

4. What physical problems were faced by the early land-colonizing plants?

5. Why are nonvascular land plants always small?

6. What group of living land plants most closely resembles the first land colonizers? Why?

## Multiple-Choice Review

In the following sentences, fill in any blanks. Complete each statement by circling the correct response.

1. Sori are found on:

   a. algae.
   b. ferns.
   c. bryophytes.
   d. club mosses.
   e. quillworts.

2. Plants that have _____ without guard cells are the:

   a. mosses.
   b. ferns.
   c. bryophytes.
   d. algae.
   e. liverworts.

3. Red algae:

   a. store carbohydrates as floridean starch.
   b. can build coral reefs.
   c. possess phycobilins.
   d. both a and c
   e. all of the above

4. Long stipes in phaeophytes:

   a. serve as a place to store starch.
   b. permit the existence of different pigment types.
   c. allow the plant to reach the water surface.
   d. provide increased area for photosynthesis.
   e. permit the presence of true phloem.

5. Gemma cups are found in:

   a. liverworts.
   b. algae.
   c. bryophytes.
   d. quillworts.
   e. hornworts.

## Exercise

Complete the following chart.

|  | Division/Class | Seed/Seedless | Vascular/Nonvascular |
|---|---|---|---|
| Fern |  |  |  |
| Green alga |  |  |  |
| Tulip |  |  |  |
| Moss |  |  |  |
| Red alga |  |  |  |
| Fir tree |  |  |  |
| Horsetail |  |  |  |
| Liverwort |  |  |  |

# 24

# THE SEED PLANTS

---

## CHAPTER AT A GLANCE

---

## CHAPTER PREVIEW

Chapter 24 presents a large number of new terms and taxonomic categories, but the chapter content—the characteristics of Earth's seed plants—is straightforward. The chapter begins by explaining the basis for seed plant classification: the features that seed plants possess that set them apart from other plant groups. Against this background you will then see the characters of various subgroups that, in turn, set them apart from each other.

The second main section discusses seed plant evolution. As you will discover, seed plants probably arose during the Permian period as descendants of the so-called seed ferns. Then the text turns to the gymnosperms, or "naked seed" plants, which still exist in large numbers today, but were at their glory during the Mesozoic "age of reptiles." Note that, in this group, the life cycle always involves delayed fertilization (Figure 24-8).

The final section of this chapter moves on to flowering plants, the angiosperms—the plant group that is at its apex today and that comprises about 95 percent of all living seed plant species. You should be impressed with the fact that many angiosperms are inextricably linked with their animal pollinators—insects, bats, birds. Table 24-1 outlines the basic differences between the two main subcategories of flowering plants, the monocots and dicots.

# LEARNING OBJECTIVES

When you have mastered the concepts of this chapter, you will be able to:

1. List the five divisions of seed plants, and give the main characteristics of the members of each.
2. List the taxonomic characteristics that link all seed plants.
3. Tell when the first seed plants appeared, and explain why, in evolutionary terms, they are often described as analogous to reptiles.
4. Name the four major groups of gymnosperms, and outline the reproductive process in each.
5. Name the two major subgroups of angiosperms and the features that distinguish them.
6. List the special reproductive adaptations of angiosperms, and describe the anatomy of a flower.
7. Explain why angiosperms are often compared to mammals in terms of their reproductive strategy.
8. Discuss the importance of flowers, and explain how flowers and fruits of the angiosperms are believed to have coevolved with various animals.

# CONCEPTS IN REVIEW

## Section I    Classification of the Seed Plants

The classification of seed-producing plants follows, as closely as possible, the evolutionary classification principles outlined in Chapter 19. Four of the five seed plant divisions are gymnosperms: those plants that bear "naked seeds" in conelike structures. The second major group of seed plants, the angiosperms, produces a protective endosperm around the seed, carries out "double fertilization," and has flowers.

The characters that unite all of the seed plants (both gymnosperms and angiosperms) as a natural group are: (1) the production of seeds, which contain a developing embryo and the necessary store of nutrients to support it; (2) a female gametophyte that is larger than the male gametophyte, and is contained in a megasporangium surrounded by an integument; (3) a male gametophyte that is very small, and male and female gametophytes that are wholly dependent on the sporophyte; and (4) the fact that water is not necessary for reproduction as it is in lower plants—that is, sperm do not have to swim through a film of water to reach the egg. Other features that unite the group are structures that serve as adaptations for terrestrial life: stems or trunks for support, vascular tissue, large leaves, and others (text p. 550).

## Section II    Evolution of the Seed Plants

Sometime during the Carboniferous period certain groups of ferns acquired the ability to produce seeds. These "seed ferns" thus were the earliest members of a new taxon of plants: the gymnosperms. During the Permian, these new gymnosperms diverged and gave rise to a number of different lineages, including the cycads, conifers, and ginkgoes. The early reptiles arose about the same time, and it is interesting to note that the "age of reptiles" (the Mesozoic) could just as well have been called the "age of gymnosperms" since both groups were in their zenith at that time. Gymnosperms are also analogous to the reptiles in that neither is tied to water for reproduction. Both the gymnosperm seed and the shelled, cleidoic egg of reptiles are adapted to withstand the harsh conditions of dry land (text p. 551).

Toward the end of the Mesozoic era angiosperms appeared in the higher and dryer environments. As they diversified, they pushed out most of the gymnosperms. The heyday of the angiosperms coincides with the Cenozoic, the "age of mammals." The evolution of angiosperms is closely linked to the evolution of numerous insects, small mammals, bats, and birds—all of which act as pollinators.

## Section III    Gymnosperms: "Naked Seed" Plants

Although the gymnosperms are a small group today, there are a number of primitive surviving groups. The Cycadophyta, or cycads, are found throughout tropical and subtropical lands, and are dioecious—that is, the male and female plants are separate. They have a typical gymnosperm life cycle: inside the strobili of female plants, megaspores are produced within the megasporangium, and these develop into the female gametophytes (megagametophytes). Egg cells then form within the many ovules. In male plants microspores are produced in microsporangia, and microgametophytes develop, to be released as pollen before they mature (text p. 553). Pollen grains mature and fertilize eggs in the female strobilus.

The division Ginkgophyta contains the ginkgoes or maidenhair trees, which were abundant in North America, Europe, and Asia during the Mesozoic. Ginkgoes are all dioecious, and although they are gymnosperms they drop their leaves in the fall (are deciduous). The Gnetophyta are a small group of broad-leaved tropical gymnosperms.

By far the most important gymnosperm order (both in numbers and economic importance) is the Coniferophyta, or conifers. Although conifers were once widespread, even in tropical areas (they were a staple of herbivorous dinosaurs), they are now largely confined to high latitudes and colder environments. Most are monoecious, with both male and female cones on a single tree. Male cones have microsporangia, which produce haploid microspores, which in turn produce pollen grains by mitotic division. The pollen is wind-dispersed. Megasporangia on female cones produce haploid megaspores, which give rise to multicellular gametophytes. Pollination occurs when a pollen grain is trapped by a sticky female ovule. Actual fertilization is delayed, however, as the pollen usually matures over a 15-month period before the sperm nuclei actually fuse with the egg nucleus (text p. 557).

## Section IV    Angiosperms: The Apex of Plant Diversity

All angiosperms are grouped into a single division (the Anthophyta) and all produce seeds surrounded by an enclosed ovule. The two large classes of angiosperms, the dicots and monocots, are separated according to the numbers of seed leaves present in the embryo. There are many differences between these classes, involving such characters as placement of the vascular bundle, root structure, leaf shape, and flower structure (text p. 558).

The flowers of angiosperms are a completely new adaptation, not seen in any other type of plant. Each flower is composed of four whorls of parts, much like a wheel-within-a-wheel arrangement. These are—from the outside in—the sepals (usually green and leaflike), the colored petals, the male stamens, and the female carpels. Angiosperm fruits, part of the protective seed "container," are the mature ovary or group of ovaries. Fruits show a number of adaptations for dispersion: they may float long distances on wind currents (cottonwood, milkweed), flutter or glide short distances (maple wings), float on salt water (coconut) or fresh water, stick to the fur of mammals or feathers of birds (numerous cockleburs or sticktights). A fruit may also be highly nutritious and be eaten whole by animals, in which case the seeds may be excreted intact (text pp. 558–559). Co-evolution of flowers and their pollinators has produced specialized physical characteristics in both (text pp. 561–562).

## KEY TERMS

| | | |
|---|---|---|
| angiosperm   *text page 557* | conifer   *554* | cotyledon   *558* |
| Anthophyta   *557* | Coniferophyta   *554* | cycad   *552* |

Cycadophyta  552

deciduous  553

dicotyledon  558

dioecious  553

flower  558

fruit  558

germination  557

ginkgo  553

Ginkgophyta  553

gnetina  553

Gnetophyta  553

gymnosperm  552

integument  553

megagametophyte  553

megasporangium  553

megaspore  553

megaspore mother cell  557

microgametophyte  553

micropyle  557

microsporangium  553

microspore mother cell  557

monocotyledon  558

monoecious  556

ovule  553

pollen grain  553

pollen tube  557

pollination  557

pollinator  559

seed  550

# SELF-QUIZ: TESTING WHAT YOU HAVE LEARNED

## Matching Key Terms

Match each term on the left with the most appropriate description on the right.

1. monocotyledon
2. pollinator
3. fruit
4. deciduous
5. angiosperm
6. ginkgo
7. cotyledon
8. pollen grain
9. dioecious
10. micropyle
11. pollen tube
12. flower
13. integument
14. dicotyledon
15. monoecious

a. maidenhair tree
b. yearly leaf loss
c. male and female cones on same tree
d. opening
e. grows from male cell to female
f. pollen vector
g. flowering plants
h. mature ovary
i. consists of carpels, sepals, petals, stamens
j. seed coat
k. two seed leaves on embryo
l. one seed leaf on embryo
m. a seed leaf
n. male gamete
o. separate sexes on separate plants

## True or False?

1. _____ Seed ferns were true gymnosperms.

2. _____ Pollen grains are the mature male gametes.

3. _____ Large leaves or tiny leaves (needles) can be equally effective at increasing surface area.

4. _____ Deciduous trees always have separate sexes.

5. _____ The micropyle does not pass through the integument.

6. _____ Most human agricultural plants are angiosperms.

7. _____ Gymnosperm reproduction depends on insect pollinators.

8. _____ Gymnosperms are more wasteful of reproductive energy than are angiosperms.

9. _____ Early angiosperm flowers were radially symmetrical.

10. _____ Honey guides are a result of coevolution.

11. _____ Annual plants disperse seeds over great distances.

## Completion

1. The megasporangium is protected by one or two tissue layers, or _____.

2. _____ are scientists who study the evolutionary history of plants.

3. The ovules and seeds of gymnosperms lie "naked" or exposed on the surface of the

   _____.

4. _____ trees are those that shed their leaves periodically (usually yearly).

5. _____ is delayed in pines.

6. An angiosperm fruit is actually a ripened _____.

7. The endosperm nutrient store of angiosperms is only produced by the ovary after

   _____.

8. A tree that produces both male and female gametes is said to be _____.

## Short Answer

1. How were the first seed plants analogous to the reptiles?

2. Why are flowers and insects believed to be the products of coevolution dating to the late Cretaceous?

3. Which groups of plants dominated during the Paleozoic? The Mesozoic? The Cenozoic?

4. Outline the life cycle of a typical pine.

5. What adaptations were necessary for plants to succeed at life on dry land?

6. What is the function of resin?

## Multiple-Choice Review

In the following sentences, fill in any blanks. Complete each statement by circling the correct response.

1. The major key to angiosperm success is:

   a. rapid growth.
   b. a vascular system.
   c. the flower.
   d. a waxy cutin.
   e. seeds.

2. In gymnosperms, most of the food reserves needed to support a growing _____ are created and stored:

   a. in the fruit.
   b. after fertilization.
   c. in the endosperm.
   d. in the integuments.
   e. before fertilization.

3. Many angiosperm species have large leaf surfaces. This feature is probably tied to:

   a. the rapid growth of angiosperms.
   b. a slow rate of growth.
   c. the limited angiosperm growing season.
   d. the need to ensure enough leaf surface after predation.
   e. the need to deflect wind and rain.

4. The dominant vegetation of high elevation, _____ forests, are:

   a. angiosperms.
   b. grasses.

    c. cycads.
    d. ginkgoes.
    e. gymnosperms.

5. The largest division (number of species) of the seed plants consists of the:

    a. gymnosperms.
    b. angiosperms.
    c. seed ferns.
    d. ginkgoes.
    e. cycads.

## Exercise

Complete the following chart.

| | Distinguishing Characteristics |
|---|---|
| Ginkgo | |
| Apple | |
| Pine | |
| Cycad | |

For an apple and a pine tree, what are the unique aspects of each plant's life cycle?

# 25

# INVERTEBRATES: EVOLUTIONARY DIVERSITY

---

## CHAPTER AT A GLANCE

# CHAPTER PREVIEW

Chapter 25 is a fascinating survey of the diverse group of animals that lack a vertebral column—and are thus lumped into the catchall category, "invertebrate." The main theme of this chapter is the increase in organizational complexity that has occurred during animal evolution, and the chapter begins with a look at the advantages and requirements of the first stage in that process—multicellularity. Next come straightforward sections on the basic anatomy and life histories of the phylum Porifera (sponges) and some radial marine animals (the Cnidaria and Ctenophora). Once you have learned something of these extremely simple creatures, you will be prepared for the following discussion on the colonial and syncytial theories of the origin of multicellularity. A small marine worm called the plakula is presented as an example of how some biologists believe primitive multicellular species may have looked.

The next section introduces animals in which more complex structure—organs and bilateral symmetry—appears. Although there are many new anatomical and taxonomic terms, you will find them well organized and summarized in Table 25-4. The final section of the chapter compares and contrasts the two major animal lineages: protostomes such as mollusks and arthropods, and deuterostomes such as echinoderms and chordates. This is extremely important material, so be sure you understand the differences in cleavage, development, and the origins of mesoderm and coelomic cavities, as well as the eventual fate of the blastopore in both lineages.

# LEARNING OBJECTIVES

When you have mastered the concepts of this chapter, you will be able to:

1. List the opportunities and consequences associated with being a multicellular organism.
2. Describe the basic anatomy of a sponge, and explain why it represents tissue-level organization.
3. Describe the basic anatomy of the cnidaria, and explain how a nematocyst works.
4. List the three classes of cnidarians, and explain how their modes of life differ.
5. Compare and contrast the cnidarians with the ctenophores, and explain why each is included in a separate group.
6. Discuss the two major theories of the origin of multicellularity, and give the evidence for each.
7. List the two major characteristics of bilaterally symmetrical animals.
8. List the three major groups of nonsegmented worms, and briefly compare their anatomy and life styles.
9. Name the two major developmental groups of animals, and list the four differences in development that are used to distinguish them.
10. Describe the anatomy of an annelid worm, and compare it with nonsegmented worms.
11. Describe the anatomy and life styles of the arthropods, and explain why they have been so successful.
12. Name the two developmental strategies of arthropods, and list the advantages of each.
13. Describe the basic anatomy of a mollusk, and describe the various molluscan life styles and adaptations.
14. Describe the basic echinoderm anatomy, and outline the functioning of the water vascular system.
15. Describe the basic anatomy of the hemichordates, and discuss the evidence for hemichordate–echinoderm–chordate relationships.

# CONCEPTS IN REVIEW

## Section I    Multicellularity: Opportunities and Consequences

Multicellularity gives organisms the potential for large size; greater mobility; a stable, controlled internal environment; and relative independence from the changeable internal environment. Along with multicellularity came structural adaptations to meet needs associated with larger, more complex animals. These tissues and organs include: muscles; circulatory, respiratory, nervous, and other systems; and sense organs. Another result of increased complexity was the evolution of a variety of simple and complex behaviors (text p. 569).

## Section II    Porifera: The Simplest Invertebrates

Sponges, the members of the phylum Porifera, are hollow, sessile filter feeders. They show the tissue level of organization, and each organism consists of three layers: epithelium, mesenchyme, and flagellated choanocytes. Sponges reproduce sexually and asexually (by budding); among the specialized cells that are found in sponges are some that may function like nerves (text pp. 569–571).

## Section III    Radial Symmetry in the Sea

Cnidarians and ctenophores—aquatic animals that include jelly fishes, corals, sea anemones, and comb jellies—also show the tissue level of organization, as well as radial symmetry. Many cnidarian species show an alternation of generations in the life cycle: a sessile polyp stage later forms a pelagic (free-floating) medusa. Both forms have a three-layered body wall and a central gut, the coelenteron, that opens onto a mouth circled by tentacles. Stinging cells on the tentacles house nematocysts, which contain harpoonlike filaments that can capture prey (text p. 571).

Cnidarians carry out extracellular digestion in the coelenteron. They are able to capture and swallow prey because they have evolved cells analogous to muscles, and a supporting hydroskeleton. They also possess nerve cells arrayed in a nerve net (text p. 573). A peculiar feature of cnidarians such as hydras is the continual replacement of body cells. Members of the phylum reproduce asexually by budding from polyps, or sexually in the medusa stage. Embryos that result from the union of haploid eggs and sperm each develop into a planula—a small, ciliated larva. A planula eventually attaches to a substrate and converts into a polyp, thus completing the alternation of generations (text pp. 573–574).

While cnidarians are likely the simplest animals near the main evolutionary line leading toward higher animals, ctenophores (comb jellies) probably arose independently. A unique ctenophore feature is the eight rows of heavily ciliated cells (combs) that enable the animal to locomote. Ctenophores also have a slightly more complex body design, including a gut with discrete parts, and do not show alternation of generations (text p. 576).

## Section IV    The Origins of Multicellularity

According to the colonial theory, multicellular organisms arose when flagellated protistan cells aggregated into hollow colonies, or formed solid organisms with a ciliated surface resembling the planula of sponges and cnidarians. In contrast, the syncytial theory suggests that ciliated, single-celled protists containing multiple nuclei became "cellularized" when plasma membranes walled off those nuclei into separate cells (text p. 576). Characteristics of the plakula, a three-layered worm that is the most primitive free-living multicellular animal known, may shed light on the postulated early, planula stage of multicellularity. Placed in its own phylum, the Placozoa, the plakula has no head or other axes of symmetry and has many features that resemble those of the solid, ciliated planula larva (text pp. 576–577).

## Section V    Organs and Bilateral Symmetry

Bilaterally symmetrical animals have a head or headlike region, right and left sides that are mirror images of each other, and move predominantly in a single direction. The first organisms that show cephalization and bilateral body symmetry are the flatworms (phylum Platyhelminthes). Flatworms are also the first animals to show the organ level of organization. Their body parts and those of all other, more complex animals develop from three embryonic layers: the ectoderm, mesoderm, and endoderm (text p. 577).

Members of the platyhelminth class Turbellaria, such as planaria, are the most primitive flatworms. Even so, they have a nervous system that includes two eyespots near the anterior end of the body for sensing light, and an organ for sensing gravity. Both connect to aggregations of nerve cells, called ganglia, that function like a primitive brain. All turbellarians are hermaphrodites (text p. 579).

Trematodes and cestodes are all parasitic flatworms. A well-known cestode is the beef tapeworm, which in the course of evolution has lost its gut and consists mainly of proglottids—segments containing gonads with sperm and eggs.

A number of separate evolutionary lines arose from simple flatworm ancestors. Members of the phylum Nemertina (ribbonworms) show a major innovation—a one-way gut with mouth and anal openings at opposite ends. The gut also has specialized regions where different phases of extracellular digestion and nutrient absorption take place. Finally, ribbonworms show an internal transport system in which a clear fluid carries nutrients in "blood" vessels (text p. 582).

The first body cavity is seen in nematodes, or roundworms—the most abundant animals on Earth. This cavity, the pseudocoelom, is a "false coelom" because it is bounded only partly by mesoderm, but it functions basically like the true body cavity of all higher animal phyla (text p. 583). Disease-causing parasitic roundworms that infect humans include pinworms and hookworms and the muscle-boring worm that causes the disease trichinosis.

## Section VI    The Two Major Animal Lineages: Protostomes and Deuterostomes

Different lines of ancient flatworms probably gave rise to the two main lineages of all higher animals: protostomes and deuterostomes. To fully grasp this section, you may want to review the material on gastrulation and cleavage in animal embryos presented in Chapter 16. Also, keep in mind that there are many variations in and exceptions to the general "rules" stated next, and that not all biologists agree about the usefulness of the protostome–deuterostome distinction.

Protostomes have the embryonic developmental pattern in which the blastopore becomes the mouth, and a separate orifice breaks through to become the anus. The reverse takes place in deuterostomes: the blastopore becomes the anus, and a second opening yields the mouth. Protostomes, including segmented worms, arthropods, and mollusks, tend to show spiral cleavage, while deuterostomes, including starfish and their relatives and all vertebrates, have radial cleavage. Also, development in protostomes is determinate, while in deuterostomes it is regulative (text pp. 584–585).

The first protostome phylum Chapter 25 considers is Annelida—the segmented worms. This phylum represents more than 8,000 species in three classes, which include earthworms and leeches. Division of the body into segments was an adaptation related to increasing size and specialization of body parts for particular functions. Annelids also have the first true coelom, which can serve functions as varied as that of a hydroskeleton to that of a site where gametes are discharged before leaving the body (text p. 587). Annelids also have organs for waste removal and a closed circulatory system in which blood circulates within tubelike vessels—a crucial adaptation that permits larger size.

Segmented worms reproduce by both asexual and sexual means. Aquatic polychaetes produce a ciliated larva called a trochophore.

The most successful protostomes—many biologists would say the most successful *animals*—are the nearly one million species of the phylum Arthropoda. This huge group of animals with jointed legs includes crustaceans, spiders, mites, and an estimated 900,000 species of insects. Three basic characteristics that distinguish arthropods from annelids include: (1) segments highly modified for different functions; (2) segmentation lost internally, and the coelom replaced by a blood-filled hemocoel; and (3) an outer, rigid exoskeleton that provides protection and serves as a site for muscle attachment. Arthropods have evolved various systems for respiratory gas exchange, all of which support the high rate of arthropod metabolism (text pp. 590–591).

The attributes of the basic arthropod body plan have allowed a great diversification of forms in the course of evolution. Subphyla include the extinct trilobites; chelicerates such as horseshoe crabs and spiders; crustaceans such as shrimps, crabs, and lobsters; and the subphylum Uniramia, which encompasses centipedes, millipedes, and insects (class Insecta). Major orders of insects and their characteristics are shown in Table 25-7 (text p. 595). Most undergo complete metamorphosis from larva to adult; insects that live in complex societies include some types of termites, ants, wasps, and bees (text p. 597).

The final group of protostomes your text considers is the phylum Mollusca. Mollusks show variations on a four-part body plan: the foot, the visceral mass, the head, and the mantle. They have a true circulatory system, with a heart and an incompressible blood-filled sinus that serves as a hemoskeleton. The major classes of mollusks include Gastropoda—snails and their relatives; Bivalvia—shelled animals in which the shell is secreted in two opposing halves; and Cephalopoda—squids, octopuses, and their relatives. In cephalopods the molluscan foot is modified and terminates in the funnel, and the head is large, central, and gives rise to eight to ten arms or tentacles. Octopuses and squids have camera eyes and show the most complex behavior of all invertebrates (text pp. 600–601).

Deuterostomes include the echinoderms (starfish and their relatives), acorn worms (Hemichordata), and the chordates—animals with notochords—that you will study in Chapter 26. In echinoderms, adults are radially symmetrical but begin life as bilaterally symmetrical larvae. Feathery sea lilies, spiny sea urchins, and cylindrical sea cucumbers reflect this phylum's varied forms. However, all possess a unique water vascular system that is involved in feeding, locomotion, and other processes. In sea stars the system ends in water-filled tube feet through which suction can be applied (text pp. 601–602).

An ancient hemichordate may have been the creature that gave rise to both echinoderms and chordates. All hemichordates are marine worms and have paired sets of gill slits that operate in feeding. Acorn worms (enteropneusts) also have a dorsally situated nerve chord that arises during early development by infolding—just as occurs in vertebrates (text p. 604).

# KEY TERMS

| | | |
|---|---|---|
| Annelida  *text page 585* | Cnidaria  *571* | ganglion  *579* |
| Anthozoa  *574* | coelom  *587* | Gastropoda  *600* |
| Arthropoda  *589* | Crustacea  *592* | Hemichordata  *600* |
| bilateral symmetry  *577* | Ctenophora  *576* | hemocoel  *589* |
| Bivalvia  *600* | deuterostome  *584* | hemoskeleton  *599* |
| cephalization  *577* | Echinodermata  *601* | Hirudinea  *585* |
| Cephalopoda  *600* | exoskeleton  *589* | hydroskeleton  *572* |
| Cestoda  *580* | extracellular digestion  *571* | Hydrozoa  *574* |
| Chelicerata  *591* | filter feeder  *569* | Insecta  *594* |

invertebrate  567

medusa  571

Mollusca  597

Monoplacophora  600

Nematoda  582

Nemertina  581

Oligochaeta  585

Onychophora  590

pelagic  571

Placozoa  576

plakula  576

planula  574

Platyhelminthes  577

Polychaeta  585

polyp  571

Polyplacophora  600

Porifera  569

protostome  584

pseudocoelom  583

radial symmetry  571

Scyphozoa  574

segmentation  587

sessile  569

social insect  597

Trematoda  580

Trilobita  591

trochophore  589

Turbellaria  577

Uniramia  592

---

# SELF-QUIZ: TESTING WHAT YOU HAVE LEARNED

## Matching Key Terms

Match each term on the left with the most appropriate description on the right.

| | | | |
|---|---|---|---|
| 1. segmentation | a. | attached and nonmoving |
| 2. ganglion | b. | sponges |
| 3. pelagic | c. | two-shelled mollusk |
| 4. polyp | d. | free-floating |
| 5. hemocoel | e. | radial jellyfish |
| 6. Bivalvia | f. | forms a medusa |
| 7. medusa | g. | formed by trapping water in coelenteron |
| 8. Trilobita | h. | larvae |
| 9. coelom | i. | three-layered marine worm |
| 10. Cestoda | j. | nerve cells |
| 11. hydroskeleton | k. | tapeworms |
| 12. planula | l. | repeated, similar body parts |
| 13. sessile | m. | filled with blood |
| 14. plakula | n. | extinct arthropods |
| 15. Porifera | o. | body cavity |

## True or False?

1. _____ Sponge amoebocytes are found in the mesenchyme.

3. _____ Sponge spicules are composed of cellulose.

3. _____ Nematocyst response is stimulus-specific.

4. _____ Cnidarian hydroskeletons are incapable of changing volume.

5. _____ Anthozoan life cycles are dominated by a medusa stage.

6. _____ Aggregations of protistan cells are a central theme of the syncytial theory.

7. _____ Flame cells are present in sponges.

8. _____ Proglottids are an example of segmentation.

9. _____ Protostome development is determinate.

10. _____ Humans are deuterostomes.

11. _____ Nymphs are associated with complete metamorphosis.

## Completion

1. A rough family tree of the invertebrates would have two main branches: _____ and _____.

2. The amoebocyte cells of sponges produce the _____, which are composed of calcium carbonate or silica.

3. Cnidarians are split into three classes: the _____, which contain hydras; the _____, which include free-living common jellyfish; and the _____, which are sea anemones and corals.

4. Organ systems and bilateral symmetry are features usually associated with _____ locomotion, and _____, or development of a head region.

5. Nematodes possess a false body cavity, or _____.

6. The presence of very similar eyes in humans, octopuses, and squids is a result of _____.

7. The first gill slits appear in the members of the phylum _____.

8. The most primitive animals with complete, one-way digestive systems are members of the class Nemertina, or commonly called _____.

9. The true worms, or annelids, are characterized by _____ of their bodies.

## Short Answer

1. How is the annelid hydroskeleton used in locomotion?

2. Why are the hemichordates considered close relatives of vertebrates?

3. How does the water vascular system of the echinoderms work? Use a starfish as an example.

4. How do insects "breathe"?

5. What are the developmental differences between protostomes and deuterostomes?

6. Why is the plakula considered a "link" between flatworms and the higher phyla?

7. How do sponges filter feed?

## Multiple-Choice Review

In the following sentences, fill in any blanks. Complete each statement by circling the correct response.

1. Sponges reproduce through:

    a. amoebalike cells.
    b. spicules.
    c. a theatrium.
    d. an osculum.
    e. incurrent pores.

2. Waste disposal by flame cells and ducts is found among:

    a. nematodes.
    b. cnidarians.
    c. flatworms.
    d. sponges.
    e. hemichordates.

3. Asexual reproduction can be found among:

    a. jellyfish.
    b. sponges.
    c. flatworms.
    d. Nemertina.
    e. all of the above

4. In deuterostomes the _____ becomes the anus.

   a. blastomere
   b. outpouch
   c. second opening
   d. first mouth
   e. blastopore

5. Leeches are members of the class:

   a. Hirudinea.
   b. Polychaeta.
   c. Oligochaeta.
   d. Cestoda.
   e. Trematoda.

## Exercise

Identify the three tissue layers in this schematic rendering of a coelomate animal. Color each layer a different color.

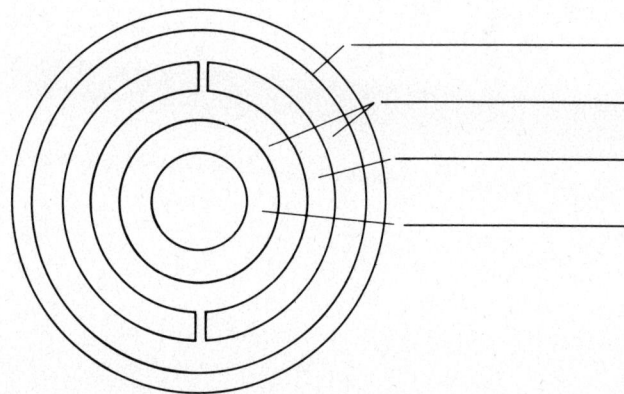

Now, do the same for this acoelomate body plan.

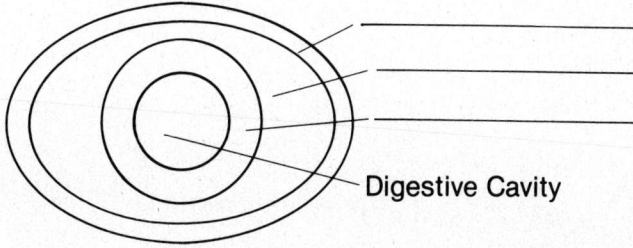

What is a typical acoelomate animal?

# 26
# VERTEBRATES AND OTHER CHORDATES

---

## CHAPTER AT A GLANCE

# CHAPTER PREVIEW

If you are like most people, you are already familiar with chordates, especially vertebrates. In general, it's much easier for us humans to relate to dogs, cats, birds, or fish than to slime molds or flukes. As you'll see in this chapter, vertebrates are extremely successful and diverse, and have adapted spectacularly to life in water, on land, and in the air. Chapter 26 completes our survey of plant and animal diversity. The remainder of the text examines the structural detail, life processes, and interactions of species.

# LEARNING OBJECTIVES

When you have mastered the concepts of this chapter, you will be able to:

1. List the four chordate features, and explain how they are distributed among the three chordate subphyla.
2. Discuss the early vertebrates, including when they first appeared, and explain why they were so small.
3. List the differences between cartilaginous and bony fish, and tell which group of bony fish gave rise to all later land vertebrates.
4. State the two innovations seen in acanthodians, and explain why these changes revolutionized vertebrate history.
5. Describe the origins of amphibians, and list this group's adaptations to life on land.
6. List the reptilian adaptations to life on dry land.
7. Explain the evolutionary origin of birds, and list their adaptations for flight.
8. Point out times of massive extinctions in vertebrate history, list some possible causes, and explain the probable effects of the extinctions on vertebrate evolution.
9. Explain the major reproductive differences among the three subclasses of mammals.
10. List the basic characteristics of mammals, and the major trends of mammalian evolution.

# CONCEPTS IN REVIEW

## Section I    Characteristics of the Chordates

At some point in their life cycle—as adults, juveniles, or embryos—all chordates possess four characteristics. The first is a stiff, rodlike notochord that functions to strengthen the animal body, prevent body shortening, and provide an axis for movement. (In higher vertebrates a vertebral column develops around the notochord.) The second characteristic is gill slits, or openings into the pharynx. These serve as exits for feeding currents in early chordates (and primitive surviving chordates), serve respiratory functions in fishes and some amphibians, and persist as embryological structures in higher vertebrates. Third, all chordates have a dorsal, hollow nerve cord that governs muscle control. Finally, chordates have a block of muscles or myotomes for movement of a tail, which extends beyond the anus (text pp. 609–611).

## Section II    The Nonvertebrate Chordates

Chordates comprise three subphyla: Urochordata (sea squirts), Cephalochordata (the lancelet), and Vertebrata (the backboned animals). Urochordates are marine species that spend most of their life cycle as sessile adults. As adults the only chordate characteristic they show is ciliated gill slits. The tadpolelike larvae, however, possess a notochord, hollow nerve cord, and myotomes. Cephalo-chordates, represented only by the lancelet, are mobile and have all four chordate characteristics as adults.

## Section III    Jawless Fishes: The First Vertebrates

The earliest vertebrates were the ostracoderms: "agnaths," or fish without jaws. Bone was confined to plates of bone lying within the skin; the internal skeleton was composed of cartilage. A definite notochord was present, but vertebrae are unknown in the fossils. Modern agnathans, or cyclo-stomes, the lampreys and hagfishes, are all specialized parasites or scavengers. They have lost the dermal bones, and have skeletons composed entirely of cartilage with only traces of primitively developed vertebrae (text pp. 613–614). Bone may have evolved first as a reservoir for phosphate and calcium storage.

## Section IV    Jawed Fishes: An Evolutionary Milestone

One of the most significant changes in chordate evolutionary history was the development of hinged jaws out of the anterior gill arches (or branchial arches). This adaptation allowed vertebrates to shift from filter feeding to feeding on larger-sized food particles. Jaws first appeared in the acanthodians about 435 million years ago. These fish also developed the first paired pectoral and pelvic fins, and the girdles associated with them. Paired fins allowed much more precise control of body movements, and paved the way for limbs in terrestrial vertebrates (text p. 616). Gill slits in jawed fishes are associated with respiration; the large amount of water moving over their surfaces makes these body sites a "logical" place for gas exchange to take place.

The four large groups of fishes discussed in the text arose in this evolutionary order: acanthodians, then the bony fishes (Osteichthyes) and placoderms, and finally Chondrichthyes—aquatic verte-brates with skeletons of cartilage, including modern sharks, skates, and rays.

From the original group of acanthodians we can derive two lineages of fishes: the Osteichthyes (and later the terrestrial vertebrates), and then, on a completely separate line, the placoderms, which gave rise to the Chondrichthyes. On the Osteichthyes line two major groups of fishes devel-oped: the Actinopterygii, or spiny-finned fish, and the Sarcopterygii, or fleshy-finned fish. Within the Actinopterygii, the teleost fishes are the best known and include over 30,000 familiar species, such as trout and bass. Teleosts owe much of their success to their gas-filled swim bladder, which allows the animal to be neutrally buoyant at nearly any depth (text p. 617). The other group of osteichthyans, the Sarcopterygii, developed internal nostrils, primitive lungs, and—in contrast to the actinopterygian spiny fins—fins that are lobe-shaped and fleshy. Inside each lobed fin is an internal bony skeleton surrounded by muscles. Sarcopterygians include fresh-water lungfish (with paired swim bladders that serve in respiration), the marine coelacanths, and the extinct marine rhipidistians (text pp. 616–617).

## Section V    Amphibians: Vertebrates Invade the Land

The rhipidistians gave rise to the Amphibia in the Devonian, some 350 million years ago. Since paired air bladders that function in respiration and fleshy fins that function as appendages had already appeared in certain sarcopterygians, the stage was set for vertebrates to move onto land. The oldest amphibians are the ichthyostegids, which shared with the rhipidistians certain tooth morphology and the presence of a lateral line sensory system. The vertebral column becomes prominent at this stage (for support on land), and the pectoral girdles, limb bones, and their muscular systems are all strengthened. Early amphibians are sometimes compared to the first land plants because both were still tied to the water for reproduction. Amphibian eggs are laid in water, undergo external fertilization, and develop into water-based larvae. Similarly, the skin still serves a respiratory function in amphibians, and must therefore be kept moist. One bone of the jaw is modified to conduct sound and moves into the inner ear cavity. Some modern amphibians (limited to frogs, salamanders, and legless salamanders) are able to inhabit dry environments by exploiting moist subhabitats, but they must still find water to reproduce (text pp. 619–621).

## Section VI    Reptiles: Adaptations for Dry Environments

About 310 million years ago amphibians gave rise to reptiles through cotylosaurs, the ancestors of the dinosaurs and of the members of the modern class Reptilia. Unlike amphibians, reptiles do not require water for reproduction. Females produce a sealed, cleidoic egg, which in effect carries its own water supply, along with food for the growing embryo. The eggs are enclosed in a hard protective shell, which also prevents dessication; gas exchange for the growing embryo takes place through pores in the shell.

Other reptilian adaptations for a dry environment include a dry skin covered with scales, and efficient lungs made possible by an expandable rib cage. Although internal lungs still must have a moist surface where gas exchange takes place, they are a much more effective adaptation for minimizing water loss than is the amphibian's moist, respiratory skin (text p. 623). Reptiles also concentrate waste products for excretion—another adaptation that reduces the loss of water from body tissues.

Reptiles are often considered analogous to the first seed plants because they could inhabit drier environments and because the cleidoic egg (seed) freed them from water-based reproduction. Reptiles also are the first group of vertebrates to have internal fertilization occurring in all members of the group. The Mesozoic era is often called the "age of reptiles" because of the diversity of reptile lines it gave rise to. Most famous were the dinosaurs, but numerous other large reptiles flourished and occupied niches on land, in water, and in the air. Surviving reptiles include turtles, lizards, snakes, crocodilians, and some primitive lizardlike species (text pp. 623–627).

## Section VII    Birds: Vertebrates Take to the Air

The earliest fossil bird is the Jurassic *Archaeopteryx*. *Archaeopteryx* shares dozens of skeletal features with the carnivorous dinosaurs, so—in a sense—biologists can think of birds as feathered dinosaurs. An older view that birds evolved from crocodiles (because they share four-chambered hearts and a certain arrangement of ear bones) has few modern adherents.

Bird ancestors came from among the fast-running, smaller bipedal carnivores. Feathers are the hallmark of birds (class Aves). They serve as insulation, they streamline and contour flight surfaces, and, in modern birds, they play a role as communication and courtship display structures. The skeleton is extensively modified for lightness, rigidity, and strength. Many of the bones in a bird's body are hollow and filled with air, while others are intricately braced with tiny girders of bone, or reduced to a bare minimum. Fusion between bones also takes place, except for the neck, which remains free to allow head and beak movement.

Birds are warm-blooded, and may have evolved feathers first for insulation rather than for flight. Flight muscles, especially of the shoulder and chest, are well developed (text pp. 628–630).

## Section VIII    Mammals: Warm Blood, Hair, Mammary Glands, and a Large Brain

The mammal-like reptiles, or therapsids, gave rise to mammals in the Triassic, about 200 million years ago. Interestingly enough, this is only about 25 million years after the first dinosaur, so even though dinosaurs and mammals initially appeared close together, reptiles had enough of a "head start" to be the dominant vertebrates throughout the Mesozoic. Mammals showed two major adaptive radiations. During the first, in the Mesozoic, species remained small and inconspicuous in comparison with dinosaurs. The second and more extensive radiation began 65 million years ago at the close of the Mesozoic and beginning of the Cenozoic, or "age of mammals." Mammal characteristics include hair for insulation, milk produced by mammary glands, warm blood (or homeothermy), and large brains. Evolutionary trends among the mammals include: (1) increased specialization, particularly in temperature regulation; (2) increased body size, although this is not a steadfast rule (some mammals specialized by becoming small); (3) diversification of tooth shapes for the different functions of cutting, slicing, and grinding; (4) elongation and specialization of limbs for flight, running, swimming, killing prey, and so on; and (5) increased brain size, the largest brain/body size ratio of all vertebrates (text pp. 631–634).

The three major subclasses of mammals—prototherians, metatherians, and eutherians—are distinguished by mode of reproduction. Surviving prototherians (monotremes) still bear young in eggs. Metatherians (marsupials) bear live embryos, but these must be nurtured in a marsupium (pouch) for some time. The ten orders of placental mammals have a much more efficient early nurturing system (the placenta), and thus can produce larger and better developed youngsters at birth (text p. 635).

## KEY TERMS

| | | |
|---|---|---|
| Acanthodii   *text page 615* | gill slits   *609* | Osteichthyes   *616* |
| adaptive radiation   *609* | Mammalia   *631* | pharynx   *610* |
| Agnatha   *612* | marsupial   *632* | placental mammal   *632* |
| Amphibia   *619* | monotreme   *632* | Reptilia   *622* |
| Aves   *628* | myotome   *610* | Urochordata   *611* |
| Cephalochordata   *611* | nerve cord   *611* | Vertebrata   *612* |
| Chondrichthyes   *616* | notochord   *610* | vertebrate   *609* |
| Chordata   *609* | | |

# SELF-QUIZ: TESTING WHAT YOU HAVE LEARNED

## Matching Key Terms

Match each term on the left with the most appropriate description on the right.

1. Acanthodii
2. Agnatha
3. Aves
4. Cephalochordata
5. Chondrichthyes
6. gill slits
7. marsupial
8. monotreme
9. myotome
10. nerve cord
11. notochord
12. Osteichthyes
13. pharynx
14. Urochordata
15. Vertebrata

a. lancelet
b. animals with backbones
c. bony fishes
d. first jaws, paired fins
e. sea squirt
f. buoyant livers
g. hydroskeleton
h. *Archaeopteryx*
i. pouched mammal
j. surrounded by gill slits
k. lamprey
l. feeding current exits
m. permitted first vertebrates to swim
n. always dorsal and hollow
o. egg-laying mammal

## True or False?

1. _____ The vertebral column develops around the notochord.

2. _____ The Cyclostomata were early land plants.

3. _____ Acanthodians were all small fishes.

4. _____ Ostracoderms are the only vertebrates lacking vertebrae.

5. _____ Lateral lines are fossil tracks left by early vertebrates.

6. _____ Reptile eggs have pores to lighten their weight.

7. _____ Mass extinctions may be cyclic.

8. _____ Scientists agree that dinosaurs were warm-blooded.

9. _____ Birds evolved from pterosaurs.

10. _____ The placenta allows eutherians to be born large.

11. _____ The lagomaorpha were slowly evolving reptiles.

12. _____ The first birds were bipeds.

## Completion

1. An important vertebrate endocrine gland, the _____, originated in jawless fishes.

2. The first jaws and paired fins appeared during the _____ Period, about 435 million years ago.

3. The _____ is a gas-filled organ that allows buoyancy at various depths in the water.

4. _____ were the earliest amphibians.

5. A major innovation in reptiles was the _____ egg.

6. The _____ mammals evolved into dozens of species in Australia and South America.

7. Modern lizards, snakes, and iguanas are placed in the order _____.

## Short Answer

1. Why are early reptiles considered analogous to early seed plants?

2. Why are early amphibians considered analogous to the first land plants?

3. Define *adaptive radiation* and give two examples from the text.

4. What environmental factor might have "encouraged" the vertebrate transition to life on land?

5. Why were all mammals so small and inconspicuous during the age of reptiles?

6. Why are marsupials born so small?

## Multiple-Choice Review

In the following sentences, fill in any blanks. Complete each statement by circling the correct response(s). Note that some questions may have more than one answer.

1. The most likely ancestors of land vertebrates are the _____ fish because they:

   a. lived at the right time in geologic history.
   b. have true jaws.
   c. have fleshy fins with internal skeletons.
   d. have the ability to gulp air.
   e. have four-chambered hearts.

2. Adults of the tiny lancelet (*Amphioxus*) are placed in the subphylum _____; they are in the phylum Chordata because they have:

   a. a notochord.
   b. hollow nerve cords.
   c. gill slits.
   d. myotomes.
   e. fins.

3. Members of the subphylum Urochordata are commonly called _____. They are chordates even though the adults lack:

   a. gill slits.
   b. a nerve cord.
   c. vertebrae.
   d. myotomes.
   e. a notochord.

4. The agnathan fish called _____ were the first vertebrates. They showed two major innovations over lower chordates, namely:

   a. a differentiated head region.
   b. plates of bone in the skin.
   c. muscle-powered intake of a feeding current.
   d. gill slits in the adult.
   e. a hollow nerve cord.

5. The first fish with true jaws were members of the class _____. Their jaws were derived by evolution of:

   a. the ostracoderm skull.
   b. ostracoderm bony plates.
   c. anterior gill arches.
   d. anterior segments of notochord.
   e. ear bones.

## Exercise

Label the four basic chordate features in the diagram below. You may use text Figure 26-1 to check your work.

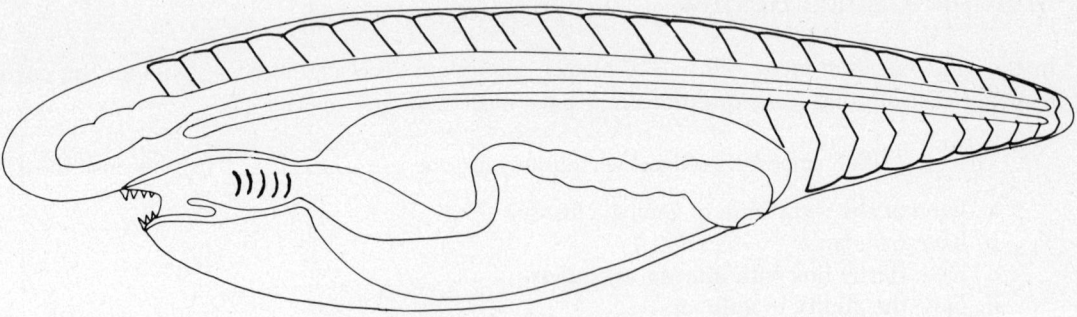

# 27
# THE ARCHITECTURE OF PLANTS

---

## CHAPTER AT A GLANCE

---

## CHAPTER PREVIEW

It is difficult to imagine our Earth spinning in space for over 200 million years with its land masses wholly barren of plants. Yet the colonization of dry environments by water-based life is certainly one of the most significant events in Earth's history. Chapter 27 opens with a discussion of the adaptations plants needed to evolve in order to inhabit dry land, including leaves, roots, stems, and seeds.

The following sections discuss each of these adaptations, beginning with leaves. As you will discover, the efficiency of leaves in collecting sunlight depends on a whole suite of characteristics, including size, shape, position, and waterproofing adaptations. Stems, considered next, are the main plant structures for support and transport, and the pipelines that ultimately connect leaves to roots. Roots can be classed into wide varieties of gross anatomical types, depending on what they look like and do: taproots for reaching deep water, fibrous roots for increased absorption, and prop roots that aid stems in providing support.

# LEARNING OBJECTIVES

When you have mastered the concepts of this chapter, you will be able to:

1. List the structures and adaptations that plants have evolved to meet the demands of life on land.
2. Explain why analogies are often drawn between the first land plants and amphibians, between the first seed plants and reptiles, and between the first angiosperms and mammals.
3. State the morphological evidence used to divide angiosperms into two broad groups: the monocots and dicots.
4. Draw a generalized sketch of a leaf, labeling and explaining the function of all photosynthetic and transport structures.
5. Draw and explain the functioning of a stoma and its associated structures, and list the adaptations leaves have for limiting water loss.
6. Draw a generalized cross section of a stem, and label and explain the functions of all structures in both monocots and dicots.
7. Draw and label a generalized cross section of a root.
8. List the functional and morphological differences between taproots and branching (or fibrous) root systems.
9. Explain why roots do not produce a cuticle, and explain the function of root hairs.

# CONCEPTS IN REVIEW

## *Section I*   Structures to Meet the Challenges of Life on Land

Vascular seed plants today grow in all areas of the Earth because they evolved structures that allowed them to meet the challenge of functioning on land. Leaves, stems, and roots, plus a vascular system and seeds were the five basic innovations that enabled plants to overcome the dangers of dessication, substantial daily and seasonal shifts in temperature, a limited water supply, and other problems of terrestrial existence (text p. 643).

Almost 400 million years ago club mosses and horsetails had already evolved all of these adaptations except seeds. Later, the gymnosperms evolved the ability to produce naked seeds, which eliminated the need for standing water during fertilization; and later still, the angiosperms, or flowering plants, produced protected seeds that develop inside flowers and fruit. With this innovation, angiosperms became the most abundant and diverse plants on Earth. They are divided into two groups: monocotyledons (monocots) and dicotyledons (dicots; Figure 27-1). The leaves, stems, and roots of the two groups differ in large and small ways (text p. 645).

## Section II   Leaves: Living Collectors of Solar Energy

To accommodate three conflicting needs, leaves have a large light-absorbing surface for collecting sunlight, a waterproof coating for preventing water loss, and pores in the watertight seal for gas exchange. Higher plants also have a transport system to carry water and nutrients between the leaves and the rest of the plant. The actual mechanisms that accomplish transport operate at the cellular level. The epidermis, a layer of cells usually one cell thick, protects upper and lower leaf surfaces and reduces water loss by secreting a transparent, waxy covering, or cuticle. Scattered throughout the epidermis are thousands of tiny pores called stomata, each one bounded by a pair of guard cells. The only epidermal cells that contain chlorophyll are the guard cells, which regulate the opening and closing of the stomatal pores and thus affect the movement of gases into and out of the leaf. The leaf's upper surface has many more stomata than the lower surface, and 90 percent of the water lost by a plant exits through the stomata.

Just inside the epidermis lies the mesophyll—the leaf's main photosynthetic tissue, composed of parenchyma cells that house many green, chlorophyll-containing chloroplasts. There are two types of parenchyma cells in dicots, one adapted for photosynthesis and the other for gas exchange (text p. 649). Within the mesophyll, veins composed of xylem and phloem cells help maintain leaf shape and also serve as a transport system that is continuous with the vascular system of the rest of the plant (text p. 650).

Leaves come in a range of shapes, colors, and sizes that represent adaptations to particular environments. A leaf's size, shape, and position on the plant determine how efficient it is at collecting solar energy. The leaves of most dicots have two regions: an enlarged, flattened blade and a stemlike petiole. The broader and flatter the blade, the more light it captures and the more radiant energy it makes available for conversion to chemical energy by photosynthesis. Many plants can modify their leaves to adjust to changing environmental conditions (text pp. 650–651).

## Section III   The Stem: Support and Transport

The stem of a typical dicot has four concentric zones of tissue. Moving from the outside inward, these zones are epidermis, cortex, vascular tissue, and pith. Together they carry out the critical functions of transport and support. The stems of some plants are modified for storage or protection (text p. 655).

Stem epidermis is similar to leaf epidermis and is always present in young plants, though not in many older ones. Just inside the epidermis is the cortex, composed of collenchyma and sclerenchyma cells specialized for strength and support (text pp. 651–652). The next smaller concentric zone consists of vascular tissue made up of xylem and phloem, which occur together in bundles. At the center of the dicot stem is the pith, composed of parenchyma cells specialized for storage. Monocots have the same four stem structures, but they are arranged differently (text p. 653).

Transport tubes are a mainstay of vascular plants. Xylem contains long, tapering tracheids (found in all vascular plants) and shorter, broader vessel elements (found only in angiosperms). Xylem transports water only after the xylem cells have died and no longer contain cytoplasm. Phloem, by contrast, must be alive to carry out its function of carbohydrate transport. It is composed of sieve tube elements stacked end to end to form sieve tubes. Perforated sieve plates occur at the ends of the sieve tube elements, and each element is associated with a living companion cell (text pp. 653–654). A layer of cells called the procambium separates the xylem from the phloem in young stems; the procambium becomes the vascular cambium in older stems.

## Section IV   The Root: Anchorage and Uptake

Roots absorb water, minerals, and oxygen from the soil and serve as a plant's anchor. The concentric zones of a root's cross section are similar to those of the stem, but the few important modifications help clarify root function (Figure 27-14).

Since its principal challenge is efficient water uptake, not the prevention of evaporation, the root epidermis does not usually excrete cutin. Instead, hairlike extensions of epidermal cells increase the root's absorptive surface, and water moves into the epidermal cells by osmosis across the plasma membrane of these root hairs. Just inside the epidermis lies the root's cortex, which is composed of nonphotosynthetic parenchyma cells. Water and minerals traverse the cortex by either the apoplastic (outside cells) or symplastic (inside cells) pathway to move from the root to other parts of the plant (text p. 657).

The innermost layer of the root cortex is the endodermis, which acts as a barrier that regulates the flow of water and nutrients into the central vascular cylinder (text pp. 657–658). With its suberin-impregnated Casparian strips, the endodermis also keeps ions from flowing back out of the xylem to the rest of the root and the soil. At the center of the root is the stele, which consists of pericycle, xylem, phloem, and (in monocots) pith; the core of dicots also has a vascular cambium. The pericycle, just inside the endodermis, generates new, lateral roots in a process that distinguishes roots from both stems and leaves (text p. 658). The arrangement of xylem and phloem in vascular bundles differs from plant to plant (text p. 658).

There are two basic types of root systems: the taproot system derived from the embryonic root, or radicle; and the fibrous root system derived from stem tissue. Prop roots are one kind of root specialization that enables plants to live in unusual places. Carbohydrate storage is another specialized root function. The amount of air (and thus oxygen) in soil determines how well a root system, and therefore the whole plant, grows. Farmers cultivate soil to aerate it; earthworms do a better job naturally.

## KEY TERMS

adventitious root   *text page* 659

apoplastic pathway   657

blade   645

Casparian strip   657

collenchyma   651

companion cell   653

cortex   651

cuticle   648

cutin   647

endodermis   657

epidermis   647

epiphyte   660

fiber   652

fibrous root system   659

ground parenchyma   653

guard cell   648

herbaceous   651

lateral root   658

leaf   645

mesophyll   648

midrib   650

palisade parenchyma   649

parenchyma cell   649

pericycle   658

petiole   645

phloem   653

pith   653

procambium   653

prop root   660

radicle   659

root   655

root hair   656

sclereid   652

sclerenchyma   651

sessile blade   646

sieve plate   653

sieve tube   653

sieve tube element   653

spongy parenchyma   649

stele   658

stem   651

stoma   648

suberin   657

symplastic pathway   657

taproot system   659

tracheid   653

vascular bundle   653

vascular cambium   653

vascular tissue   653

vein   649

vessel element   653

xylem   653

# SELF-QUIZ: TESTING WHAT YOU HAVE LEARNED

## Matching Key Terms

Match each term on the left with the most appropriate description on the right.

| | | |
|---|---|---|
| 1. epiphyte | a. | regulate opening and closing |
| 2. procambium | b. | leaf pore |
| 3. tracheid | c. | water pipe |
| 4. stoma | d. | long, tapering, and tubelike |
| 5. blade | e. | storage tissue |
| 6. phloem | f. | air plant |
| 7. xylem | g. | deep, subsurface water |
| 8. pith | h. | carbohydrate transport |
| 9. taproot system | i. | lies between the xylem and phloem |
| 10. guard cell | j. | flattened region |

## True or False?

1. _____ Guard cells have the same basic shape in monocots and dicots.

2. _____ Most monocots have a taproot system.

3. _____ Sclerenchyma cells are derived from the parenchyma.

4. _____ Procambium gives rise to the vascular cambium.

5. _____ Vessel elements are approximately twice the diameter of tracheids.

6. _____ Sieve tube elements are always associated with a companion cell.

7. _____ Orchids are epiphytes.

8. _____ Apoplastic water lies between cell walls.

## Completion

1. Angiosperms first appeared in the _____ period, about 120 million years ago.

2. Palisade parenchyma cells are _____-shaped, and are arranged in a layer below the _____ .

3. _____ occurs at the center of a dicot stem, and functions as a storage tissue.

4. The _____ lies on the inside of the endodermis, and surrounds the xylem and phloem.

5. Trees that live in habitats with soft substrata often show a number of _____ that help support their weight and hold them upright.

6. In monocots the vascular bundles are found scattered throughout the _____ .

## Short Answer

1.  What are the anatomical characteristics that distinguish the monocots and dicots?

2.  How do stomata open and close?

3.  What specific adaptations do leaves have to prevent water loss?

4.  How is the uptake of water, oxygen, and minerals accomplished by roots? What are the types of roots? And how is their structure related to their functions?

5.  What major advances were made by land plants during the Triassic and Jurassic periods, or about 230 to 120 million years ago?

6.  What mechanical and physiological functions are served by stems?

## Multiple-Choice Review

In the following sentences, fill in any blanks. Complete each statement by circling the correct response.

1. Pith is a _____ tissue found in:

    a. monocot stems.
    b. monocot roots.
    c. dicot roots.
    d. dicot stems.
    e. dicot bark.

2. In vascular bundles the xylem is almost always:

    a. facing the epidermis.
    b. toward the center of the stem or root.
    c. on the inside of the epidermis.
    d. exterior to the phloem.
    e. all of the above

3. In monocots:

    a. leaf veins tend to be branching.
    b. root systems tend to be deep taproots.
    c. there are two seed leaves.
    d. there is no vascular cambium.
    e. there is no pith.

4. Phloem transport cells:

    a. are living.
    b. have several nuclei.
    c. are never associated with sieve tube elements.
    d. never transport carbohydrates.
    e. are nonliving.

5. Cortical cells are separated from vascular tissues by the:

    a. stele.
    b. epidermis.
    c. endodermis.
    d. bark.
    e. pith.

## Exercise

In the diagram below, fill in the parts of a leaf, including the upper and lower epidermis, the palisade parenchyma, spongy parenchyma, stoma, mesophyll region, guard cells, and any veins.

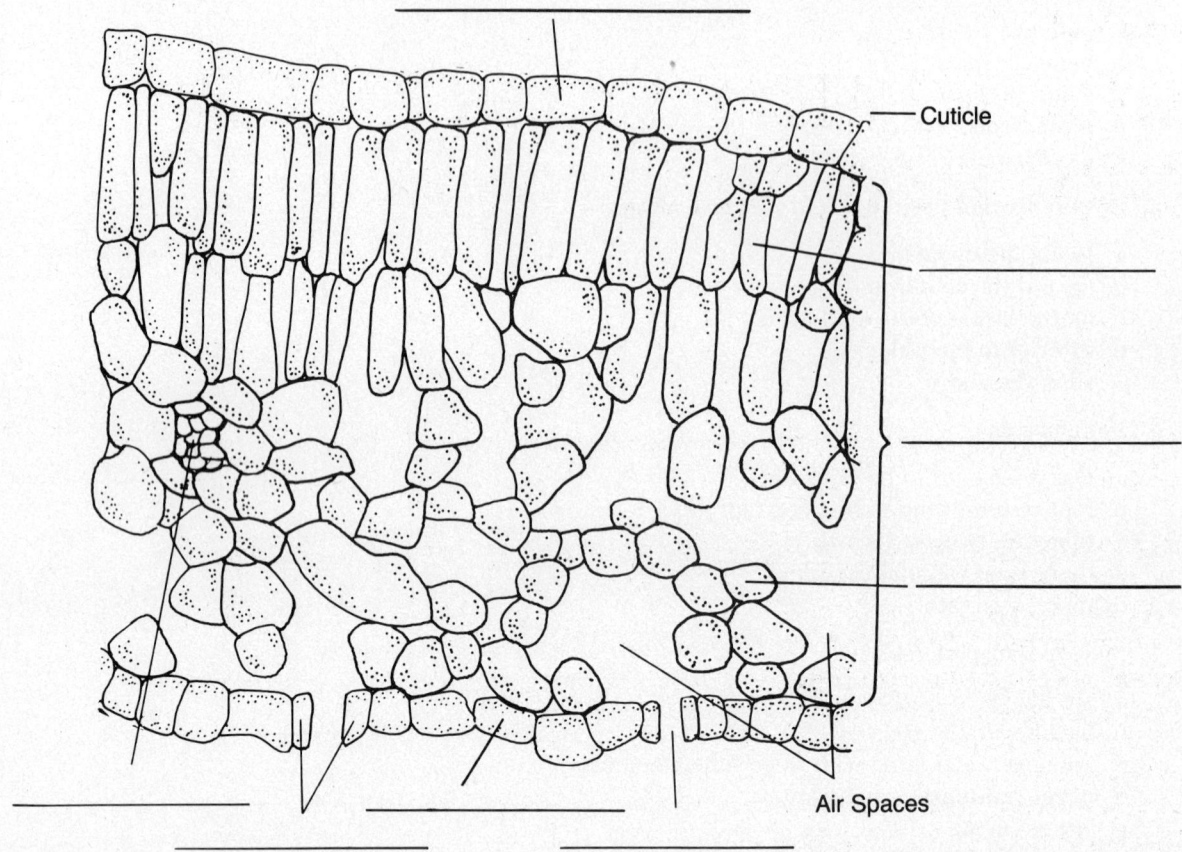

Now, briefly describe the function of each structure you identified.

# 28
# HOW PLANTS REPRODUCE, DEVELOP, AND GROW

## CHAPTER AT A GLANCE

## CHAPTER PREVIEW

If you have ever mowed a lawn, planted and weeded a garden, or trimmed around a sidewalk and remarked how the plants were all "in a string," then you will be well prepared for Chapter 28.

The first section reviews the important concept of alternation of generations presented earlier, and then looks at vegetative reproduction—the plant reproductive strategy accomplished via rhizomes, stolons, bulbs, tubers, and "stringlike" runners. Next, the text considers the gross anatomy of flowers, and the reproductive processes of microsporogenesis and megasporogenesis. These latter topics tend to be complicated, so be sure to become familiar with Figures 28-5 and 28-8. In the following section, take care to note the differences between the embryos of monocot and dicot plants. Other important variations occur in germination and seedling development.

With the basics of plant reproduction and development as background, Chapter 28 briefly discusses an intriguing subject—the comparative development of flowering plants and mammals. As you will see, in these two highly complex groups the similarities and differences are striking. The following two chapter sections cover primary and secondary plant growth; you will find the text illustrations useful in clarifying the features of each process. Finally, the chapter closes with a description of the three types of plant life spans: annual (a single year or growing season), biennial (two years), and perennial (many years).

## LEARNING OBJECTIVES

When you have mastered the concepts of this chapter, you will be able to:

1. Explain how vegetative reproduction produces plants that are genetically identical, and list the advantages of this type of reproduction.
2. List and define the various structures involved in vegetative reproduction.
3. Draw and label the parts of a typical angiosperm flower.
4. Outline the gamete-producing processes in angiosperms.
5. Outline the sequence of events that takes place in pollination, germination, and fertilization.
6. Draw and label the parts of a typical seed, trace its development from fertilization, and explain how it is equipped for survival and dispersal.
7. Define primary growth and explain where it occurs.
8. Draw two freshly sprouted seed plants (a monocot and dicot) and label all parts.
9. Draw a cross section of a growing root and label its three major zones.
10. Explain how leaves and branches are produced by the apical meristem.
11. Describe secondary growth, and list the tissues produced by this process.
12. Define the term *wood*, and list the major wood types.
13. Define the terms *annual*, *biennial*, and *perennial*, and describe the conditions for which each of these strategies is adapted.

## CONCEPTS IN REVIEW

### Section I  Vegetative Reproduction: Multiplication Through Cloning

The emergence from the parent's body of new plants genetically identical to the parent and each other is known as vegetative reproduction. If a plant is well suited to its environment, vegetative reproduction—a kind of cloning—can be a distinct advantage (text p. 665).

While modified leaves, stems, and roots can all give rise to new individuals, the most familiar vegetative reproduction occurs via stem structures: rhizomes, stolons, runners, bulbs, corms, and tubers (text pp. 666–667). Farmers and gardeners hasten the natural process of asexual reproduction by cutting corms and tubers and planting the pieces, and, in woody plants, by grafting.

There is one great disadvantage to vegetative propagation: the lack of genetic variation makes whole clones of plants susceptible to disease, unfavorable changes in the environment, or other negative factors (text pp. 668–669).

## Section II    Sexual Reproduction: Multiplication and Diversity

Flowers are the overall structures of sexual reproduction in angiosperms and are an example of the diversity fostered by the process. A group of modified leaves, flowers consist of four whorls of parts: an outer ring of green, photosynthetic sepals; colored petals that attract pollinators; stamens that produce microspores which develop into pollen; and a carpel or set of fused carpels in which eggs develop and fertilization takes place (text pp. 671–674).

Pollination occurs when molecules that make up the surface of a pollen grain interact with proteins and sugars on the outer surface of the carpel's upper region. The specificity of this interaction helps ensure that plants mate with others of the same species. After contacting a suitable target, the pollen grain germinates. A pollen tube then grows down to the ovary at the base of the carpel, and it is here that fertilization occurs. Angiosperms undergo double fertilization, forming a diploid zygote and a triploid endosperm cell that will generate the zygote's first source of nutrition (text pp. 674–675).

## Section III    The Development of Plant Embryos

After fertilization the diploid zygote begins to form the embryo, while the triploid endosperm cell undergoes rapid mitotic divisions to create storage tissue that takes up nutrients from the parent plant. Mitosis of the zygote forms cells that will develop into the embryo and the columnlike suspensor that connects the embryo to the ovule wall. Inside cells of the globular embryo give rise to vascular tissue, while outside cells give rise to protoderm, the forerunner of epidermis (text p. 676). In most dicots, the embryo consumes the endosperm while developing, and nutrients are stored in two embryonic leaves, called cotyledons. In grasses and many other monocots, endosperm forms the main component of the mature seed. There is one cylindrical cotyledon, called the scutellum (text p. 676).

Two other precursor tissues also begin to form during early plant development. One, the apical meristem, is an organizing center of actively dividing, undifferentiated cells. In dicots it is wedged between the two cotyledons, while in monocots it occurs at one side of the scutellum. A second precursor is the root meristem, which arises at the opposite end of the embryo. Meristems, unique to plants, are zones where cells for new organs can be generated throughout the life of the plant (text pp. 676–677).

## Section IV    Seeds: Protection and Dispersal of the New Generation

As the embryo grows, an outer seed coat forms and serves as a protective seal around the embryo. At the same time, the ovary at the base of the flower enlarges to surround and protect the seed(s). Most fruits consist of tissue derived from the wall of the ovary.

Seeds disperse only after the embryo they enclose has gained weight; the endosperm and cotyledons have stored food and replaced their water with dry weight; and the embryo has dehydrated (text pp. 678–679). Once dispersed by wind, animals, or other agents, seeds can overwinter or survive a drought by entering a resting state called dormancy. Later, when exposed to the right conditions, a seed germinates (text p. 680).

## Section V    Germination and Seedling Development

Rehydration and increased metabolism contribute to germination—a rupturing of the seed coat and elongation of the embryonic root to form the radicle. Soon after this, the initial stem (hypocotyl) and the future stem (epicotyl) emerge and begin to grow.

For a review of the differences in germination and early development between monocots and dicots, reread text pages 681–682. Both kinds of angiosperms establish their first photosynthetic leaves as soon as possible, and meristematic centers in young seedlings' roots and shoots start generating new cells that grow and differentiate.

## Sections VI and VII    Primary Growth: From Seedling to Mature Plant and Angiosperm and Mammalian Development Compared

New tissues in a young seedling, and a corresponding increase in size, result from primary growth of the rapidly growing root and shoot meristems. In the young root, there are three zones: the meristematic region, the region of elongation, and the region of differentiation. Cell divisions at the front of the meristem produce the root cap. New cells behind the cap plus the lengthening of these cells in the region of elongation cause vertical root growth. Development through cell specialization takes place in the third region, which is marked by root hairs.

The shoot grows upward from the apical meristem, which gives rise to the plant's stem, leaves, and bud primordia (text pp. 683–685). If bud primordia grow into branches, these branches may develop their own apical meristems and thus perpetuate growth.

A comparison of similarities and differences between angiosperm and mammalian development can help clarify how plants have adapted their forms to the functional requirements of their restricted habitats. For example, both groups of organisms protect and nourish their embryos with maternal tissue. Both also share certain details of development (text p. 685). Striking differences are apparent, too. For instance, whereas animal tissues and organs are formed in part by the movements and rearrangements of cells, plant structures such as roots, leaves, and flowers are generated by totipotent meristems in different regions of the plant (text p. 685). And, unlike animals, healthy plants are always growing new organs.

## Section VIII    Secondary Growth: The Development of Wood and Bark

In nonherbaceous plants, a thickening and strengthening of the stem comes from secondary growth. This growth arises from the two regions of lateral meristems: the vascular cambium and the cork cambium.

The vascular cambium is a cylinder of actively dividing cells separating xylem from phloem in both stem and root. Inner cambium cells differentiate to secondary xylem, or wood; outer cambium cells become secondary phloem, a part of bark (text p. 686). The youngest xylem, near the periphery of a trunk or stem, forms sapwood; the older, more central xylem dies, stops conducting, and becomes heartwood. The vascular cambium also contains ray initial cells that differentiate to form a system of lateral transport tubes within the expanding trunk.

The cork cambium develops from the cortex in stems and from the pericycle in roots (text pp. 651 and 658) and lies just beneath the epidermis. It produces cork to replace an epidermis split by an expanding trunk or root. The bark of woody stems includes cork, cork cambium, cortex, and carbohydrate-conducting phloem. This is why girdling (removal of a ring of bark all the way around a tree) can be fatal to the plant (text pp. 686–687).

## Section IX    Plant Life Spans and Life Styles

Plants live one, two, or many years. Annuals germinate, grow, develop, produce flowers and seeds, and die within one growing season. Most of their energy goes toward flowering and seed production.

Less common are biennials, which go through the same cycle in two years. Both annuals and biennials stop growing when the apical meristem is converted to a flower.

Perennials live for many years and often require several years of growth before they flower and produce seeds (text pp. 688–689).

## KEY TERMS

aleurone    *text page 681*

annual plant    *688*

apical meristem    *677*

axillary bud    *684*

bark    *687*

biennial plant    *688*

bud primordium    *684*

bulb    *667*

calyx    *671*

cleavage division    *677*

coleoptile    *681*

cork    *687*

cork cambium    *687*

corm    *667*

corolla    *671*

cotyledon    *677*

dormancy    *680*

double fertilization    *676*

early wood    *686*

epicotyl    *681*

flower    *671*

flower bud    *685*

fruit    *678*

fusiform initial cell    *687*

germination    *674*

heartwood    *687*

hypocotyl    *681*

inside cell    *677*

internode    *684*

late wood    *686*

lateral meristem    *686*

leaf primordium    *683*

lenticel    *687*

megagametogenesis    *673*

megasporogenesis    *673*

microgametogenesis    *672*

microsporogenesis    *672*

node    *684*

outside cell    *677*

perennial plant    *689*

perianth    *671*

polar nuclei    *674*

primary growth    *682*

protoderm    *677*

radial transport    *687*

radicle    *681*

ray initial cell    *687*

region of differentiation    *682*

region of elongation    *682*

rhizome    *666*

root cap cell    *682*

root meristem    *678*

root sucker    *667*

runner    *667*

sapwood    *687*

scutellum    *677*

secondary growth    *686*

seed    *678*

stolon    *667*

suspensor    *677*

terminal bud    *684*

tuber    *667*

vascular ray    *687*

vegetative reproduction    *665*

vertical transport    *687*

wood    *686*

# SELF-QUIZ: TESTING WHAT YOU HAVE LEARNED

## Matching Key Terms

Match each term on the left with the most appropriate description on the right.

1. biennial plant    a. all sepals
2. cotyledon    b. yearly
3. fruit    c. ovary wall

4. stolon              d. secondary xylem
5. root cap cell       e. two-year flower
6. annual plant        f. initial stem
7. wood                g. all petals
8. corolla             h. "crash helmet"
9. hypocotyl           i. aerial stem root
10. calyx              j. seed leaf

## True or False?

1. _____ Corms are underground structures involved in vegetative reproduction.

2. _____ All flower parts are derived from modified leaves.

3. _____ Double fertilization is a normal process in gymnosperm reproduction.

4. _____ The monocot cotyledon is called the scutellum.

5. _____ "Late wood" is derived from spring xylem cells.

6. _____ Biennial plants produce two generations each year.

7. _____ Wood is secondary xylem.

8. _____ Heartwood is nonconducting xylem.

## Completion

1. Rhizomes, runners, stolons, bulbs, corms, and tubers are all structures plants utilize to accomplish _____.

2. Flowers are an angiosperm adaptation to ensure _____ and _____ via a variety of vectors.

3. Microspores differentiate into functional pollen grains during _____.

4. The pollen tube must pass through the _____ at one end of the ovule.

5. In most monocots _____ forms the largest portion of the mature seeds.

6. Most fruits are formed from the mature _____ wall.

## Short Answer

1. What adaptations do angiosperm seeds have for dispersal?

2. What are the dormancy mechanisms that enable seeds to survive dehydration?

3. What are the three major zones in a growing root?

4. What structures are derived from the apical meristem?

5. What structures are derived from the cork cambium?

6. What are the three major classifications of plants, based on their growing seasons or life spans?

## Multiple-Choice Review

In the following sentences, fill in any blanks. Complete each statement by circling the correct response.

1. Vegetative _____ produces offspring that are:

   a. less variable than their parents.
   b. more variable than their parents.
   c. genetically identical to the parents.
   d. nonfertile.
   e. nonviable.

2. The perianth of a flower is composed of:

   a. the calyx and corolla.
   b. the sepals and calyx.
   c. the corolla and sepals.
   d. the receptacle and sepals.
   e. the petals and the receptacle.

3. For pollen to fertilize an egg, the route the _____ must follow is:

   a. ovary-stigma-style.
   b. stigma-style-ovary.
   c. stigma-ovary-style.
   d. style-stigma-ovary.
   e. style-ovary-stigma.

4. The ovule contains:

   a. two diploid megaspore mother cells.
   b. two haploid megaspore mother cells.
   c. a single tetraploid megaspore mother cell.
   d. a single diploid megaspore mother cell.
   e. a single haploid megaspore mother cell.

5. The endosperm is formed from:

   a. a haploid cell.
   b. a diploid cell.
   c. two diploid cells.
   d. two polar nuclei and two sperm.
   e. a triploid cell.

6. At the globular embryo stage, the two types of cells are:

   a. inside and outside cells.
   b. monocotyledons and dicotyledons.
   c. dividing and nondividing cells.
   d. suspensor and scutellum cells.
   e. apical and root meristem cells.

# 29
# EXCHANGE AND TRANSPORT IN PLANTS

---

## CHAPTER AT A GLANCE

---

## CHAPTER PREVIEW

The precise mechanisms by which plants—from diminutive garden pansies to 300-foot redwood trees—are able to draw water from roots to the tips of leaves have puzzled biologists for centuries. Chapter 29 explores the modern explanations of exchange and transport in plants, beginning with an explanation of "passive" transport processes—those that operate by physical rather than chemical laws of nature.

The following section considers xylem-based water transport and some concepts that can be confusing. Take special care to learn the differences between hypo- and hypertonic solutions, and solute and water potential.

The chapter then focuses on translocation, or solution transport in phloem, including a fascinating account of how researchers use aphid stylets to study sieve tube fluids. You'll find that, just as with xylem transport, there are several theories about the mechanism of translocation. You may want to review basic xylem and phloem anatomy (Chapter 27) before studying this material in detail.

The next section discusses how plants absorb nutrients via roots. It also "recycles" and provides more information on nitrogen fixation, a process touched on in earlier chapters. The final section of Chapter 28 surveys the features of two of the most curious plant types: carnivorous plants and "sensitive" plants, those capable of rapid tissue movement.

# LEARNING OBJECTIVES

When you have mastered the concepts of this chapter, you will be able to:

1. List the various anatomical locations where gas exchange takes place in plants.
2. Explain how plants solve the problems of dealing with metabolic waste products.
3. Tell how osmosis and water potential move water through the xylem.
4. Outline and discuss the two basic theories (root pressure and capillary action) of how plants move water upward, against gravity.
5. Explain how water may rise to great heights in large plants, using the transpiration-pull theory.
6. Explain how water loss is regulated by the stomata, and list at least three factors that affect gas movement through stoma.
7. Trace the route that sugars travel in the phloem.
8. Use the mass flow theory to explain how solutions move through the phloem.
9. Distinguish between active and passive transport in plants.
10. List the factors that influence the rate at which plants obtain minerals.
11. Explain why clay soils are more favorable to mineral procurement than sandy soils.
12. Outline in detail the process of nitrogen fixation in plants.
13. Explain how carnivorous plants obtain nitrogen.

# CONCEPTS IN REVIEW

## Section I   Plant Strategies for Meeting Basic Needs

To exchange gases, acquire water and nutrients, and dispose of wastes, plants expend as little energy as possible. They conserve energy by relying on passive physical processes instead of chemical reactions driven by high-energy compounds like ATP.

Gases—mainly carbon dioxide and oxygen—enter and leave the plant by diffusion. Inside the plant, they diffuse short distances to reach each and every cell. Fluids move within the vascular system of xylem and phloem tubes by passive physical processes. Most plant wastes are stored in cells or broken down and recycled (text p. 694).

## *Section II*    Transport of Water in the Xylem

Water potential is the overall water status of a cell; it is determined by adding the cell's turgor pressure (pressure exerted by water in the cell) to its solute potential (concentration of solutes in the cell). The principles of turgor pressure and solute potential are the basis for understanding how water moves in plants (text pp. 695–696). Botanists have devised three theories to account for that movement.

In the root pressure theory, adherents maintain that water pressure builds up in roots because of the active uptake of mineral ions from the soil. This root pressure forces water into and up the stem xylem (text p. 696). Root pressure is not sufficient, however, to move water in the stems of most plants.

Another theory involves capillary action, which is based on the cohesion of water molecules to each other and on their adhesion to the sides of a tube. Like root pressure, capillary action produces some upward movement within the xylem. However, capillary action is also not sufficient to explain how water moves to the tops of the tallest trees.

The best hypothesis for how water makes its upward climb involves transpiration. According to the transpiration-pull theory, evaporative water loss through stomata in the leaves and stem pulls water up by a long liquid "chain" of water molecules held together by cohesion (text p. 697). The opening and closing of stomata, which depend on ion (mainly $K^+$) flow and changes in turgor pressure in surrounding guard cells, help regulate evaporative water loss. Contributing to the process are the concentration gradients and diffusion properties of water vapor, $O_2$, and $CO_2$, as well as wind blowing across the leaf's surface (text pp. 698–699).

## *Section III*    Transport of Solutions in the Phloem

Phloem is composed of living sieve tube elements that have cytoplasm but no nuclei, and companion cells that have nuclei. Using aphids, plant physiologists have shown that phloem transports a concentrated solution of mainly sucrose and nitrogen- and sulfur-containing amino acids.

The organic compounds in the phloem solution move up and down the plant from areas of highest concentration (sources) to areas of lowest concentration (sinks). In the mass flow theory, researchers propose that this translocation of solutes results from a building and release of turgor pressure in phloem cells, and experiments show that the solutes are transported down a turgor pressure gradient (text pp. 700–701). Although there are minor problems with the concept of mass flow, the theory is the best available explanation of the transport of solutions in phloem.

## *Section IV*    How Roots Obtain Nutrients from the Soil

To survive and thrive, plants need an array of minerals—inorganic substances such as potassium and phosphorus that originate in Earth's rocks. The table on text page 702 summarizes the functions of essential minerals required in large (as macronutrients) and small (as micronutrients) amounts.

Minerals are dissolved in soil water. From there, they enter a plant's roots in one of two ways: by passive diffusion, or by active uptake requiring an expenditure of energy (text p. 702). The minerals are transferred to the xylem and then move throughout the plant by passive diffusion. A soil's concentration of minerals, pH, and structure all affect the rate at which plants take up minerals (text pp. 703–704).

Nitrogen, the most important factor limiting plant growth, is derived from the atmosphere, but plants cannot directly assimilate the molecular nitrogen ($N_2$) found in air. They rely instead on fixation (conversion of molecular nitrogen to nitrogen compounds) by nonliving agents such as lightning and, more importantly, by bacteria such as *Azobacter, Clostridium,* and *Rhizobium* (text pp. 704–706). A symbiotic association with certain fungi (called mycorrhiza) enables plant roots to take up water and minerals more readily.

## Section V   Carnivorous and "Sensitive" Plants

The Venus's-flytrap and other carnivorous plants get some of their nitrogen the hard way—from trapping and digesting insects. Action potentials enable them to respond to touch quickly enough to snare prey. Such changes in electrical potential are similar to the nerve impulses of animals. Even some noncarnivorous plants, such as mimosa, respond visibly to touch (text p. 707).

The rapid movements of plants differ significantly from those of animals. Plants move their parts by rapid alteration of turgor pressure in critically placed cells; in animals the specialized proteins of muscle tissue move legs, wings, and jaws.

## KEY TERMS

bacteroid   *text page 704*

capillary action   *696*

carnivorous plant   *706*

cohesion-adhesion-tension
theory   *697*

guttation   *696*

mass flow theory   *700*

mineral   *701*

nitrification   *704*

nitrogen fixation   *704*

root nodule   *704*

root pressure theory   *696*

solute potential   *695*

translocation   *699*

transpiration   *697*

transpiration-pull
theory   *697*

water potential   *695*

# SELF-QUIZ: TESTING WHAT YOU HAVE LEARNED

## Matching Key Terms

Match each term on the left with the most appropriate description on the right.

| | |
|---|---|
| 1. nitrification | a. leaf water loss |
| 2. bacteroid | b. droplet formation |
| 3. translocation | c. cannot be made by photosynthesis |
| 4. water potential | d. live in nitrogen-poor soils |
| 5. root nodule | e. oxidation of $NH_3$ |
| 6. capillary action | f. legumes |
| 7. transpiration | g. lives in root nodules |
| 8. guttation | h. equal to turgor pressure + solute potential |
| 9. mineral | i. solute transport |
| 10. carnivorous plant | j. thin tubes required |

## True or False?

1. _____ Electrical action potentials are important to carnivorous plants.

2. _____ Minerals are transported in the xylem.

3. _____ Wind speed has no effect on stomata.

4. _____ The root pressure theory explains how water is brought to the top of a very tall tree.

5. _____ Companion cells have no visible nuclei.

6. _____ Meristems are strong carbon sinks.

7. _____ Movement in plant cells is based on the proteins actin and myosin.

8. _____ A hypotonic medium results in plasmolysis.

## Completion

1. Water potential − turgor pressure = _____.

2. Cells placed in a _____ solution will take up liquid by osmosis.

3. Water will neither leave nor enter a plant's cells if the water potential is _____.

4. The xylem is composed of _____ and _____.

5. Protoplasts are prepared from isolated _____ cells.

6. The fluid transported by phloem may contain up to _____ percent (by volume) of dissolved minerals.

7. Translocation occurs down a gradient of _____.

8. Minerals that are required in very large amounts are lumped under the heading of

_____.

## Short Answer

1. What two factors control the rate at which plants take up minerals?

2. What mechanisms oxidize atmospheric nitrogen in the soil?

3. Why are carnivorous plants so rare and unusual?

4. How does the mimosa close its leaves?

5. How does capillary action move water?

6. What causes guttation?

## Multiple-Choice Review

In the following sentences, fill in any blanks. Complete each statement by circling the correct response.

1. The basic needs of plants, such as gas exchange and transport of water and nutrients, depend mainly on:

   a. large amounts of ATP.
   b. wind speed.
   c. chemical processes.
   d. passive processes.
   e. symbiosis with bacteria.

2. The folding leaf response of a mimosa plant may be:

   a. an adaptation to prevent water loss.
   b. an intermediate form, on the way to carnivory.
   c. a method of generating heat.
   d. an adaptation to avoid being eaten.
   e. an adaptation for avoiding strong sunlight.

3. The best explanation for how _____ can travel upward hundreds of feet to the tops of the tallest trees is:

   a. the root pressure theory.
   b. guttation.
   c. osmosis.
   d. simple capillary action.
   e. the transpiration-pull theory.

4. Water moving upward in the xylem is under:

    a. tension.
    b. compression.
    c. tremendous pressure.
    d. osmotic pressure.
    e. shear force.

5. Guard cells that are subjected to _____ acid:

    a. plasmolize.
    b. close their stomata.
    c. rapidly divide.
    d. burst.
    e. none of the above

## Exercise

In the figure below, use arrows to show which structures enable a plant to exchange gases and to transport and distribute water and nutrients. Be sure to include internal structures.

# 30
# PLANT HORMONES

---

## CHAPTER AT A GLANCE

---

## CHAPTER PREVIEW

Chapter 30 is concerned with the hormone-based regulating mechanisms of higher plants. The text begins by pointing out the basic differences between plant and animal hormones, and then considers, in turn, four classes of plant hormones. The first, auxins, play roles in plant tropisms (movement toward gravity and light), and act as inhibitors of cell growth. The next group, gibberellins, are probably the largest class of plant hormones and are the main promoters of growth in cells, stems, and fruit. The third group, the cytokinins, are compared and contrasted in their role of increasing size, not by elongation, but by promoting cell division by stimulating mitosis. Then you will consider abscisic acid, the growth-slowing agent that is so important in leaf, fruit, and flower separation, and in maintaining seed dormancy.

Next the chapter looks at the role of the gaseous hormone ethylene in fruit ripening, and how the various classes of hormones may interact in complex ways to bring about various effects. Plant scientists have several different hypotheses about the hormone mechanism that brings about flowering, but as yet there is no general theory that fully explains the trigger for this marvelous event. Other hormone-mediated phenomena in plants include leaf-color change and the separation of leaf from stem that makes autumn "fall." The chapter closes with a survey of photoperiodism, and the characteristics of short-day and long-day plants.

# LEARNING OBJECTIVES

When you have mastered the concepts of this chapter, you will be able to:

1. Give a general description of plant hormones, and compare and contrast their effects with those of animal hormones.
2. Name the basic types of plant tropisms and explain how they are accomplished.
3. Explain what is meant by apical dominance.
4. List the five major classes of plant hormones.
5. Explain how auxins are involved in cell elongation.
6. Explain the roles of gibberellins in plant growth.
7. Discuss the roles of cytokinins, and tell how they promote cell division.
8. Explain the roles of abscisic acid in plant growth, and tell what is meant by the term *abscission*.
9. List and explain the sequence of events that takes place when plants lose leaves.
10. Discuss the effects of photoperiod on plant growth and flower production.
11. Discuss the chemical and ecological (external) factors that control the production of flowers.

# CONCEPTS IN REVIEW

## Section I    Auxins: Cell Elongation and Plant Movements

Hormones are substances made in very small amounts in one part of an organism that produce effects in another part of the same organism. The five major classes of plant hormones help control growth and internal processes.

The auxins, a group of related hormones, are generally produced in a plant's stem and leaves, and act in seemingly contradictory ways. Charles and Francis Darwin and F. W. Went showed that auxins cause plants to bend toward the light (phototropism) by diffusing away from a light source and triggering elongation in nonilluminated cells. Auxins can also suppress cell enlargement in the lower branches of a tree and produce apical dominance, speed the maturation of fruit tissue, and orient a shoot (or root) in relation to gravity (text pp. 714–715).

The mechanism of phototropism is only partially understood. When pigment molecules receive light (particularly blue light of the shorter wavelengths) on one side of a stem, they cause auxins to travel down the darker, opposite side in a type of movement known as polar transport. On that darker side, the auxins cause cells to expand by promoting the secretion of protons ($H^+$), which set in motion a chain reaction that results in cell wall loosening and cell elongation (text p. 716).

## Section II    Gibberellins: Growth Promoters

The gibberellins are a group of more than seventy related compounds. Like the auxins, they are made in the young, growing regions of the shoot, and they cause plant growth by increasing the size and (unlike the auxins) the number of a stem's internodal cells.

In rosette plants, gibberellin application can induce bolting, and in many fruiting plants, treatment with gibberellins results in fruit enlargement. Although gibberellins and auxins have similar effects, different cells in the plant bind and respond to them, making both hormones necessary for normal growth (text pp. 717–718).

## Section III   Cytokinins: Cell-division Hormones

In another instance of hormonal "cooperation," the cytokinins promote cell division, but only in the presence of auxins. The ratio of cytokinin to auxin determines the rate of that cell division. Cytokinins are structurally related to adenine and are usually produced in the roots.

A high proportion of cytokinin causes cells to divide rapidly and differentiate to shoot or leaf tissue; if more auxin is present, roots or disorganized callus tissue forms. Cytokinins also delay aging in leaves (text p. 719), possibly by making the leaf a sink for nutrients.

Because they promote growth and development, cytokinins and auxins are useful in plant–tissue–culture techniques. Plant cells and tissues grow much better in culture than do animal cells. With some plants, one cell (a protoplast) can generate an entire organism, reflecting the totipotency that is common to plant cells. In fact, the growth of new organs from constantly proliferating meristems is a key feature of plants, one of the many that distinguish them from animals (text p. 721).

## Section IV   Abscisic Acid: The Growth-slowing Hormone

Found in dormant bulbs and seeds, and in some fruits, leaves, and other tissues, abscisic acid suppresses a plant's natural tendency to grow. It works in a delicate balance with the growth-promoting hormones to enable plants to respond to stress or to enter dormancy; it also accelerates senescence and promotes abscission. Abscisic acid is generated mainly by cells experiencing stress. Although abscisic acid's specific interaction with other hormones is complex, it is not difficult to understand the overall concept of give and take between growth-promoting and growth-suppressing hormones (text pp. 721–722).

## Section V   Ethylene: The Gaseous Hormone

The only gaseous plant hormone is ethylene, which accelerates and regulates fruit ripening. Like the cytokinins, it has a complex association with auxin and may be produced wherever there is a high concentration of auxin.

Ethylene inhibits the transport of auxin across stem or root, and the ratio of ethylene to auxin seems to determine whether cells elongate or expand radially in response to auxin. The presence of auxin, in turn, stimulates the formation of ethylene.

## Section VI   Interactions of Plant Hormones

According to the generally accepted principle of hormonal interaction in plants, two or more positive hormonal signals are required for growth, but one negative signal can prevent further growth. For every instance of inhibition by abscisic acid, the proper combination of auxin, gibberellin, and cytokinin can reverse the effect (text p. 723).

Falling leaves and germinating seeds demonstrate how slight shifts in the subtle balance of complex hormonal interactions can terminate or initiate growth (text p. 723).

## Section VII   Control of Flowering

How does a plant keep track of the seasons so that it can flower at an appropriate time? The answer is that its developmental program and responses to environmental cues result in hormonal activity that can trigger flowering.

Night length is one environmental cue that determines whether a plant will flower. Short-day plants require a long, uninterrupted night (the flip side of a short photoperiod, or day length) to flower. Long-day plants require a short, uninterrupted night (the natural consequence of a day length of more than some critical number of hours). Day-neutral plants appear to be unaffected by night or day length (text pp. 725–726).

Plants detect day and night length with phytochrome, a pigment that is active in leaves. Phytochrome's interconversion between two forms registers the effects of sunlight by day and no sunlight by night. But phytochrome's activity does not *measure* day length. For that, plants probably have some sort of internal clock. Phytochrome also plays a role in seed germination (text p. 727).

Flowering is induced by a suitable photoperiod but is controlled by several factors including perhaps gibberellins, the hypothetical florigen hormone, a plant's developmental status, the temperature, and the availability of nutrients and water (text pp. 727–728).

## KEY TERMS

abscisic acid    *text page 721*

abscission   722

abscission layer   723

apical dominance   714

auxin   714

callus tissue   719

cytokinin   719

ethylene   722

florigen   727

gibberellin   716

gravitropism   714

hormone   712

photoperiod   724

phototropism   712

phytochrome   726

tropism   712

vivipary   721

# SELF-QUIZ: TESTING WHAT YOU HAVE LEARNED

## Matching Key Terms

Match each term on the left with the most appropriate description on the right.

1. gravitropism
2. photoperiod
3. hormones
4. tropism
5. auxins
6. florigen
7. phototropism
8. cytokinins
9. vivipary
10. abscission

a. regulate tropisms
b. a hypothetical hormone
c. the fate of leaves in the fall
d. cell division
e. toward light
f. lethal mutation
g. length of the day
h. "down"
i. less specific in plants than in animals
j. movements

## True or False?

1. _____ Charles Darwin and his son Francis published "The Power of Movement in Plants."

2. _____ Auxins work by causing cells to expand.

3. _____ Gibberellins increase both size and number of internodal cells.

4. _____ Cytokinins speed up cell division; but they also speed aging.

5. _____ Totipotency is common in both plants and animals.

6. _____ Plant hormones are more specific than animal hormones.

7. _____ Abscisic acid is a prime factor in seed dormancy.

8. _____ Plants measure day length with phytochrome.

## Completion

1. _____, or a plant growth movement due to gravity, is caused by a build-up of _____ on the plant's lower surface.

2. _____ and _____ promote cell growth and elongation, while the _____ stimulate division and senescence.

3. A lethal mutation called _____, or "live birth," causes root and shoot elongation in the embryo while the embryo is still in the flower.

4. The hormone _____ is produced by ripening fruits.

5. The length of day, or _____, is important in determining when angiosperms _____.

6. The pigment that plants use to determine day length actually has two chemical forms: _____ and _____.

## Short Answer

1. What were the three groups of plants described in the photoperiod work of Garner and Allard?

2. What are the two conformations of phytochrome; when and how does each work?

3. What were the early Darwin family experiments that probed the phenomenon of tropisms?

4. What is the evidence that plant hormones are less specific than many animal hormones?

5. How do plants and animals differ in their abilities to produce new clones from individual cells?

6. What causes parts (leaves, dead flowers, ripe fruits) to drop off of plants?

## Multiple-Choice Review

In the following sentences, fill in any blanks. Complete each statement by circling the correct response.

1. Cell elongation requires:

   a. auxins and gibberellins.
   b. auxins and ethylene.
   c. gibberellins and abscisic acid.
   d. gibberellins.
   e. cytokinins.

2. The "flower-making" hormone is:

   a. ethylene.
   b. abscisic acid.
   c. an auxin.
   d. florigen.
   e. gibberellin.

3. The "simple" procedure of removing weeds from agricultural crops increases crop yield by:

   a. 10 percent.
   b. 15 percent.
   c. 25–50 percent.
   d. 50–60 percent.
   e. 50–75 percent.

4. Auxin may be able to move _____ through plant tissues that already contain much auxin. This is called:

   a. expansion.
   b. polar transport.
   c. gradient transport.
   d. auxatropism.
   e. none of the above

5. Ethylene—the fruit-ripening hormone—can also affect plant growth by:

   a. causing cell elongation.
   b. increasing gibberellin effects.
   c. increasing cytokinins.
   d. destroying abscisic acid.
   e. inhibiting auxin transport.

# 31
# THE CIRCULATORY AND TRANSPORT SYSTEMS

---

## CHAPTER AT A GLANCE

---

## CHAPTER PREVIEW

Circulation and transport are vital for every type of animal. This chapter begins with a general survey of the simplest systems, seen in lower animals, and then progresses to the larger and more complex systems in higher species. In single-celled organisms the transport method is, of course, simple diffusion. Diffusion also meets the needs of small, flat-bodied multicellular species.

Next the focus shifts to the concepts of "open" and "closed" circulatory systems, and the story of William Harvey's discovery of blood circulation in humans. With this material as background, the chapter takes a comprehensive look at the structures and evolution of circulatory systems in representative vertebrates: fish, birds, and the mammals. The hierarchical nature of blood vessels (from the largest to smallest) is described, along with a detailed explanation of blood pressure.

At this point you will examine the self-generated (myogenic) beat of the vertebrate heart, and explore the vasomotor center of the brain. The next section outlines the components of blood, and introduces a number of new terms. Finally the chapter presents what may be the most mysterious component of circulation, the lymphatic system. You will encounter some of this again in a later chapter on the immune system.

# LEARNING OBJECTIVES

When you have mastered the concepts of this chapter, you will be able to:

1. Explain the relationship between body size and type of circulatory system, and compare and contrast open and closed circulatory systems.
2. Explain why blood flows in a circuit, and list the four types of muscles that propel it through the system.
3. Outline the methods that William Harvey used in his discovery of circulation.
4. Draw and label a cross section of a mammalian heart, and trace the route of blood flow through it.
5. Trace the various stages of development in the evolution of the vertebrate heart.
6. Draw and label generalized cross sections of an artery and a vein.
7. Trace the complete circular route taken by a blood cell as it passes from the heart to arteries, capillaries, veins, the lungs, and back through the heart.
8. Describe the anatomy of a capillary bed, and list four routes by which capillary fluids can reach body cells.
9. Define blood pressure, and explain how it differs among different types of vertebrates and at different body locations.
10. List and discuss the four major categories of blood flow regulation.
11. List, describe, and discuss the function of the blood's fluid and solid components.
12. Explain the functioning of the lymphatic system.

# CONCEPTS IN REVIEW

## Section I  Strategies for Transporting Materials in Animals

A transport system that carries nutrients, oxygen, and other essential raw materials to and from all body cells is essential to the physiological activities of animals. Single-celled organisms and flat, porous, or thin-walled multicellular animals depend for such transport on simple diffusion down a concentration gradient. But diffusion is a slow process and can only service cells that are a very short distance from the source.

All multicellular animals that are not flat, porous, or thin-walled have a circulatory system in which fluid circles by bulk flow. Insects, some mollusks, and several other types of invertebrates have an open circulatory system in which a fluid such as hemolymph cycles partly in enclosed vessels and partly in open spaces called sinuses. In the closed circulatory system of earthworms, squids, octopuses, and all vertebrates, blood moves through a continuous set of interconnected arteries (leading to the tissues), capillaries (in the tissues), and veins (exiting the tissues).

The fluid in all circulatory systems flows in one direction for four reasons: (1) there is a heart that acts as a pump; (2) its contraction (along with contractions of some vessel walls) creates blood pressure; (3) sphincter muscles shut off or open up vessels in local tissues; and (4) valves inside some of the vessels prevent backflow (text pp. 736–738).

Even in animals with circulatory systems, it is by diffusion that materials enter and leave cells. Such diffusion moves materials back and forth between the extracellular fluid and the intracellular fluid. In a closed circulatory system, the materials also must diffuse into and out of the blood (text p. 738). William Harvey was the first to deduce how the vertebrate circulatory system actually works (text pp. 738–739).

## Section II    The Vertebrate Circulatory System

The human heart is a double pump, divided in half longitudinally, that propels blood through two separate circuits: the pulmonary, or lung, circulation (in which deoxygenated blood leaves the right heart, goes to the lungs, and returns to the left heart); and the systemic, or body, circulation (in which oxygen-rich blood leaves the left heart, cycles through the body, and returns to the right heart). The mammalian heart and circulation evolved from the simpler pumping system of the fish through the progressively more powerful systems of amphibians and reptiles (text pp. 740–745).

Differences in the structure of arteries, veins, and capillaries suit these vessels for their specific functions. Arteries have relatively thick, three-layer walls, the outer layer containing collagen and springy elastin fibers; these fibers give arteries their ability to contract vigorously and remain elastic. In contrast, veins are more pliable and can increase in volume as they accumulate blood. The minute capillaries, which permeate every body tissue, have extremely thin walls composed of one layer. Through this layer, materials diffuse into and out of the extracellular fluid (text pp. 745–748).

As blood moves through the pipes of the vertebrate circulatory system, it exerts a force called blood pressure against the vessels' walls. Differences in pressure between the two sides of the heart (and within the capillary beds) are what cause blood to flow.

In humans, blood pressure is highest as blood leaves the heart and lowest during relaxation of the heart muscle. Substances move between blood and extracellular fluid in the capillaries because of changes in blood pressure and the effects of colloidal osmotic pressure (text p. 749).

## Section III    Regulation of Blood Flow

The rate at which blood flows through an animal's body must change with the animal's varying needs for material transport, because the speed of transport depends on the rate of flow. Four factors determine the flow rate: vessel size; frequency of heartbeat; volume of blood pumped per beat; and the activity of sphincter muscles in capillaries.

The speed of blood flow to any spot in the circulatory system depends on the cross-sectional total of the vessels at that point (text p. 750). Heartbeats are triggered at regular intervals by electrical impulses originating in the heart itself: in the sinoatrial (S-A) node—the pacemaker; and in the atrioventricular (A-V) node. Nerve signals can alter the rate of heart contraction to meet the body's changing demands (text pp. 751–752). Nerve impulses can also alter the stroke volume, or amount of blood pumped at each heartbeat, and they can regulate blood flow to different capillary beds by vasoconstriction (contraction of capillary sphincters and vein walls) and vasodilation (relaxation of sphincters and veins). The vasomotor center in the brain stem coordinates the nervous system's regulation of heart and blood vessel activity (text p. 753).

## Section IV    Blood: The Fluid of Life

The life-sustaining solution we call blood consists of a liquid plasma portion and a solid cellular fraction. The plasma is water that contains dissolved gases such as oxygen and nitrogen; ions and nutrient molecules such as $Na^+$, $K^+$, $Cl^-$, glucose, and vitamins; the wastes of cellular metabolism; and blood proteins that function as hormones, constituents of the immune system, or in blood clotting.

The blood's solid components include red blood cells (erythrocytes), white blood cells (leukocytes), and platelets (text p. 754). Platelets help initiate the clotting that prevents excessive bleeding from a cut. Blood clotting is an example of biological amplification in which a small event triggers a cascade of increasingly larger events (text pp. 754–756).

## Section V    The Lymphatic System and Tissue Drainage

Since every second of every day there is a net movement of fluid out of capillaries, why doesn't all the fluid leave the blood and accumulate as extracellular fluid? Without a lymphatic system it would. But this second system of vessels running roughly parallel to the circulatory system's veins (Figure 31-16) collects excess extracellular fluid and returns it to the blood circulation. (It also functions as part of the immune system; for a discussion of this role, see text Chapter 32.) In addition to mammals, fish, amphibians, and reptiles also have a lymphatic system (text p. 756).

Lymph is propelled in mammals by body muscles that knead the vessels while valves prevent backflow, and by smooth muscles in the walls of larger lymphatic vessels. Vertebrates other than mammals have lymph hearts (text p. 757).

## KEY TERMS

aorta  *text page 743*

aortic arch   742

arteriole   743

artery   736

atrioventricular (A-V)
   node   751

blood pressure   737

bulk flow   736

capillary   736

cardiac output   752

circulatory system   735

closed circulatory
   system   736

colloidal osmotic
   pressure   749

diastole   749

erythrocyte   754

granulocyte   754

heart   737

leukocyte   754

lymphatic system   756

lymphocyte   754

monocyte   754

open circulatory system   736

plasma   753

platelet   754

pulmonary (lung)
   circulation   742

sinoatrial (S-A) node   751

Starling's law   741

systemic (body)
   circulation   742

systole   749

vasoconstriction   752

vasodilation   752

vasomotor center   753

vein   736

venule   744

# SELF-QUIZ: TESTING WHAT YOU HAVE LEARNED

## Matching Key Terms

Match each term on the left with the most appropriate description on the right.

| | | | |
|---|---|---|---|
| 1. leukocyte | | a. | heart contracted |
| 2. platelet | | b. | controls nerves |
| 3. vasoconstriction | | c. | heart relaxed |
| 4. monocyte | | d. | carries ions |
| 5. vasomotor center | | e. | immune response |
| 6. diastole | | f. | clotting |
| 7. systole | | g. | white blood cell |
| 8. open circulatory system | | h. | increases blood pressure |
| 9. plasma | | i. | red blood cells |
| 10. erythrocytes | | j. | most invertebrates |

## True or False?

1. _____ Diffusion is feasible only when the distances traveled are very small.

2. _____ The one-way flow of blood is maintained by sphincters.

3. _____ Extracellular fluid is not required in animals with closed circulatory systems.

4. _____ Veins never carry highly oxygenated blood.

5. _____ Hypercholesterolemia is a genetic disease that is less severe in heterozygotes than in homozygotes.

6. _____ Arteries are the volume reservoir of the circulatory system.

7. _____ Heartbeat is self-generated by the heart.

8. _____ The fight-or-flight response is triggered by epinephrine.

## Completion

1. The circular movement of blood through a system is often called _____.

2. Blood is helped along its route by muscle contractions not only in the heart, but also by _____ contractions from the blood vessel walls.

3. Blood flow into and out of specific regions is also regulated by _____ muscles that control a vessel's diameter.

4. _____ is the watery component of blood.

5. _____ states that the more that cardiac muscle is stretched, the greater the response, and the larger the amounts of blood that are pumped per contraction.

6. Capillaries have a single layer of _____ cells.

## Short Answer

1.  What are the pressure differences between systole and diastole blood pressures?

2.  What is meant by the term *colloidal osmotic pressure?*

3.  What problems of transport limit the body size of invertebrates? And how do some invertebrates (octopus and giant squid) overcome this?

4.  What are the advantages of a closed circulation system?

5.  What are the components of blood?

6.  What path is followed by blood in the pulmonary and the systemic routes?

## Multiple-Choice Review

Complete the following statements by circling the correct response.

1. The largest (diameter) artery is the:

   a. carotid artery.
   b. pulmonary artery.
   c. aorta.
   d. aortic arch.
   e. none of the above

2. The heart's pacemaker is the:

   a. bundle of His.
   b. lymphatic system.
   c. A-V node.
   d. sinoatrial node.
   e. atrioventricular node.

3. The vessel that serves as the main pressure reservoir is the:

   a. pulmonary artery.
   b. coronary artery.
   c. carotid artery.
   d. superior vena cava.
   e. aorta.

4. The first complete separation of the heart into a right and left half is seen in:

   a. fish.
   b. amphibians.
   c. reptiles.
   d. birds.
   e. mammals.

5. A blood cell, entering a fish heart, would pass through the following structures (posterior to anterior):

   a. conus arteriosus, atrium, ventricle, sinus venosis.
   b. sinus venosis, atrium, ventricle, conus arteriosus.
   c. sinus venosis, conus arteriosus, atrium, ventricle.
   d. sinus venosis, ventricle, atrium, conus arteriosus.
   e. conus arteriosus, ventricle, atrium, sinus venosis.

## Exercise

Explain the significance of varying pressures within the arteriole and venule ends of a capillary. What role does the lymphatic system play? Use the diagram below as a reference.

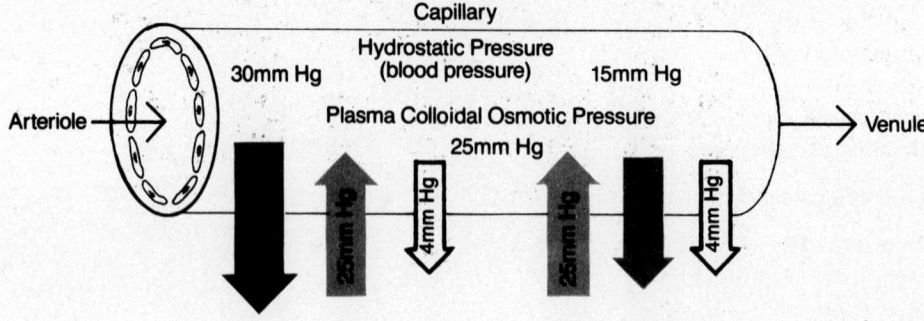

# 32
# THE IMMUNE SYSTEM

## CHAPTER AT A GLANCE

## CHAPTER PREVIEW

This chapter is one of the most timely chapters in your text, in part because it contains much material that is related to the condition AIDS. It begins with a general survey of the immune system, including the various types of lymphocytes and the organs from which they arise. Then, the lines of immune defense are divided into the nonspecific (a "shotgun" approach) and the specific (tailor-made) defenses.

The next several sections describe the two kinds of immunity—humoral and cell-mediated—and the third class of lymphocytes: the "natural killers." The chapter then turns to concepts that are sometimes more difficult to grasp. One of these is the phenomenon of tolerance, or the ability of the body to recognize its own tissues as "self" instead of as an invader. Breakdown of such tolerance results in autoimmune disease. Other topics follow: the difference between active and passive immunity, and, finally, the phenomenon of allergic reaction.

# LEARNING OBJECTIVES

When you have mastered the concepts of this chapter, you will be able to:

1. List the organs, tissues, and molecules that comprise the immune system.
2. List the various types of white blood cells, tell where they are found, and state the function of each.
3. Describe the structure of lymph nodes, and list the functions they serve.
4. List the functions of the spleen.
5. List the functions of the thymus.
6. Distinguish between specific and nonspecific defense systems, and give examples of each.
7. Explain what antibodies are and how they work.
8. Describe the chemical structure of antibodies, and explain how they work.
9. Outline the clonal selection theory.
10. Explain how monoclonal antibodies are produced and how they function.
11. Name the three kinds of T cells, and describe the function of each.
12. Describe the functioning of "natural killers."
13. Explain the nature of autoimmune diseases and how they are created.
14. Distinguish between active and passive immunity.
15. Define the term *allergy,* and explain why an allergy may grow progressively worse.
16. Explain what histocompatibility antigens are, and describe how they work.

# CONCEPTS IN REVIEW

## *Section I*   Components of the Immune System

The body's immune system is a "standing army" that holds potentially harmful microbes and molecules at bay. It is composed of: several classes of white blood cells; the specialized molecules (such as antibodies) those cells secrete; and the organs in which the cells arise, differentiate, and carry out their activities. Its smooth functioning depends on a few simple principles: specific recognition by the binding of molecules with reciprocal shapes; gene shuffling that generates a great diversity of molecular binding sites; and natural selection at the cellular level.

Of the several types of white blood cells in the immune system, macrophages and lymphocytes are the principal agents of immune responses. Macrophages engulf and digest foreign particles and help initiate immune responses. Lymphocytes attack invaders directly or indirectly. There are three kinds of lymphocytes: T cells, B cells, and natural killer (NK) cells. All of these cells arise in the bone marrow and then migrate to other organs for differentiation and action. The organs of the immune system include the lymph nodes, spleen, and thymus (text pp. 762–763).

## Section II    Lines of Defense: Nonspecific and Specific

Nonspecific defense mechanisms work the same against all foreign substances and include the barrier formed by skin and mucous membranes, as well as the inflammatory response (text p. 764).

Specific defense mechanisms occur only in the immune system, which reacts in a unique way to each species of bacterium or virus. The specificity of immune responses works in tandem with their diversity, memory, and tolerance to achieve strong, versatile protection. Despite the diversity of foreign invaders, the immune system, because of the diversity of its own elements, can usually respond with a precisely tailored defense. With its memory, the immune system can mount secondary responses that are stronger and faster than its primary responses. And the system normally tolerates (does not attack) the body's own tissues because it can distinguish "self" from "nonself."

There are two broad categories of immune responses: humoral, carried out by B cells and antibodies; and cellular, carried out by T cells.

## Section III    Antibodies and Humoral Immunity

Antibodies are Y-shaped globular protein molecules that recognize (bind to) antigenic determinants, or clusters of atoms on antigens. An antigen is any virus, bacterium, fungus, or other substance that elicits an immune response. In humoral immunity, antigens stimulate B cells to produce antibodies.

Each antibody is composed of two light polypeptide chains and two heavy chains, and each chain has variable and constant regions. The amino acid variations in the most variable (hypervariable) regions determine the specificity of antibody binding sites.

There are five classes of antibodies: IgG, IgM, IgA, IgD, and IgE. While all antibodies bind antigen, different classes go different places in the body and carry out different functions. The binding of IgG or IgM antibodies to a cell may activate the complement response (text pp. 767–768).

The clonal selection theory explains how antigens "select" from a large pool of B cells those clones that make antibodies specific for the antigen's determinants. After selection by antigen, the B cells divide and redivide to form a clone of antibody-secreting plasma cells and long-lived memory cells (text pp. 769–770). A unique process of gene shuffling takes place in the nuclei of immature B cells, and as a result the cells become committed to producing antibodies with only one shape of binding site. For the details of this gene shuffling, review text pages 770–773.

Researchers have used their understanding of clonal selection and antibody specificity to make monoclonal antibodies that can help diagnose cancer and other diseases (text pp. 773–776).

## Section IV    T Cells and Cell-mediated Immunity

Though T cells do not make and secrete antibodies, they mature and undergo clonal selection just as B cells do (text p. 776). Three types of T cells cooperate to establish cell-mediated immunity. Killer T cells directly attack foreign cells or virus-infected self cells. Helper T cells assist B cells and other T cells in starting and building immune responses. Suppressor T cells play a role in turning off immune responses or in limiting their magnitude and duration, and in tolerance. Macrophages help initiate a T cell attack by bringing antigens and immature T cells together and by secreting lymphokines (text p. 777).

## Section V    Natural Killers: The Third Class of Lymphocytes

Like other lymphocytes, natural killer (NK) cells arise in the bone marrow. But unlike B cells and T cells, they appear to lack binding sites for antigens on their surfaces. Nevertheless, NK cells can lyse tumor cells while leaving normal cells unharmed. NK cells also directly attack viruses and virus-infected cells.

## Sections VI and VII   Autoimmune Disorders: A Breakdown in Tolerance and Immune Deficiency

When normal tolerance breaks down, an autoimmune response can result in diseases such as arthritis or lupus erythematosus. Tolerance most likely develops by several mechanisms. In clonal deletion, clones of lymphocytes with reactivity to self-antigens are destroyed or inactivated. Suppressor T cell activity may also help bring about tolerance (text p. 778). It is not yet clear what triggers an autoimmune attack by T or B cells.

People born without components of the immune system and people who lose the ability to mount an immune response suffer from immune deficiency. Acquired immune deficiency syndrome, or AIDS, is a threatening new immune deficiency disease that was first detected in the early 1980s. It is the result of a viral infection that apparently destroys the function of helper T cells; without these helper cells, killer T cells and B cells cannot mount an attack, and AIDS patients fall prey to many kinds of infections other people overcome easily (text p. 779).

## Section VIII   Immune Phenomena and Human Medicine

It is possible to stimulate the immune system to provide broader protection. Active immunity results from recovery from a disease or from vaccination (text p. 780), both of which activate a specific immune response and leave behind a pool of memory cells. In passive immunity, a person receives antibodies made by another person or other kind of animal. Fetuses and newborn infants are protected by antibodies they receive from maternal blood (text pp. 780–781).

Allergies are the result of immune reactions in which an abundance of IgE antibodies activate the mast cells to which they are bound. This activation triggers the release of histamines and other chemicals. The most serious allergic reactions produce anaphylactic shock, a life-threatening state during which mast cells in many tissues discharge the contents of their granules simultaneously.

Histocompatibility genes encode proteins that appear on the surfaces of most cells throughout the body. It is because a person's tissues are labeled with histocompatibility antigens that the immune system often rejects an organ transplanted from another person. Drugs can help suppress such rejection responses.

In addition to playing an important role in immune responses, histocompatibility genes and the proteins they encode also contribute to sexual acceptance in mice and reproductive success in humans (text pp. 782–783).

## KEY TERMS

active immunity   *text page 780*

allergy   782

anaphylactic shock   782

antibodies   765

antigen   765

autoimmune response   778

B cell (B lymphocyte)   762

cell-mediated immunity   776

clonal selection theory   769

complement response   768

constant region   766

heavy chain   766

humoral immunity   765

hypervariable region   767

immune deficiency   779

immune system   761

immunoglobulin   767

inflammatory response   764

light chain   766

lymph node   762

macrophage   761

major histocompatibility complex   782

memory cell   770

monoclonal antibody   773

naive B cell   769

natural killer (NK) cell   762

passive immunity   780

peripheral lymphoid tissue   762

plasma cell   770

spleen   762

T cell (T lymphocyte)   762

thymus   763

variable region   767

# SELF-QUIZ: TESTING WHAT YOU HAVE LEARNED

## Matching Key Terms

Match each term on the left with the most appropriate description on the right.

| | | | |
|---|---|---|---|
| 1. | antigen | a. | stores lymphocytes and red blood cells |
| 2. | macrophage | b. | in a "resting state" |
| 3. | naive B cells | c. | large B cells |
| 4. | allergy | d. | attack tumor cells and viruses |
| 5. | plasma cell | e. | allow a fast secondary immune response |
| 6. | memory cells | f. | a foreign substance |
| 7. | thymus | g. | engulfs and digests |
| 8. | lymph node | h. | T lymphocyte differentiation |
| 9. | spleen | i. | reaction of immune system to foreign substance |
| 10. | NK cell | j. | small mass of tissue |

## True or False?

1. _____ Viruses can be attacked directly by NK cells.

2. _____ The thymus becomes more active with age.

3. _____ Arthritis is the most common autoimmune disease.

4. _____ The AIDS virus works by killing helper T cells.

5. _____ Anaphylactic shock may be counteracted with epinephrine.

6. _____ T lymphocyte differentiation takes place in the adenoids.

7. _____ Passive immunity is created by exposure to a pathogen.

8. _____ Autoimmune diseases result from a breakdown of tolerance.

## Completion

1. The two major types of lymphocytes are _____ and _____.
2. The basic structure (polypeptide chains) of an antibody molecule is held together by

   _____.

3. The clonal selection theory was developed by _____ and _____.

4. The three major kinds of T lymphocytes are _____, _____, and

   _____.

5. _____ is immunity based on a previous exposure to a pathogen.

6. Most mild allergies can be treated with _____, which counteracts the effects of

   mast cell _____.

## Short Answer

1. What are the differences between macrophages, T cells, and B cells?

2. Why must cancer sufferers sometimes have their lymph nodes removed?

3. What is immune system "memory"?

4. Describe the difference between the constant and variable regions of an antibody.

5. What is the relationship between B cells and plasma cells?

6. How do biologists account for the remarkable diversity of antibodies?

## Multiple-Choice Review

In the following sentences, fill in any blanks. Complete each statement by circling the correct response.

1. The first vaccination was accomplished by:

   a. Roux.
   b. Karl Landsteiner.
   c. Niles Jerne.
   d. Edward Jenner.
   e. Yersin.

2. Cells that can engulf and _____ foreign particles are:

   a. macrophages.
   b. antibodies.
   c. NK cells.
   d. B cells.
   e. T cells.

3. Invading cells can be burst by a series of _____, or the:

    a. complement response.
    b. clonal selection response.
    c. clonal expansion.
    d. inflammatory response.
    e. autoimmune response.

4. Rattlesnake victims are best treated with:

    a. gamma globulin.
    b. antivenins.
    c. vaccination.
    d. natural killer cells.
    e. T cells.

5. Autoimmune disease is caused by a breakdown in:

    a. T lymphocytes.
    b. tolerance.
    c. macrophages.
    d. immunoglobulin.
    e. heavy chains.

## Exercise

The diagram below shows the inflammatory response. Describe what is taking place, including the roles of granulocytes and macrophages.

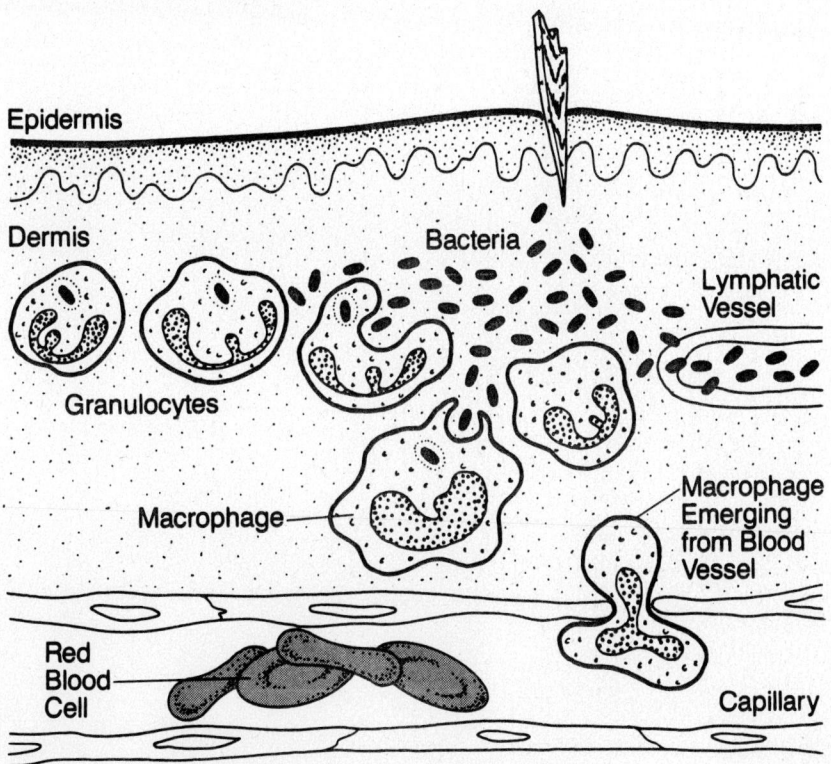

# 33

# RESPIRATION: THE BREATH OF LIFE

---

## CHAPTER AT A GLANCE

---

## CHAPTER PREVIEW

Chapter 33 begins with a fairly simple but important lesson on the physics of gases—the properties and interactions that, in part, determine whether an air-breathing animal such as yourself will be able to take in sufficient oxygen to meet bodily needs. Next the text surveys the various respiratory systems found in the animal kingdom (body surface, gills, tracheae, and lungs).

The logical outcome of this survey is a detailed look at the human respiratory system, the best studied example of lung-based respiration. Next, you will explore some of the specialized functions of structures seen in other vertebrates, including the air sacs of birds and the lung-derived swim bladder of fish. You may find the following section on respiratory pigments and gas transport the most challenging part of this chapter because it involves some chemistry and chemical formulas.

Mechanisms that control breathing come next, followed by basic facts on the functioning of the brain's respiratory center. A final, fascinating section considers some unusual adaptations to oxygen-poor environments, including the huge arteries of some high mountain-dwelling humans, and the vertebrate diving reflex.

# LEARNING OBJECTIVES

When you have mastered the concepts of this chapter, you will be able to:

1.  Explain the physical factors that affect the behavior of gases, and describe their consequences for biological gas movements.
2.  Explain how simple diffusion of respiratory gases takes place.
3.  List the types of internal and external respiratory organs found among animals, and discuss the advantages and limitations of each.
4.  Describe the countercurrent exchange system of fish gills.
5.  Explain why air-breathing animals expend less energy to obtain oxygen than do aquatic, gill-using species.
6.  Describe the gas exchange system of insects and the limits it sets on body size.
7.  Make a simple sketch of a vertebrate lung, name its structures, and explain how gas exchange takes place.
8.  Explain why *all* types of respiratory organs have large surface areas.
9.  Sketch the upper respiratory pathway of mammals (from mouth and nostrils to the lungs), and label all structures.
10. Sketch the microstructure of a mammalian lung, and explain where gas exchange takes place.
11. Explain how air is drawn into the lungs of mammals.
12. Define the terms *vital capacity* and *tidal volume*, and explain how each may vary with level of activity.
13. Describe the air sac system of birds, tell how it works, and explain how air is drawn into the lungs of birds.
14. Explain what a swim bladder is, and how it functions.
15. Explain the role of hemoglobin in respiration.
16. Explain the oxygen dissociation curve and the Bohr effect.
17. Tell how myoglobin functions in respiratory storage.
18. Explain how respiration is regulated by the brain.
19. List five adaptations animals have for coping with low oxygen environments.

# CONCEPTS IN REVIEW

## Section I    Respiration and the Physics of Gases

Respiration is the process by which whole organisms exchange gases with the environment. It includes the intake of oxygen, its transport and delivery to cells, and the removal of carbon dioxide.

The pressure exerted on the Earth by the weight of gases in the air is called atmospheric pressure. Since air is composed of several gases, each one of which makes up only part of the total, each gas exerts only a partial pressure. The partial pressure of nitrogen is abbreviated $PN_2$; of oxygen, $PO_2$; and of carbon dioxide, $PCO_2$. Partial pressure helps determine what quantity of a gas will travel in blood and through extracellular fluid to reach the tissues; the higher the partial pressure of a gas, the more of that gas enters a liquid. Other factors also play a role, including the solubility of the gas in water (for most gases, it is quite low), the temperature (colder liquids hold more gas), and the diffusion rate of the gas (which is much slower in liquids than in air; text pp. 789–791). By breathing into lungs or pulling water across gills, animals set up a bulk flow that compensates for water's limited capacity for carrying gases.

## *Section II*  Respiratory Organs: Structures for Efficient Gas Exchange

Throughout evolution, the organs that make up an animal's respiratory system have overcome the physical limitations on the movement and content of gases in liquids by expanding their surface area, maintaining a wet surface for gas exchange, and developing efficient ventilating and pumping mechanisms. Single-celled organisms and some multicelled animals with a slow metabolism exchange gases with the environment by diffusion across moist membranes that are not special respiratory membranes (text p. 791). In many aquatic animals, external respiratory organs called gills provide a large surface area and a short distance for gases to diffuse between the surrounding water and the circulation. Countercurrent flow in and across the gills of fish improves the efficiency of oxygen uptake (text pp. 791–794).

Insects and most terrestrial arthropods have tracheae, branching hollow tubes that deliver air directly to tissues throughout the body. Spiracles are the closable vents that lead into the tracheae (text p. 794).

Terrestrial vertebrates, many aquatic vertebrates, and some invertebrates have lungs, with varying surface areas for gas exchange. In all lungs, the actual exchange occurs in the watery layer covering the innermost layer of lung cells.

Animals with lungs consume less energy in respiration than do gilled animals because air holds much more oxygen than water, and the animals have ventilatory mechanisms to bring large quantities of air to their lungs (text pp. 794–796).

## *Section III*  The Human Respiratory System: The Hollow Tree of Life

Air enters a person's upper respiratory tract through the nostrils and moves through other structures to the trachea (text pp. 796–797). The trachea branches into two bronchi, which in turn branch to bronchioli that end in blind cavities called alveoli. Through the thin-walled alveoli, oxygen reaches the blood and carbon dioxide leaves it. Once in the blood plasma, most oxygen enters red blood cells and binds to hemoglobin.

Ventilation—the filling and emptying of internal respiratory organs—operates by a force pump system in lungfish and amphibians. Land vertebrates evolved a suction pump action based on the reptilian innovation of an expandable rib cage. Forcible exhalation in mammals increases inhalation capacity from 10 percent (tidal volume) to 80 percent (vital capacity) of the trachea's and lungs' total volume (text pp. 798–799).

## *Section IV*  Respiration in Birds: A Special System

Birds have a highly efficient respiratory system that includes air sacs in addition to lungs. The system functions by continuous flow as well as a crosscurrent exchange system (text pp. 799–800).

## *Section V*   Swim Bladders in Fish: Buoyancy Organs Derived From Lungs

Analogous to the air sacs of birds is the swim bladder of fish, which enables these aquatic vertebrates to adjust their buoyancy. Air sacs and swim bladders probably evolved in parallel from the lungs of ancient air-breathing fish.

## *Section VI*   Respiratory Pigments and the Transport of Blood Gases

Special pigment molecules, circulating freely in blood or contained in red blood cells, do most of the actual transporting of gases in complex respiratory systems. In some invertebrates, the pigment is hemocyanin; in other animals, the pigment is hemoglobin.

Hemoglobin picks up oxygen in regions where $PO_2$ is high and releases it where $PO_2$ is low (text pp. 802–803). A decrease in blood pH (increase in acidity) or increase in blood temperature lowers hemoglobin's affinity for oxygen, as does the binding to hemoglobin of ATP (text pp. 803–805). When hemoglobin binds carbon monoxide, it is unable to release its oxygen cargo in the tissues. In muscles, the respiratory pigment myoglobin stores oxygen, which it releases during heavy exercise (text p. 805).

Carbon dioxide diffuses out of cells and into the venous end of capillaries because $PCO_2$ is higher in the tissues than in the blood of those capillaries. The $CO_2$ then moves to the lungs either dissolved and unchanged in the blood, as a dissolved ion of carbonic acid, or inside red blood cells. In the lungs, $CO_2$ passes into the alveoli, from which it is expelled from the body (text pp. 806–807).

## *Section VII*   Control of Breathing

The respiratory center, a group of nerve cells in the brain's medulla, regulates the muscle contractions that control breathing. In aquatic vertebrates, changes in the level of $O_2$ affect the respiratory center, but in terrestrial vertebrates, it is the level of $CO_2$ that propels respiration. The mechanoreceptor reflex triggers a change in breathing rate to fit the level of exercise and not simply in response to changes in blood chemistry (text p. 807).

Improper brain control of breathing may cause death in older people and in infants.

## *Section VIII*   How Animals Adapt to Oxygen-poor Environments

Animals adjust to oxygen-poor environments through structural and physiologic modifications. These adaptations enable them to pick up more oxygen and deliver it more efficiently to the tissues, with the smallest expenditure of energy.

High-flying birds and mountain-dwelling mammals have hemoglobin with a high affinity for oxygen. Indians of the Andes have enlarged chests and lungs. And deep-diving marine mammals have a large amount of myoglobin in their muscles; during a dive, they shunt blood from other tissues to heart, brain, and skeletal muscles (text pp. 808–810).

## KEY TERMS

air sac   *text page* 799

alveolus   797

atmospheric pressure   787

Bohr effect   *803*

bronchus   797

countercurrent exchange system   *793*

crosscurrent exchange system   799

diaphragm   798

diving reflex   *809*

DPG (2,3-diphospho-glycerate)   *804*

external intercostal  *798*

gill  *791*

internal intercostal  *798*

larynx  *797*

lung  *794*

mechanoreceptor reflex  *807*

myoglobin  *805*

oxygen debt  *809*

oxygen dissociation
   curve  *802*

partial pressure  *788*

pharynx  *797*

respiration  *787*

respiratory pigment  *801*

solubility  *790*

swim bladder  *801*

tidal volume  *798*

trachea  *797*

tracheae  *794*

vital capacity  *798*

# SELF-QUIZ: TESTING WHAT YOU HAVE LEARNED

## Matching Key Terms

Match each term on the left with the most appropriate description on the right.

1. air sac
2. diving reflex
3. bronchus
4. tidal volume
5. myoglobin
6. vital capacity
7. alveolus
8. gills
9. diaphragm
10. Bohr effect

a. birds
b. pH and $O_2$
c. can store oxygen
d. marine mammals
e. 80 percent
f. 10 percent
g. only found in mammals
h. blind-end cavities
i. major branch of trachea
j. aquatic animals

## True or False?

1. _____ Atmospheric pressure is about 14.7 pounds per square inch at sea level.

2. _____ Gas solubility in liquids is very high.

3. _____ Water flow over fish gills is a countercurrent exchange system.

4. _____ Lungs use much more energy in respiration than do gills.

5. _____ Humans (and all mammals) possess a secondary palate.

6. _____ Swim bladders are controlled by the red and oval glands.

7. _____ Crosscurrent exchange systems interlace at 90 degrees.

8. _____ Bradycardia consists of a rapid speed-up in heart rate.

## Completion

1. The actual site of lung-based gas exchange is a _____ layer covering the
   _____ layer of lung cells.

2. Within the lungs, fluid surface tension is lowered by _____.

3. Although the main muscle involved in mammalian breathing is the _____, the
   _____ muscles also assist.

4. The muscles of diving mammals (whales and others) contain abundant amounts of

_____, which allow for oxidative metabolism when diving.

5. The automatic breathing cycle is controlled by the _____ of the medulla.

6. Breathing rate can increase to meet the demands of exercise through the _____,
which is based not on blood chemistry but on mechanical feedback information from tendons
and joints.

## Short Answer

1. Why is atmospheric pressure important for oxygen-breathing organisms?

2. How do tracheae-based respiration systems adjust for increased activity?

3. What mechanical adaptations do mammals have to keep food, air, and liquids separate as they
pass through the oral cavity and on their way to lungs or stomach?

4. How is air drawn into the lungs in reptiles? Birds? Mammals?

5. How do respiratory pigments work?

6. What effects does pH have on the blood's ability to carry oxygen?

## Multiple-Choice Review

In the following sentences, fill in any blanks. Complete each statement by circling the correct response.

1. The proper sequence is:

    a. spiracles, tracheae, tracheoles, air capillary.
    b. spiracles, tracheae, air capillary, tracheoles.
    c. spiracles, tracheoles, tracheae, air capillary.
    d. tracheae, tracheoles, spiracles, air capillary.
    e. tracheae, spiracles, tracheoles, air capillary.

2. In mammals, the nasal and mouth cavities are kept separate by a _____, which makes it possible to chew food and breathe at the same time.

    a. larynx
    b. nostril
    c. trachea
    d. epiglottis
    e. secondary palate

3. Human speech, especially the vowel sounds, is made possible by the:

    a. pharynx.
    b. trachea.
    c. supra-laryngeal pharynx.
    d. secondary palate.
    e. larynx.

4. The function of a swim bladder is:

    a. to prevent urine contamination of public pools.
    b. to act as a hydrostatic organ.
    c. to act as a lung.
    d. to increase solubility of oxygen.
    e. none of the above.

5. The air sac system of _____ functions to:

    a. lighten bones.
    b. assist in respiration.
    c. move air into the lungs.
    d. all of the above
    e. a and b

# 34
# DIGESTION AND NUTRITION

---

## CHAPTER AT A GLANCE

---

## CHAPTER PREVIEW

As heterotrophs, animals must take in food and obtain from it the nutrients that sustain life. In this chapter you will begin with a survey of animal strategies for ingestion and digestion, starting with the simple systems of protozoans, and progressing to the complex digestive systems of mammals.

Many single-celled species obtain nutrients directly through simple diffusion, while in protozoans and many parasites the strategy is intracellular digestion of materials taken in directly through cell walls. As you see in the first section, most of the information about invertebrates is easy to understand, but once an animal's anatomy becomes more complex, so too does digestion. This becomes evident as the chapter considers the digestive systems of large herbivores, which are the most complex in the animal kingdom.

The human digestive system serves in this chapter as the main, most detailed example. Most of the anatomy is familiar and should present no problems—in fact, many students are intrigued to learn what actually happens to the food they eat. The chemical breakdown of food molecules by enzymes is the basis for digestion, and is the subject of the chapter's third major section. Table 34-1 presents a clear summary of the enzymes' uses, their source(s), what they act upon, and the end products.

Coordination of digestion—including subsections on control of hunger, feeding, and body fat—is described next, followed by a discussion of the roles of molecules that qualify as macro- and micronutrients, and vitamins and minerals. You will find summaries of much of this information in Tables 34-2 and 34-3.

# LEARNING OBJECTIVES

When you have mastered the concepts of this chapter, you will be able to:

1. Distinguish between ingestion and digestion.
2. Distinguish between the feeding strategies of herbivores, carnivores, and omnivores.
3. List the parts of an alimentary canal, and list the alimentary canal organs found in humans.
4. Explain what happens to food in the buccal cavity (mouth).
5. Trace the movement of food through the route of ingestion (from the mouth to the stomach).
6. Explain what happens to food in the stomach.
7. Trace the passage of food through the small intestine, explain what happens to it there, and define the term *peristalsis*.
8. Outline the digestive roles of the pancreas and liver.
9. List and discuss the functions of the large intestine.
10. Explain and discuss the symbiotic relationships (related to digestion) between vertebrates and bacteria and protozoans.
11. State the principle of chemical digestion.
12. Explain how proteins, carbohydrates, and fats are broken down chemically.
13. Name the hormones that control the secretion of pancreatic enzymes and liver bile.
14. Explain what macronutrients are, and list at least three.
15. Explain what micronutrients are, and list four functions that depend on them.

# CONCEPTS IN REVIEW

## Section I    Animal Strategies for Ingestion and Digestion

To secure the nutrients they need for survival, animals take in (ingest) and chemically break down (digest) food. The organs of the digestive system enable them to do this, and these organs vary in shape and size with the type and amount of food consumed. In fact, animals have evolved basic body plans and behaviors largely designed around the needs of finding and consuming food.

In small or very flat animals like protozoans, food is absorbed or engulfed by body cells and broken down by intracellular digestion. Such animals must live where there is a continuous supply of food (text pp. 814–815).

Animals with larger, more complex bodies have a hollow gut chamber (lumen) in which extracellular (outside body cells) digestion takes place. Although flatworms and cnidarians have a blind-ended lumen with a single opening, most animals have a two-ended gut called the alimentary canal, or digestive tract. As food makes a one-way passage through this canal, it encounters discrete regions—such as the earthworm's pharynx, esophagus, crop, and gizzard—specialized for different tasks. The digestive tracts of insects show variations on a basic plan (text pp. 815–816).

Meat-eating (carnivorous) vertebrates have sharp teeth with which to tear their food to ingestible size; plant-eating (herbivorous) vertebrates have teeth adapted for slicing, crushing, and grinding plant material. Land animals that don't generate their own body heat eat less often and have a much shorter gut than animals that do generate their own body heat, such as humans or cows (text pp. 817–818). Symbiotic microbes living in extra stomach chambers and the long intestinal tract of some herbivores enable these plant eaters to harvest calories from plant cellulose (text pp. 819–820).

## Section II   The Human Digestive System

Functioning as omnivores, most humans consume a wide variety of food; as a result, their digestive system has the ability to process many kinds of substances.

Food enters an oral cavity—guarded by lips, containing teeth and a tongue, and bathed by saliva. The teeth break down food mechanically; the tongue moves the broken-down food along, and saliva moistens the food and begins the process of chemical digestion (text pp. 820–822). From the mouth, food passes through the pharynx and esophagus (text pp. 822–824) to the J-shaped bag known as the stomach. The stomach contains gastric juice, the most acidic substance in the body, which helps turn chunks of food to chyme (text p. 825). Sugars, salts, and any alcohol present in the stomach are absorbed across its wall into the circulation.

Chyme passes through the pyloric sphincter to the small intestine—the main site of chemical digestion and subsequent absorption into the body of usable nutrients. The small intestine has an enormous surface area due in part to villi and microvilli. The pancreas manufactures some of the enzymes active in the small intestine. (For details of intestinal structure and function, review text pp. 825–827.) The final stretch of the human alimentary canal is the large intestine, where water and minerals are absorbed and the feces forms. The large intestine houses symbiotic bacteria that secrete amino acids and vitamin K, which can make their way into the body.

The liver, the largest organ in the body, regulates the nutrient content of blood and thus functions as an intermediary between digestion and the body's metabolic needs. It also produces enzymes that break down toxins and an agent that controls bone growth.

## Section III   The Chemistry of Digestion

The general strategy of chemical digestion is the breaking of bonds that link monomers of protein, carbohydrate, and lipid polymers. Hydrolysis and the coupling of enzymatic tasks are basic to the process.

Amylase, which is both in saliva and secreted by the pancreas into the small intestine, breaks the amylose bonds in starch molecules. A battery of enzymes in the small intestine cleaves disaccharides to glucose, which is absorbed through the intestinal wall (text p. 828).

The enzymatic breakdown of fats is similar to that of carbohydrates, but because fats are insoluble in water, their breakdown requires the emulsifying activity of bile salts.

Endopeptidases, which cleave specific bonds, and exopeptidases split the bonds of proteins. Several well-coordinated mechanisms prevent these enzymes from attacking the cells of the body (text pp. 829–830).

## Section IV  Coordination of Ingestion and Digestion

Animals possess a well-controlled system of ingestion and digestion. Food is taken in when or before cells need nutrients, and digestive mechanisms begin as soon as food enters the alimentary canal.

The sight or smell of food elicits brain impulses that stimulate the secretion of saliva and gastric juices. The presence of food in the stomach triggers more secretions, as does the arrival of chyme in the small intestine (text pp. 830–833).

Parts of the brain also control the impulse to eat or to stop eating. The amount of food in the stomach and levels of nutrients in the liver and circulation appear to be the triggers of these controls. Hunger, instinct, and nutrient needs help determine what an animal eats, while a set point in conjunction with the proportion of stored fat seems to determine each individual's body weight (text p. 834). An increase in exercise or environmental stress can shift the set point.

## Section V  Nutrition

In addition to water, heterotrophs need large amounts of four macronutrients: proteins, carbohydrates, fats, and some minerals. They also need smaller amounts of vitamins and trace minerals, the so-called micronutrients.

The energy value of a food is measured by the number of calories produced when the food is completely oxidized. One calorie is the amount of heat required to raise the temperature of 1 gram of water by 1°C. Nutritionists use the kilocalorie (1,000 calories, or 1 Calorie) as their unit of measure. One gram of protein or carbohydrate yields 4 kilocalories of energy; 1 gram of fat yields 9 kilocalories.

Energy yield is only one measure of a food's value. Chemical makeup is another. Animals, including humans, cannot synthesize eight of the twenty amino acids needed to build proteins, and thus must ingest these essential amino acids in their food. People on a vegetarian diet must eat the proper mix of foods at every meal to ensure that their bodies have a full set of amino acids for building proteins, because the body does not store protein (text p. 835). Another macronutrient people cannot synthesize is linoleic acid; an essential component of cell membranes, it is found in vegetable oils.

Among the micronutrients, vitamins are organic molecules that function as coenzymes or cofactors of enzymes. Minerals are inorganic molecules that provide ions critical to the functioning of many cellular enzymes and proteins like hemoglobin. Tables 34-2 and 34-3 (text pp. 838–839) summarize the sources and functions of the main vitamins and minerals we need.

Vitamin and mineral deficiencies are rare in the North American population. And although malnutrition is a problem among the very poor, obesity resulting from too much dietary fat, sugar, and alcohol and too little exercise affects many more people. Aerobic exercise produces several health benefits.

## KEY TERMS

alimentary canal  *text page 815*

bile  *826*

bolus  *821*

caecum  *816*

calorie  *835*

carnivore  *817*

chyle  *826*

chyme  *825*

crop  *816*

digestion  *814*

enterogastric reflex  *833*

esophagus  *816*

essential amino acid  *835*

fat-soluble vitamin  *836*

gall bladder  *826*

gastrin  *831*

gizzard  *816*

herbivore  *818*

ingestion  *814*

intestine  *816*

large intestine  *827*

liver   *827*

macronutrient   *834*

micronutrient   *835*

microvillus   *816*

mineral   *836*

omnivore   *820*

pancreas   *826*

pepsin   *824*

peristalsis   *824*

pharynx   *816*

salivary gland   *821*

small intestine   *825*

stomach   *817*

vasoactive intestinal peptide
  (VIP)   *833*

vitamin   *836*

water-soluble vitamin   *836*

zymogen   *830*

# SELF-QUIZ: TESTING WHAT YOU HAVE LEARNED

## Matching Key Terms

Match each term on the left with the most appropriate description on the right.

| | | |
|---|---|---|
| 1. omnivores | a. | lions, tigers, and bears |
| 2. vitamin | b. | liver |
| 3. chyme | c. | causes gastric glands to release HCl |
| 4. bolus | d. | coenzyme |
| 5. calorie | e. | a unit of heat |
| 6. peristalsis | f. | semiliquid mass |
| 7. gastrin | g. | waves of contraction |
| 8. herbivores | h. | moistened lump of food |
| 9. bile | i. | plant eaters |
| 10. carnivores | j. | can eat anything (plants or meat) |

## True or False?

1. _____ Ingestion starts in the stomach.

2. _____ Digestion starts in the mouth.

3. _____ Herbivores spend more time chewing than do carnivores.

4. _____ The rumen is the equivalent of a stomach.

5. _____ The epiglottis prevents aspiration of food.

6. _____ The small intestine is longest in carnivores, shortest in herbivores.

7. _____ The human appendix is a nonfunctional caecum.

8. _____ Fat-soluble vitamins include C, A, D, and E.

## Completion

1. The surface area for absorption in the small intestine is greatly increased by numerous

   _____.

2. Typical mammals have four types of teeth, which are (from front to back): _____,

   _____, _____, and _____.

3. Food is moved down the esophagus to the stomach by _____, which is a moving
   wave of muscle contraction.

4. Animals that are _____ consume both plant and animal matter.

5. The primary site for digestion in your body is the _____.

6. _____ is an important product of the liver that controls bone growth.

## Short Answer

1. What factors control the amounts of body fat in humans?

2. What functions do vitamins have?

3. How does the alimentary canal of an earthworm compare to that of a vertebrate?

4. What processes of ingestion and digestion occur in the mouth?

5. What digestive functions are performed by the pancreas?

6. What digestive functions are performed by the liver?

## Multiple-Choice Review

Complete the following statements by circling the correct response.

1. The large intestine serves mainly to:

    a. digest starch.
    b. digest protein.
    c. produce bile.
    d. remove water.
    e. absorb amino acids.

2. Protein digestion starts in the:

    a. mouth.
    b. stomach.
    c. jejunum.
    d. ileum.
    e. large intestine.

3. Which of the following amino acids can humans *not* synthesize?

    a. lysine
    b. valine
    c. threonine
    d. isoleucine
    e. None of the above can be synthesized.

4. Water-soluble vitamins include:

    a. A.
    b. C.
    c. D.
    d. E.
    e. all of the above

5. Vitamin A is important because it:

    a. is used in visual pigment production.
    b. prevents scurvy.
    c. plays a role in blood clotting.
    d. maintains red blood cells.
    e. is involved in the Krebs cycle.

## Exercise

In the diagram below, label the parts active in digestion. Briefly describe what each does.

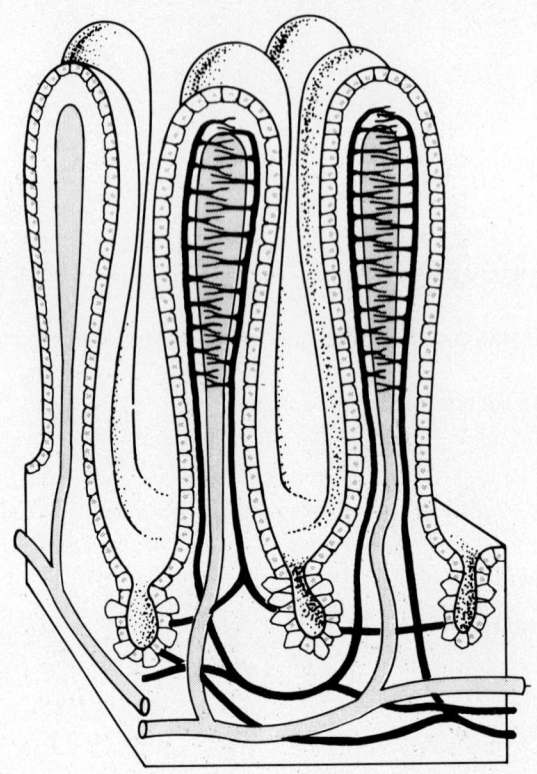

# 35

# HOMEOSTASIS: MAINTAINING BIOLOGICAL CONSTANCY

---

## CHAPTER AT A GLANCE

---

## CHAPTER PREVIEW

This chapter covers three vital systems in the animal body, all of which play major roles in maintaining a relatively constant internal environment. Because the processes are elaborate and some of the anatomy is complex, you will need to allow plenty of time to study and review each of the chapter's major sections.

The first section introduces the principles of fluid regulation and the concept of tonicity, and then traces the various strategies that enable different organisms to maintain their body's water balance while living in fresh and salt water and on dry land. Next the text examines excretion of nitrogenous wastes—the potentially toxic by-products of protein metabolism—and surveys the process in various species of invertebrates and vertebrates. The core of this section is the intricate structure and multipurpose functioning of the mammalian kidney. The anatomy in this section can be difficult, so be sure to make good use of the text figures.

The final section of Chapter 35, on mechanisms of temperature regulation, focuses on the science behind fascinating details of animal life—how insects maintain high body temperatures, why reptiles do "push-ups," why birds and mammals are able to live in habitats all over the Earth, and so forth.

# LEARNING OBJECTIVES

When you have mastered the concepts of this chapter, you will be able to:

1. Name the three general types of homeostatic mechanisms and give the function of each.
2. Cite the problems of osmoregulation faced by fresh-water animals, and describe the solutions to those problems.
3. Cite the major problem of osmoregulation faced by marine fish, and describe the solution to that problem.
4. List the osmoregulatory problems faced by birds and reptiles that inhabit desert and marine environments, and describe the solutions to those problems.
5. Give the source of nitrogenous waste products, and explain how these waste products are excreted in aquatic and land animals.
6. Sketch a cross section of a mammalian kidney, label all its structures, and explain how it works.
7. Explain how the kidneys produce urine.
8. Explain what thermoregulation is.
9. Distinguish between poikilothermy and homeothermy.
10. Distinguish between ectothermy and endothermy.
11. Explain how homeotherms regulate body temperature.
12. Explain how a rete mirabile works.
13. List the factors of thermiogenesis.
14. Explain the role of the hypothalamus in maintaining homeostasis.

# CONCEPTS IN REVIEW

## Section I    Regulation of Body Fluids

Homeostasis is the maintenance of a relatively constant internal environment despite fluctuations in the external environment. It depends on a continuous expenditure of energy. Osmoregulation involves those homeostatic operations that maintain internal fluids in a constant state. For a review of the mechanisms that determine water and salt balance, see the discussion of tonicity and osmosis in Chapter 5.

In most invertebrates and vertebrates, a compatible osmolyte strategy helps regulate water flow without a deleterious build-up of salts. Such animals also maintain a special salt balance, in which there is a higher concentration of $K^+$ inside of cells and of $Na^+$ outside of cells (text p. 844).

Fresh-water organisms must prevent excessive hydration and the outward movement of salts. To cope with these problems, they rarely drink water, excrete a large volume of water as dilute urine, and actively take up salt (text p. 845). In contrast, salt-water organisms drink water frequently, excrete some salt ions in a small volume of concentrated urine, and excrete other salt ions through salt-secreting cells on the gills. Some fish switch from pumping salt in to pumping it out as they migrate from fresh to salt water. Sharks solve their dehydration problem by a counteracting osmolyte strategy that relies on urea and TMO to raise the internal osmotic pressure (text p. 846).

Land animals have a tendency to dehydrate through the evaporation of body fluids into the air. To counteract this tendency, amphibians take up water through the skin with the help of a brain hormone, and regulate how much water is reclaimed from their bladder with another hormone. In many mammals, a countercurrent exchange operating in the nasal cavities combats water loss from the lungs; but for most terrestrial vertebrates, the main means of regulating the osmotic balance of body fluids is to control urine composition through kidney function. Animals in dry and salty environments have glands near the eye or in the tongue that remove NaCl from the blood (text pp. 847–848).

## Section II    Excretion of Nitrogenous Wastes

Nitrogenous wastes form from the breakdown of proteins, nucleic acids, and other nitrogen-containing compounds. To get rid of ammonia (or $NH_4^+$), a toxic nitrogenous waste, fish carry out deamination in the gills; from there the toxic wastes diffuse away with the respiratory current. Land animals convert ammonia and ammonium ions to the much less toxic urea or to uric acid. Concentrated uric acid crystallizes and precipitates out of solution. Table 35-1 (text p. 850) lists the major nitrogenous wastes of various vertebrates and reviews the animals' strategies of osmoregulation.

Most invertebrate and vertebrate animals use the same basic "plumbing"—a tubule able to filter and reabsorb water, ions, and molecules—to rid their body of wastes, ions, and excess water. In earthworms, organs called nephridia take care of excreting wastes and maintaining salt and water balance, while in most insects, the Malpighian tubules fulfill these functions. And in reptiles, birds, and mammals, the kidney filters blood and excretes wastes. Nephrons, the functional units of the kidney, are composed of a Bowman's capsule, a proximal convoluted tubule, a loop of Henle, and a distal convoluted tubule that feeds into a collecting duct. Together, these nephron parts produce a primary filtrate and then reabsorb from it water and other substances needed by the body. They carry out their tasks by means of filtration, reabsorption, secretion, and concentration (text pp. 852–856).

Several hormones regulate kidney function. For example, aldosterone governs the rate of salt reabsorption in the distal convoluted tubule; and ADH, secreted in response to nerve impulses from the hypothalamus, helps regulate the osmotic pressure of blood plasma by controlling the amount of urine produced (text pp. 856–858). The kidney's activities require a great deal of energy, but because its activities are essential to the maintenance of homeostasis, this energy expenditure is a price the organism must pay to survive.

## Section III    Homeostasis and Temperature Regulation

Because many biochemical reactions and biological molecules work best within a small range of temperatures, temperature has been a strong source of selective pressure. Thermoregulation, the control of internal temperature, is a complex, highly integrated process involving the organs of several physiologic systems, and it is a keystone of homeostasis.

Animals cope with the wide range of environmental temperatures by living where the temperature remains constant, by physiological processes that allow them to adapt to a range of temperatures, or by generating heat internally so that body temperature remains constant despite changes in external temperatures (text p. 859).

Poikilotherms are animals such as most fish, amphibians, reptiles, and insects that have a variable body temperature attuned to the temperature of their surroundings. Homeotherms such as birds and mammals have a relatively constant body temperature. Ectotherms derive most of their body heat from the environment, whereas endotherms generate their own body heat. In nature, many organisms do not fit neatly into these categories. And even within each category, most animals have a variety of ways—ecological, metabolic, behavioral, structural, and physiological—of regulating their responses to heat and cold.

Most aquatic organisms live in a fairly stable thermal environment and are ectothermic poikilotherms. Mako sharks and tuna are exceptions: they generate their own heat and maintain a high core temperature (text p. 861).

On land, where temperatures can fluctuate dramatically, amphibians, reptiles, and insects derive heat from the environment and have variable body temperatures that respond to changes in external temperature. Frogs (a representative of amphibians) have a low metabolic rate and become inactive in cold weather. Because of evaporative cooling problems associated with moist skin, they thrive best in the tropics. Dry-skinned reptiles, in contrast, seek out direct sunlight and other sources of heat to raise their body temperature, and they continuously adjust their behavior to achieve homeothermy during the day. Invertebrates have mainly metabolic adaptations for temperature regulation (text pp. 863–864).

Birds and mammals, the most active and behaviorally complex animals, can live all over the world because they maintain constant body temperatures of 35°C to 42°C with internally generated heat. Among the features that help them maintain a constant body temperature are insulation and piloerection, evaporation through body coverings and sweat glands, vasoconstriction and vasodilation, and behavioral mechanisms as diverse as migrating to warmer climates and changing clothes (text pp. 864–866).

Birds and mammals generate the heat they need to maintain their constant body temperatures by four mechanisms: (1) muscle contraction; (2) ATPase pump enzymes; (3) brown fat; and (4) a high basal metabolic rate controlled largely by thyroxine (text p. 866). The portion of the brain called the hypothalamus functions as a thermostat that coordinates the heat-generating mechanisms and counteracts external changes in temperature to maintain internal constancy. Normally, the thermoregulatory activity of the hypothalamus works in conjunction with rapidly responding peripheral control mechanisms to ensure an even core temperature.

Pyrogens alter the hypothalamic set point to produce fever, an important defense mechanism for fighting disease. In contrast, the set point of various homeotherms falls during hibernation, and along with this decrease, metabolism slows (text pp. 867–868). Torpor, a variation on hibernation, allows small animals like hummingbirds to function as homeotherms.

## KEY TERMS

# SELF-QUIZ: TESTING WHAT YOU HAVE LEARNED

## Matching Key Terms

Match each term on the left with the most appropriate description on the right.

| | | | |
|---|---|---|---|
| 1. medulla | a. heat from the environment |
| 2. ectotherm | b. constant body temperature |
| 3. rete mirabile | c. found in newborn mammals and polar bears |
| 4. torpor | d. in humans it is 98.6°F |
| 5. endotherm | e. contains collecting ducts and loops of Henle |
| 6. cortex | f. outer rim of kidney |
| 7. homeotherm | g. a steroid |
| 8. set point | h. miraculous net |
| 9. aldosterone | i. temporary lowering of set point |
| 10. brown fat | j. produce their own body heat |

## True or False?

1. _____ Fresh-water fish rarely drink water.

2. _____ Reptiles are good examples of ectotherms.

3. _____ Mammals and birds are good examples of endotherms.

4. _____ The set point rises during torpor.

5. _____ A rete mirabile requires blood flow in opposing directions.

6. _____ Set points are regulated by the hypothalamus.

7. _____ Fresh lake water is hypertonic to the fish that live in it.

8. _____ Sea water is hypotonic to marine fish.

## Completion

1. Sharks can get rid of excessive salt by way of a _____ gland that secretes a concentrated salt solution.

2. Animals that face water shortages often secrete their nitrogenous wastes as

   _____.

3. The simplest excretory structure of animals is the _____.

4. Insect excretory systems are based on blind-ended sacs called _____.

5. The _____ of the amphibians produce large volumes of urine, and are anatomically similar to the kidneys of fresh-water fish.

6. The kidney-regulating hormone antidiuretic hormone (or ADH), is also called

   _____.

## Short Answer

1. What are diuretics and how do they work?

2. What technique do research physiologists use to investigate water movement through kidney systems?

3. What effect does an organism's body temperature have on proteins?

4. What are some of the behavioral modifications that "cold-blooded" animals have to raise their body temperatures?

5. What are the four sources of thermiogenesis?

6. What physiological modifications take place when animals enter hibernation?

## Multiple-Choice Review

In the following sentences, fill in any blanks. Complete each statement by circling the correct response.

1. Mammals can lower their body _____ temperatures by evaporation of fluids from:

   a. loops of Henle.
   b. fresh water.
   c. saliva.
   d. urine.
   e. sweat glands.

2. Because of their small body size, some heterothermic birds and mammals can maintain high body temperatures:

   a. constantly.
   b. only in summer.
   c. only for short periods.
   d. only by eating special foods.
   e. only in special "thermal caves."

3. Amphibians must remain in _____ habitats because their kidneys lack:

   a. loops of Henle.
   b. a renal artery.
   c. nephrons.
   d. glomerular capillaries.
   e. a Bowman's capsule.

4. The color of "brown fat" comes from:

   a. blood vessels.
   b. numerous mitochondria.
   c. osmolytes.
   d. thyroxine.
   e. fatty acids.

5. Water reabsorption is regulated in the _____ ducts by:

   a. angiotensin II.
   b. salt glands.
   c. ANF.
   d. ADH.
   e. aldosterone.

## Exercise

Supply the missing labels in this diagram of a nephron.

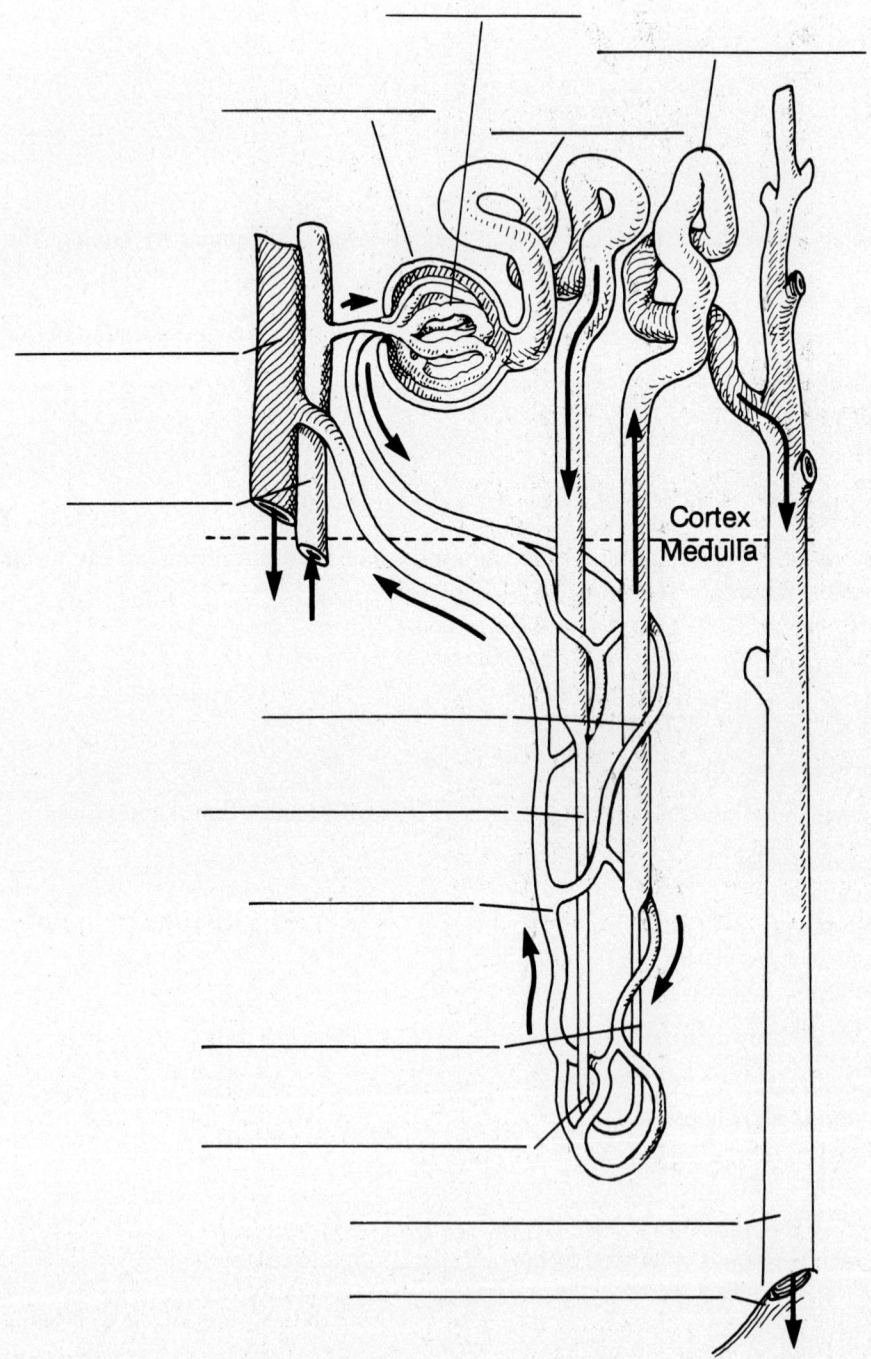

Now, briefly describe the role of each kidney structure.

# 36

# THE NERVOUS SYSTEM

---

## CHAPTER AT A GLANCE

---

## CHAPTER PREVIEW

As with Chapter 35, in this survey of the nervous system—with emphasis on the nervous system in humans—you will need to assimilate complex concepts and master some difficult terminology. You should also devote time to careful study of the chapter's illustrations and figures.

The first section presents a detailed overview of the different kinds of neurons and their anatomy. If you fully understand this material before going on, you will find it much easier in later sections to visualize the anatomical patterns associated with the reception and processing of sensory information. Next, the chapter presents the mechanisms of neuron action, and then moves into a survey of nervous systems among various species, from the simplest to the highly complex human nervous system. A final section presents the escape response of the crayfish. By the time you get this far, you should have little trouble in explaining how a series of simple sets of neural activity make this complex behavior possible.

# LEARNING OBJECTIVES

When you have mastered the concepts of this chapter, you will be able to:

1. Compare and contrast the endocrine and nervous systems.
2. Sketch a neuron, and explain the function of the dendrites, axons, synaptic terminal, and cell body.
3. List the three types of neurons and give the function of each.
4. Explain in detail the causes and effects of a nerve impulse.
5. Compare and contrast electrical and chemical synapses.
6. Explain what is meant by postsynaptic potential.
7. Describe a simple reflex arc.
8. Define the following terms: central nervous system, peripheral nervous system, somatic nervous system, autonomic nervous system, parasympathetic nervous system, and sympathetic nervous system; then arrange the terms in a simple outline that shows their hierarchical relationships.
9. Describe the crayfish escape response as an example of apparent learned behavior that is wholly instinctive.

# CONCEPTS IN REVIEW

## Section I    Neurons: The Basic Units of the Nervous System

Neurons are cells specialized for electrical switching and conducting. In most multicellular animals, they are arranged in networks whose function is to coordinate, or integrate, all of the animal's activities. The networks of neurons make up the nervous system and work via the same principles of cell-to-cell communication as the elements of the endocrine system. In addition to neurons, the nervous system contains glial cells, or glia. These provide mechanical support, insulation, and metabolic aid to neurons.

Each neuron is an energy transducer. When acted upon by chemicals, heat, pressure, or other forms of energy, it transforms the stimulus into an electrical signal known as the nerve impulse, or action potential. The neuron also conducts the information contained in nerve impulses and communicates with other cells in the nerve network (text p. 872).

Four structures characterize neurons: (1) dendrites (the receiving antennae); (2) axons (the transmitting cables); (3) synaptic terminals (the cable endings that form part of a synapse); and (4) the cell body (the maintenance site) (text p. 874).

There are three major classes of neurons. Receptor (sensory) neurons are specialized energy transducers and each one is sensitive to a particular type of stimulus. Effector (motor) neurons transmit messages to muscles and glands and thereby cause some sort of action. Interneurons convey information from sensory to motor neurons and are the sites of integration within the nervous system (text p. 875).

## Section II    How Neurons Signal

Neurons generate impulses, or action potentials, by electrochemical events that change the permeability of the cell membrane and allow ions to move in and out. The movement of ions changes the electrical polarity of the neuron and these changes initiate and propagate a nerve impulse.

The resting potential is a state of stored electrical energy in which the inside of the cell has a negative electrical charge (fewer positive ions, mostly $K^+$) relative to the outside of the cell (more positive ions, mostly $Na^+$; text p. 878). In most neurons, the resting potential is $-70$ mV.

Although all cells have resting potentials, only neurons (and a few other cell types) have a cell membrane that responds in an explosive way to chemical or electrical signals. In response to such stimuli, ion channels for $Na^+$ in the neuronal membrane open and allow some of the positive sodium ions in the extracellular fluids to flow into the cell. This influx, which transforms the negative internal electrical charge to a positive charge, depolarizes the cell and sets up an impulse. During the short refractory period that follows, the cell membrane in the immediate area of the impulse cannot react to additional stimuli (text pp. 878–881). After the refractory period ends, the cell repolarizes; that is, ion pumps in the cell membrane restore the normal ionic balance of the resting potential, once again making the intracellular charge negative relative to the extracellular charge.

Neurons initiate action potentials in an all-or-none fashion. A stimulus must reach a threshold before sodium channels will open; but once it reaches the threshold, no matter how far beyond the threshold it extends, the stimulus will initiate the same action potential (text p. 881).

Action potentials propagate because the initial depolarization event sets up local circuit currents that depolarize adjacent membrane areas in a rapid one-way succession down the axon. The greater the diameter of the axon, the greater the speed of impulse propagation. In most animals, the largest axons are in areas of the nervous system where speed of conduction contributes the most to survival. Temperature also affects the speed of conduction. The higher the temperature, the faster the propagation. Insulation by myelin sheaths, in conjunction with nodes of Ranvier, also speeds up impulse conduction by means of saltatory propagation (text pp. 882–883).

An impulse propagated in one neuron is transmitted to an adjacent neuron or effector cell across a synapse. The neuron whose impulse moves toward the synapse is the presynaptic neuron. The cell on the other side is the postsynaptic cell. In electrical synapses, transmission occurs by direct contact between cell membranes. In chemical synapses, chemical messengers called neurotransmitters diffuse across the synapse from presynaptic to postsynaptic cells (text pp. 884–885). At an excitatory synapse, neurotransmitters depolarize the postsynaptic cell, while at an inhibitory synapse, transmission prevents depolarization of the receiving cell (text pp. 886–887).

## *Section III*    The Organization of Neurons into Systems

Radially symmetrical cnidarians (jellyfish and hydra) have nerve nets composed of a fast-conduction circuit and a slow-conduction circuit. There is no centralized control center. As animals became more complex and evolved bilateral symmetry and cephalization, they also evolved reflex arcs and nervous control centers. Rare, two-neuron arcs link a sensory neuron to a motor neuron, but most reflex arcs have one or more interneurons. The more interneurons in a circuit, the greater the opportunity for complex information processing and output. In the most complex circuits, learning occurs (text p. 890).

Ganglia are groups of neuronal cell bodies that coordinate functions and behaviors in bilaterally symmetric animals. The cerebral ganglia, or brains, are the most complex ganglia. Nerve cords are bundles of axons that relay data.

The vertebrate nervous system is divided into a central nervous system (CNS) and a peripheral nervous system (PNS). The central nervous system consists of the brain and spinal cord. Together they integrate and control all functions of the body (text pp. 890–891).

The peripheral nervous system is composed of receptor cells, sensory cells that relay information inward to the CNS, and motor neurons that carry signals out to muscles and organs throughout the body. Divided into two functional subsystems, the PNS controls voluntary muscle activity through the somatic nervous system and involuntary physiologic functions through the autonomic nervous system. The autonomic nervous system, in turn, has sympathetic and parasympathetic components, which often act antagonistically (text pp. 891–894).

## Section IV   How Neurons Regulate a Simple Behavior: The Crayfish Escape Response

The crayfish escape response depends on a "hard-wired" circuit that forms in the embryo according to genetic and developmental instructions. The escape action is instinctive and does not depend on the modification of circuits that characterizes learning (text pp. 795–797).

The crayfish modifies its escape response not by purposeful behavior, but by specific neural pathways that for a moment override the escape response and lead the animal to nearby food.

## KEY TERMS

action potential (nerve impulse)   *text page 873*

autonomic nervous system   *892*

axon   *874*

central nervous system (CNS)   *890*

chemical-gated channel   *879*

dendrite   *874*

depolarization   *879*

effector (motor) neuron   *875*

excitatory synapse   *886*

ganglion   *890*

glial cell (glia)   *874*

grand postsynaptic potential (GPSP)   *887*

hyperpolarization   *879*

inhibitory synapse   *887*

interneuron   *875*

local circuit current   *881–882*

myelin sheath   *883*

nerve   *874*

nerve impulse (action potential)   *873*

nerve net   *889*

neuron   *872*

neurotransmitter   *884*

node of Ranvier   *883*

oligodendrocyte   *883*

parasympathetic nervous system   *892*

peripheral nervous system (PNS)   *890*

polarization   *878*

postsynaptic potential (PSP)   *887*

receptor (sensory) neuron   *875*

reflex arc   *890*

refractory period   *881*

resting potential   *877–878*

saltatory propagation   *883*

Schwann cell   *883*

somatic nervous system   *892*

summation   *887*

sympathetic nervous system   *892*

synapse   *884*

synaptic terminal   *874*

synaptic transmission   *884*

threshold   *881*

voltage-gated channel   *879*

## SELF-QUIZ: TESTING WHAT YOU HAVE LEARNED

### Matching Key Terms

Match each term on the left with the most appropriate description on the right.

1. dendrite
2. Schwann cell
3. neuron
4. myelin sheath
5. ganglion
6. synapse

a. transmits action potentials
b. receives information from receptors
c. insulates and supports neurons
d. a spinelike extension of cell surface
e. produces myelin
f. insulates axons

7. threshold     g. aggregations of neuron cell bodies
8. axon          h. junction
9. glial cell     i. all-or-none
10. interneuron   j. energy transducer

## True or False?

1. _____ Action potentials are thought to be generated at the axon hillock.

2. _____ A cell at its resting potential is in an unpolarized state.

3. _____ Propagation slows as temperature rises.

4. _____ The majority of human nervous system cells are interneurons.

5. _____ Peripheral nerve myelin sheaths are produced by Schwann cells.

6. _____ Reflex arcs probably evolved after nerve nets.

7. _____ The brain's surface appears gray because of billions of myelinated axons.

8. _____ The parasympathetic nervous system stimulates organ function.

## Completion

1. _____ are short, multibranched, and often spinelike extensions of the cell surface.
2. One of the important functions of glial cells is to generate layers of electrical insulating material, _____, around individual atoms.

3. _____ are very large axons near the spinal cord of fish that function in the startle response.

4. The process of impulses jumping from node to node is called _____.

5. The adding together of individual postsynaptic potentials is called _____.

6. Brains are composed primarily of _____.

## Short Answer

1. What is the structural difference between gray matter and white matter?

2. What are the two major divisions of the autonomic nervous system?

3. What are the major divisions of the peripheral nervous system?

4. Briefly, what role do sodium ions play in the propagation of an action potential?

5. What is the function of the refractory period?

6. What is the chemical/physical basis of the action potential?

## Multiple-Choice Review

In the following sentences, fill in any blanks. Complete each statement by circling the correct response.

1. The nucleus of a neuron is found in the:

   a. dendrite.
   b. synaptic terminal.
   c. axon hillock.
   d. cell body.
   e. axon.

2. The actual junction where one neuron communicates with another is called a:

   a. glial junction.
   b. synapse.
   c. dendrite.
   d. receptor.
   e. neurotransmitter.

3. When the inside of a cell is electrically negative relative to the outside of the cell, the electrical charge is said to be in:

   a. a depolarized state.
   b. a voltage-gated state.
   c. resting potential.
   d. an inaction potential.
   e. none of the above

4. In mammals, eye position is controlled by the:

   a. olfactory nerve.
   b. vagus nerve.
   c. hypoglossal nerve.
   d. facial nerve.
   e. trochlear nerve.

5. Brain and _____ myelin is produced by:

   a. Schwann cells.
   b. oligodendrocytes.
   c. saltatory propagation.
   d. dendrites.
   e. effector neurons.

## Exercise

In the graph below, use arrows and labels to show where on the action potential curve sodium flows into the cell and potassium flows out of it.

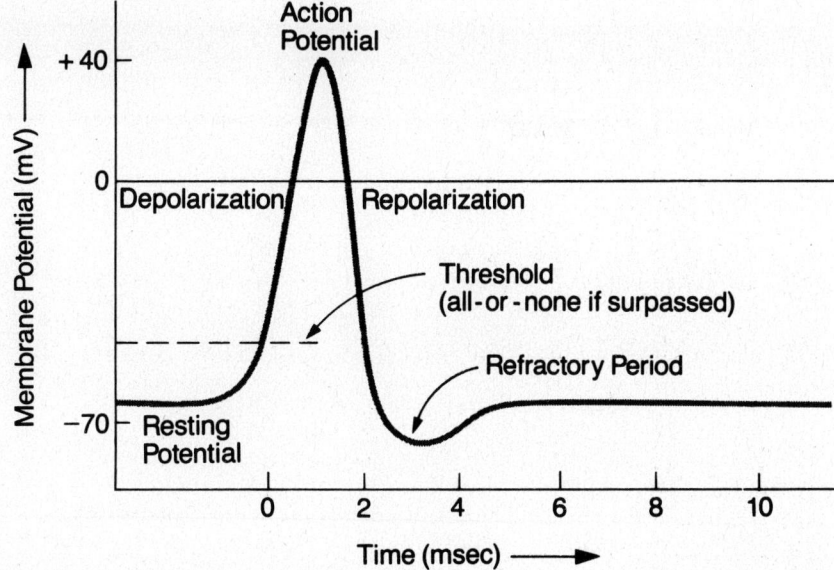

What happens during depolarization and repolarization, respectively? What happens during the refractory period?

# 37
# HORMONAL CONTROLS

## CHAPTER AT A GLANCE

## CHAPTER PREVIEW

This chapter deals with hormones—the chemical messengers produced at one site in the body that produce effects at sometimes very distant target cells. There are few difficult concepts in this chapter, and Tables 37-1 and 37-2 clearly summarize major hormones, their sources and effects.

After briefly setting the stage with the six principles of hormonal action, the text quickly traces the chemical classification of hormones as proteins, amines, and steroids. The well-studied roles of hormones in the development of invertebrates come next, followed by a detailed survey of the vertebrate endocrine system.

The latter portion of the chapter probes how hormones influence complex physiological, behavioral, and developmental processes. A final short section is devoted to recent advances in our understanding of hormones, especially prostaglandins and the brain peptides.

# LEARNING OBJECTIVES

When you have mastered the concepts of this chapter, you will be able to:

1. Create a simple, basic definition of a hormone that includes its source, effect, and target.
2. List and discuss the six principles that apply to hormone action.
3. Give a simple, chemically based definition of hormones.
4. Explain how steroid hormones work.
5. Explain the roles of cyclic AMP, cyclic GMP, inositol triphosphate, diacyl glycerol, juvenile hormone, and ecdysone.
6. Explain how blood-glucose levels are controlled, which hormones are involved, and where they are produced.
7. List the hormones secreted by the kidneys, and explain what they do.
8. Name the hormones secreted by the heart and explain their functions.
9. Name the three groups of corticosteroids and explain their functions.
10. Give the sources of epinephrine and norepinephrine, and explain what these hormones do.
11. List the hormones secreted by the thyroid and explain their functions.
12. List the pituitary hormones, tell which ones originate from the anterior pituitary and which from the posterior pituitary, and explain the role(s) of each.
13. List and discuss the hormones that affect circadian rhythms.
14. Explain the function of prostaglandins.

# CONCEPTS IN REVIEW

## Section I    Basic Characteristics of Hormones

Hormones are secreted by the organs, tissues, and cells of the endocrine system. Once they enter the circulatory system, they travel to other parts of the body, where they affect the activities of target cells. Some hormones regulate growth; others control sexual maturation or reproductive activity; still others maintain the chemical balance of blood or alter an animal's behavior with the changing seasons.

Exocrine glands secrete substances into ducts that empty into body cavities or onto body surfaces. Endocrine glands have no ducts and secrete their hormones into the extracellular space, from which hormone molecules diffuse into the bloodstream. Paracrine hormones diffuse to affect local cells and do not enter the circulation. Autocrine hormones act back on the secreting cells.

There are three chemical categories of hormones—protein, amine, and steroid (text p. 902)—but all hormones function according to the same basic principles. They affect only target cells that possess specific receptors; the receptors develop as the target cells differentiate. Different cells may respond in different ways to the same hormone. Some hormones act steadily over a long period; some appear sporadically and stimulate a brief burst of activity. Negative feedback loops govern the amount of a circulating hormone. Hormones are usually broken down rapidly (text pp. 902–903).

## Sections II and III    Cellular Mechanisms of Hormonal Control and Endocrine Functions in Invertebrates

The binding of a hormone to a receptor (either on the surface or within the cytoplasm of a cell) transmits a chemical message that can have one of three effects: (1) increase the rate at which other substances enter or leave the target cell; (2) stimulate the cell to synthesize enzymes, proteins, or other substances; (3) prompt the cell's machinery to activate or suppress existing cellular enzymes.

To transmit their chemical message, steroid hormones (such as estrogens or testosterone) must gain entry to a cell and bind to receptors in the cell's nucleus. Formation of a hormone–receptor complex results in gene activation and the production of specific proteins (text pp. 903–904).

With the exception of thyroxine, nonsteroid hormones bind to receptors in the target cells' plasma membrane. This binding results in an increase in cAMP production; and the cAMP, functioning as a second messenger, sets in motion a cascade of chemical reactions that either activate or inactivate specific enzymes or other proteins. Cyclic GMP is a second messenger that has the opposite effect from cAMP. In another second messenger system, $IP_3$ triggers the release of $Ca^{2+}$ (text pp. 905–907). All second messengers amplify the original hormonal message so that a small amount of hormone produces large effects.

Hormones control periodic molting, reproductive cycles, and metamorphosis in insects. Ecdysone, a steroid hormone that controls molting and metamorphosis, is produced by all insects and many other arthropods. Several peptides that serve as regulatory molecules in protozoans and simple multicellular invertebrates evolved to function as neurotransmitters in humans and other vertebrates (text pp. 907–908).

## Section IV    The Vertebrate Endocrine System

The organs of the vertebrate endocrine system can be divided into two categories: multipurpose organs that combine endocrine functions with other biological activities, and organs whose sole function is to secrete hormones. Control of these organs' activities is either by nerve cells in the brain or by mechanisms operating independently of the brain.

The pancreas is a dual-purpose organ. In addition to secreting digestive enzymes, it secretes hormones (insulin, glucagon, somatostatin) that regulate glucose levels in the blood and glucose absorption by the intestine (text pp. 909–911). The kidneys are also multipurpose organs. They secrete hormones that stimulate bone marrow to produce red blood cells; regulate calcium uptake in the small intestine; and help regulate the sodium-transport activity of the kidney's own tubule cells (text p. 911). The heart is another multipurpose organ with an endocrine function. It secretes atrionatriuretic factor (ANF), which helps lower blood pressure.

The adrenal glands function as endocrine tissues only and consist of a cortex that secretes steroid hormones and a medulla that secretes amine hormones. The steroid hormones of the cortex contribute to the regulation of glucose production and circulation (glucocorticoids); affect blood pressure and muscle function by controlling the flow of sodium and potassium ions (mineralocorticoids); and control sexual development, maturation, and activity (androgens and estrogens; text pp. 911–912). In the medulla, chromaffin cells secrete mainly epinephrine and norepinephrine, which produce the fight-or-flight reaction (text pp. 912–913).

Another purely endocrine gland is the thyroid. In response to a pituitary hormone, the thyroid secretes two hormones that set an animal's metabolic rate, and a third hormone that decreases the level of calcium in the blood. A fourth hormone, secreted by the parathyroid glands (which sit on the surface of the thyroid), increases blood calcium. The calcium-increasing and calcium-decreasing hormones participate in a complex feedback loop that functions independently of the nervous system or pituitary gland (text pp. 913–914).

Situated at the base of the skull, the pituitary gland is composed of an anterior and a posterior lobe that together produce at least ten hormones. Table 37-1 on text page 917 summarizes the structure and function of the hormones of the anterior pituitary. Only two hormones are secreted by the posterior pituitary: oxytocin (which stimulates contraction of smooth muscle cells) and ADH (which increases urine production). These hormones are synthesized in the nearby hypothalamus.

Although only a small part of the brain, the hypothalamus controls the pituitary gland and governs many physiological and behavioral responses including hunger, thirst, body temperature, and sexual stimulation. Table 37-2 (text p. 919) reviews the releasing and release-inhibiting hormones of the hypothalamus.

## Sections V and VI   Hormonal Control of Physiology and Hormonal Control of Behavior

By integrating and controlling an animal's physiological functions, hormones help maintain home-ostasis. For example, when glucose enters the blood plasma, insulin produced by pancreatic cells stimulates muscle, fat, and liver cells to take up the glucose. When blood glucose falls below a certain level, other pancreatic cells secrete glucagon, and this hormone causes the liver to release some of its stored glucose into the blood. As a result, the levels of blood glucose remain fairly stable.

Hormones also control the daily (circadian) and annual rhythms by which animals function. While external cues may "set" the daily or annual clocks, the internal rhythms have a genetic basis and are propelled by the brain and its hormones (text pp. 921–922).

The mating and other complex behaviors of all vertebrates and invertebrates depend on a variety of hormones and the physiologic and motor responses they produce in reaction to sensory inputs to the brain and hypothalamus.

## Section VII   Hormones and Development: Amphibian Metamorphosis

The metamorphosis of a tadpole into an adult frog illustrates the role hormones play in development. In the changeover from an aquatic herbivore to an amphibious insectivore, legs appear, the tail disappears, and the type of hemoglobin alters—all through the control of hormones secreted mainly by the thyroid and the brain (text p. 923).

The brain probably initiates metamorphosis in response to changing environmental conditions. Some conditions result in neotony in which salamander larvae develop mature gonads and repro-duce without metamorphosing to adult forms (text p. 924).

## Section VIII   Hormones: New Types, New Sites, New Modes of Synthesis

Prostaglandins and brain peptides are recently discovered molecules that function as hormones. Prostaglandins are produced by most cells and seem to appear in response to local tissue distur-bances; they stimulate smooth muscle cells.

Brain peptides have relatively long-lasting effects and most likely help maintain normal relation-ships among nerve-cell activity, blood flow, and brain-cell metabolism. Most brain peptides are also found in other vertebrates, invertebrates, single-celled organisms, and even plants. Thus, these molecules apparently had an ancient evolutionary origin and have functioned in biochemical communication for millions of years.

## KEY TERMS

adrenal gland   *text page 911*

adrenocorticotropic hormone (ACTH)   *917*

aldosterone   *911*

antidiuretic hormone (ADH)   *917*

calcitonin   *913*

chromaffin cell   *912*

corticosteroid   *911*

cyclic AMP (cAMP)   *905*

cyclic GMP (cGMP)   *906*

diacylglycerol (DG)   *906*

1,25-dihydroxy-chole-calciferol   *911*

ecdysone   *907*

endocrine system   *901*

epinephrine   *912*

erythropoietin   *911*

glucagon   *909*

glucocorticoid   *911*

goiter   *913*

growth hormone (GH)   *915*

inositol   trophosphate   (IP₃)   *906*

insulin   *909*

islets of Langerhans   *909*

juvenile hormone (JH)   *908*

lipotropin (LPH)   *917*

melanophore-stimulating hormone (MSH)   *917*

melatonin   *921*

mineralocorticoid   *911*

neotony   *924*

norepinephrine   *912*

oxytocin   *917*

parathyroid gland   *914*

parathyroid hormone (PTH)   *914*

pineal gland   *921*

pituitary gland   *914*

prolactin   *915*

prostaglandin   *924*

release-inhibiting hormone (RIH)   *918*

releasing hormone (RH)   *918*

somatomedin   *915*

somatostatin   *909*

suprachiasmatic nucleus (SCN)   *921*

thyroid gland   *913*

thyrotropic hormone (TH)   *913*

thyroxine ($T_4$)   *913*

triiodothyroxine ($T_3$)   *913*

---

# SELF-QUIZ: TESTING WHAT YOU HAVE LEARNED

## Matching Key Terms

Match each term on the left with the most appropriate description on the right.

1. prolactin
2. adrenal gland
3. ACTH
4. glucagon
5. goiter
6. MSH
7. pineal gland
8. thyroid gland
9. insulin
10. neotony

a. may cause infertility in humans
b. regulates corticosteroid production
c. releases melatonin
d. caused by lack of iodine
e. juvenile reproduction
f. stimulates conversion of glycogen
g. influences fur color in some mammals
h. derived from the endostyle
i. "toward the kidney"
j. regulates glucose

## True or False?

1. _____ Endocrine glands are ductless.

2. _____ The hormones LH, TH, and FSH are produced in the posterior pituitary.

3. _____ Gluconeogenesis takes place in the liver.

4. _____ Hormones affect only specific, target cells.

5. _____ Metamorphosis is triggered by high levels of juvenile hormone.

6. _____ The vertebrate brain cannot be considered part of the endocrine system.

7. _____ Falling levels of erythropoietin trigger red blood cell production.

8. _____ Too much ADH produces the excessive urination of diabetes insipidus.

## Completion

1. The two basic types of vertebrate glands are the ducted _____ glands and the ductless _____ glands.

2. Juvenile hormone is secreted by the _____ organs during the caterpillar instar stages.

3. Insulin and glucagon are produced in the _____ of the pancreas.

4. The pancreatic hormone _____ is secreted by delta cells and inhibits the secretion of insulin and glucagon.

5. _____ is a phenomenon wherein juveniles are fully capable of reproduction.

6. Epinephrine causes brain and skeletal blood vessels to _____, but skin and kidney blood vessels to _____.

## Short Answer

1. What are the advantages/disadvantages of the fight-or-flight response? What is the chemical trigger of the response?

2. What are the various hormones produced by the thyroid, and what are their functions?

3. What are the various hormones produced by the pituitary, and what are their functions?

4. How are rhythms and biological clocks controlled by hormones?

5. What makes hormones so specific?

6. What are the second messengers and how do they work?

## Multiple-Choice Review

In the following sentences, fill in any blanks. Complete each statement by circling the correct response.

1. The metamorphosis of a silkworm from _____ stage to the adult stage is controlled by:

   a. epinephrine.
   b. adenyl cyclase.
   c. juvenile hormone.
   d. prolactin.
   e. brain hormone.

2. The second messenger phenomenon was discovered by:

   a. John Hunter.
   b. E. U. Sutherland.
   c. E. H. Starling.
   d. W. M. Bayliss.
   e. none of the above

3. In a fasting animal, the brain can obtain energy by the _____ of:

   a. glucose.
   b. glycogen.
   c. ketone bodies.
   d. all of the above
   e. none of the above

4. Circadian rhythms take place over a period of:

   a. a year.
   b. 6 months.
   c. 12 hours.
   d. about 24 hours.
   e. no set times

5. The lack of glucocorticoid hormones in humans produces:

   a. Addison's disease.
   b. goiter.
   c. diabetes mellitus.
   d. diabetes insipidus.
   e. dwarfism.

# 38

# THE SENSES

---

## CHAPTER AT A GLANCE

---

## CHAPTER PREVIEW

Senses form the interface between an organism and its external and internal environments. This chapter surveys the four basic sensory groups, beginning with a general discussion of how animals sense their environment via specialized receptors that are usually concentrated in body areas termed *sense organs*. These organs, in turn, are classified according to the type of stimulus to which they respond: chemicals, temperature, light, or mechanical differences.

The first sensing structures you will examine are the chemoreceptors, which are involved with smell and taste. Next come mechanoreceptors, where the discussion includes pressure detection in fish, the structure of the inner ear in land vertebrates, and unusual abilities such as magnetic and electrical field detection, and echolocation. The thermoreceptors described in the following section are simple and easily understood.

There is more complexity in the chapter's final subject, photoreception—the biological basis of vision. Here you will encounter a large amount of detail, from the functions of light-detecting molecules to the structurally complex eyes of mollusks and vertebrates. Biologists now know a great deal about how the brain processes visual input, and, once again, the text diagrams will help you visualize (no pun intended) how this amazingly intricate system perceives and interprets the external world.

# LEARNING OBJECTIVES

When you have mastered the concepts of this chapter, you will be able to:

1. Outline the sequence of physical and chemical events that must take place in order to convey a sensory message from a receptor to the central nervous system.
2. Name the two basic types of chemical perception, and describe the operation of each.
3. Explain what mechanoreceptors are, and provide an example from fish or amphibians.
4. Explain how the inner ear functions to maintain balance and detect movement.
5. Draw and correctly label a cross section of the mammalian ear, and discuss the functions of its structures.
6. Describe the structure and functioning of several "specialized" senses: echolocation, electroreception, and magnetic field perception.
7. Explain how various organisms detect different wavelengths of radiant heat.
8. Draw a cross-sectional view of a mammalian eye, label its structures, and explain how each structure works.
9. Outline how visual information is processed by the brain.
10. Compare and contrast the vertebrate eye with examples from among the invertebrates.

# CONCEPTS IN REVIEW

## Section I    Sensory Receptors

Each species possesses sensory equipment attuned to its own way of life. But in every animal, sensory receptor cells in the skin, eyes, and other parts of the body respond to environmental stimuli in much the same way. The rate of ion flow through the receptor cells changes in response to chemical, mechanical, or electromagnetic energy; and in this ionic alteration lies an electrical "message" that is relayed to the central nervous system. Sensory neurons either function as receptor cells themselves or synapse with separate receptor cells. Neurons in the central nervous system translate the electrical message they receive into a perception; thus, it is information processing in the brain that determines how a stimulus is actually perceived.

Receptor cells occur singly or in clusters called sense organs, and they have a baseline level of activity. Changes in that level signal the arrival of information from the environment. The brain distinguishes between background noise and information of survival value (text pp. 929–930). Classified according to the environmental factor they detect, receptors are divided into several categories: chemoreceptors, mechanoreceptors, thermoreceptors, and photoreceptors (see Table 38-1).

## Section II    Chemoreceptors

Chemoreceptors detect specific molecules and ions, and are prominent in taste and smell. Ten thousand taste buds dot the human tongue, and in them, taste receptor cells sense sweet, bitter, sour, and salty tastes. A combination of information from many receptors plus information about odor determines a food's flavor. Taste cells die and are regenerated (text pp. 930–932).

Unlike taste receptors, olfactory receptors in vertebrates are true neurons. They detect low concentrations of odorants and transmit information that enables animals to recognize territorial boundaries or carry out the reproductive cycle. Researchers believe that the shape of a molecule determines what olfactory receptor it binds to and thus what it smells like. Binding activates a second messenger system and generates nerve impulses that travel to the olfactory bulb at the front of the brain. Olfactory receptor cells respond to stimuli in a graded way that is proportional to the number of odor molecules bound (text p. 934).

## Section III    Mechanoreceptors

Mechanoreceptors monitor pressure, stretching, and bending. To do this, they depend on a distortion of the shape of the mechanoreceptor cell and its membrane. In vertebrate skin, Meissner's and Pacinian corpuscles detect light and heavy pressure, while simple pain receptor nerve endings respond to pressure, temperature, and chemicals (text pp. 934–935).

Some mechanoreceptors, such as those in the lateral line of fish, have cilia or microvilli that, when bent, deform the cell's membrane and enable the animal to detect perturbations in the environment. Through similar deformations in sensory hair cells, other animals monitor their position in space relative to gravity and in relation to their linear and angular movements. For this spatial perception, invertebrates rely on statoliths whereas vertebrates depend in part on otoliths and the semicircular canals of the inner ear (text pp. 936–938).

Sensory hair cells are also the key to vertebrate hearing. Pressure waves generated by sound that has entered the ear displace a particular spot on the basilar membrane and cause microvilli on overlying hair cells to bend against the tectorial membrane at an angle. The precise angle determines the rate at which the hair cells release neurotransmitter, and this release ultimately results in nerve impulses that travel through the auditory nerve to the brain (text pp. 938–940).

In addition to perceiving sound, animals must detect the direction from which it comes. To do this, they use two sets of cues: the relative loudness of a sound at one ear or the other, and the difference in time between the sound's registration in one ear and the other (text pp. 940–942). Small nocturnal animals and aquatic mammals have evolved the ability to use sound to sense their way in the dark. The sound-based system they use for navigation is called echolocation.

Variations of mechanoreceptors include the electroreceptor cells of some fish and the nerve fiber endings located near the electromagnetic crystals in pigeons (text pp. 943–944).

## Section IV    Thermoreceptors

Nearly all animals are able to respond to variations in temperature. Vertebrates detect such changes through nerve endings in the skin and tongue. Snakes hunt their food by detecting temperature changes. Infrared detectors within facial pits that contain thousands of receptor nerve endings make the snakes supersensitive to the body heat of prey (text p. 944).

## Section V    Photoreceptors

Light-sensitive structures in animals are amazingly diverse. Simple multicellular animals can sense the direction as well as the absence or presence of light. In more complex animals, a lens consisting of accessory cells focuses light on underlying light receptors. These higher animals can detect light of lower intensity and sense motion as well as direction.

In the vertebrate eye, the receptor cells that actually receive and transform light to electrical energy are in the retina, and there are two types: rods and cones. The highly sensitive rods respond to low levels of light and are active in night vision; the cones react to higher light intensities only. The density of cones in the fovea determines the sharpness of vision, and the differences between three classes of cones form the basis of color vision (text pp. 945–947).

Both rods and cones contain rhodopsin molecules consisting of opsin and the light receptor retinal. When light strikes retinal, it causes changes in molecular configuration, and these changes transduce the light energy to an electrical message for the nervous system (text pp. 948–949). According to the trichromatic theory, different classes of cones have variations of opson that make them most sensitive to different wavelengths of light (text p. 950).

The wiring of rods, cones, bipolar cells, ganglion cells, and other nerve cells determines what animals see. The visual field of a ganglion cell includes the full set of rods and cones associated with it. The larger the visual field of a ganglion cell, the lower its acuity (resolving powers).

The optic nerve carries impulses from the eye via the optic chiasma to the visual cortices of the brain. The cells in each visual cortex are arrayed in precisely ordered columns and layers, and they respond to bars of light oriented at certain angles, to squares or rectangles of light, to moving edges, and to light hitting the same spot on both retinas. This cellular arrangement and specificity enable the visual cortices to transform patterns of light into perceptions of faces or moving tennis balls (text pp. 951–952).

Cephalopods' eyes are similar to those of vertebrates, but arthropods have compound eyes composed of many ommatidia (text pp. 952–954).

## KEY TERMS

auditory nerve   *text page* 938

basilar membrane   936

bipolar cell   950

chemoreceptor   930

cochlea   938

compound eye   952

cone   947

cornea   945

echolocation   942

electroreceptor cell   943

endolymph   936

fovea   945

ganglion cell   950

gustation   930

iris   945

lateral line   935

lens   945

mechanoreceptor   930

neural retina   945

olfaction   930

ommatidium   952

opsin   948

optic nerve   945

organ of Corti   938

otolith   936

oval window   938

photoreceptor   930

pigmented retina   945

proprioceptor   935

pupil   945

retinal   948

rhabdomere   952

rhodopsin   948

rod   947

round window   938

semicircular canal   936

sense organ   930

statocyst   936

statolith   936

taste bud   930

tectorial membrane   938

thermoreceptor   930

transducin   948

trichromatic theory of color vision   949

tympanic membrane   938

visual cortex   952

visual field   951

# SELF-QUIZ: TESTING WHAT YOU HAVE LEARNED

## Matching Key Terms

Match each term on the left with the most appropriate description on the right.

| | |
|---|---|
| 1. cone | a. fluid-filled sac |
| 2. otolith | b. high-resolution viewing |
| 3. fovea | c. a daytime photoreceptor |
| 4. rhodopsin | d. a nighttime photoreceptor |
| 5. statocyst | e. focuses |
| 6. pupil | f. visual pigment |
| 7. semicircular canal | g. opening in the iris |
| 8. rod | h. coiled tubes |
| 9. cochlea | i. provide a sense of balance |
| 10. lens | j. ear stones |

## True or False?

1. _____ Mechanoreception may involve detection of shear forces.

2. _____ A human tongue can detect three basic tastes.

3. _____ Otoliths detect straight-line acceleration.

4. _____ The tympanic membrane stretches across the oval window.

5. _____ The purpose of a lens is to gather and focus light.

6. _____ Animals adapted to life in dim light have many more rods than cones.

7. _____ Compound eyes are composed of many separate optic units called ommatidia.

8. _____ The sensory hair cells of mammals are found within the organ of Corti.

## Completion

1. The spontaneous level of electrical activity for receptor cells is called a _____.

2. The _____ system of fish and amphibians detects pressure variations, sound, and movement.

3. The semicircular canals of a vertebrate contain _____, a fluid that exerts a force in response to movement.

4. The tiny bones of the mammalian inner ear are, in order from the eardrum to the oval window:

   _____, _____, and _____.

5. In most vertebrates the photoreceptors of the eye are the _____ and

   _____.

6. _____ lines the nasal cavities of most vertebrates, and contains the olfactory neurons.

## Short Answer

1. What are the structural and functional differences between otoliths and statoliths?

2. How do insect eyes differ in structure and function from vertebrate eyes? How do their images compare in terms of sharpness?

3. How do the electroreceptor cells of fish work?

4. How is an image formed in a vertebrate eye?

5. How is color vision accomplished according to the trichromatic theory?

6. How do bats use echolocation to locate objects by sound?

## Multiple-Choice Review

In the following sentences, fill in any blanks. Complete each statement by circling the correct response.

1. Human taste receptors can detect:

   a. sweet.
   b. salty.
   c. sour.
   d. bitter.
   e. all of the above

2. The incus is connected to:

   a. the round window.
   b. the tympanic membrane.
   c. the oval window.
   d. the cochlea.
   e. the endolymph.

3. Statoliths are composed of:

   a. calcium carbonate.
   b. endolymph.
   c. magnitite.
   d. transducin.
   e. opsin.

4. The _____ ear chamber is connected to the pharynx by means of the:

   a. auditory nerve.
   b. tectorial membrane.
   c. Eustachian tube.
   d. organ of Corti.
   e. tympanic membrane.

5. Which of the following is not found among mammals?

   a. tympanic membrane
   b. stapes
   c. oval window
   d. lateral line
   e. incus

## Exercise

Identify the six types of cells in this cross section of the retina.

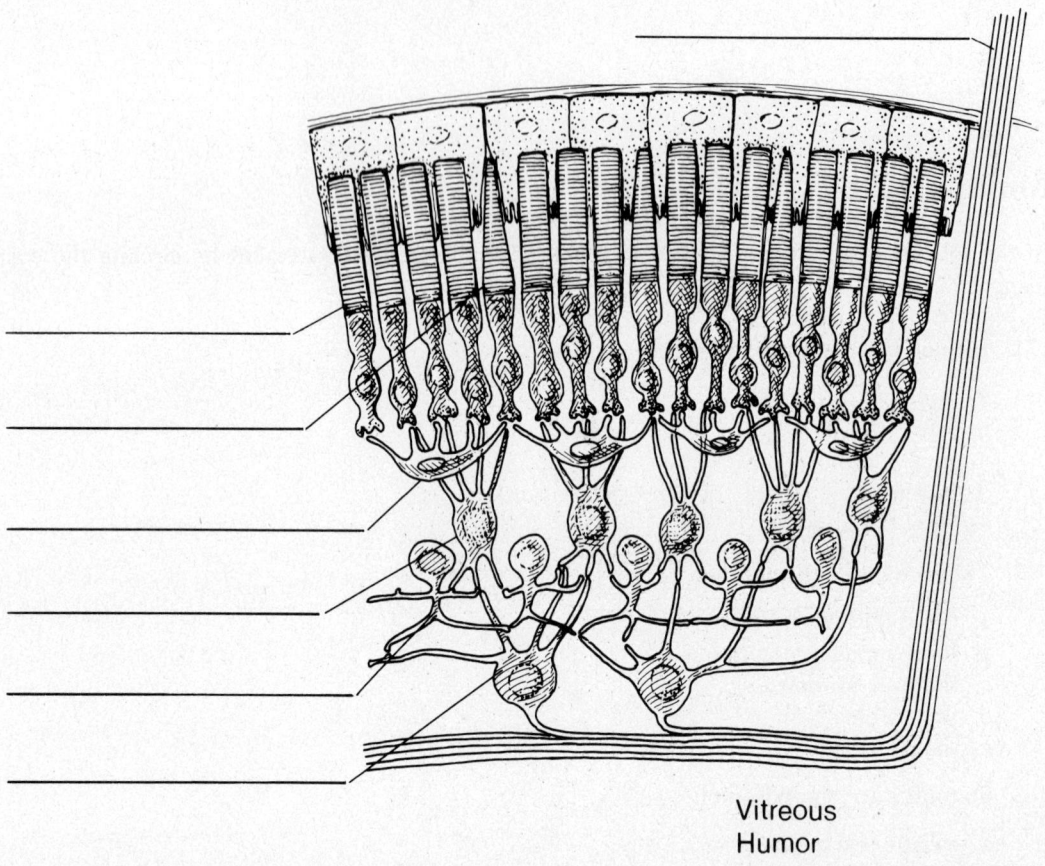

Vitreous
Humor

Now, give the function of each.

# 39
# SKELETONS AND MUSCLES

---

## CHAPTER AT A GLANCE

---

## CHAPTER PREVIEW

Skeletons and muscles, the subjects of this chapter, often work subtly. You probably are not aware that you are using dozens of muscles to read this page, hundreds to process food in your alimentary canal, a few to breathe, dozens more to hold your sitting posture, and one that is so important that, should it cease to work for just a minute or two, you would never again have to think about a biology test. Moreover, most of these muscles require a movable substrate of bones to work against.

The chapter starts by looking at the basic features of animal skeletons (both external and internal), and the inner gross structure and microstructure of bone and cartilage. Then it considers how muscles detect changes in the environment through detectors such as stretch receptors. The next section probes the classification of muscles; Table 39-1 sums up the important anatomy.

A core section of this chapter presents the rather complicated mechanism of muscle action, the sliding-filament theory. Be sure to have the microstructure of muscle fibers well in mind as you study. A description of muscle action, covered next, discusses muscle tone and the all-or-none functioning of muscles. After that, you will be prepared for a detailed look at the basic types of muscle tissue, including the abundant smooth muscles of gut and blood vessels, striated (skeletal) muscle, and the all-important cardiac muscle tissues. A brief final section considers the distribution of contractile systems among the invertebrates.

# LEARNING OBJECTIVES

When you have mastered the concepts of this chapter, you will be able to:

1. Describe an exoskeleton, including its composition, and list three of its functions.
2. Define the following: cartilage, bone, Haversian system, spongy bone, compact bone, joint, ligament, axial skeleton, appendicular skeleton, tendon.
3. Draw a cross section of a long compact bone (such as a femur or humerus) and label its components.
4. Distinguish between agonist and antagonist muscles.
5. List the three classes of muscles and explain where each is found.
6. Explain why striated muscles are often referred to as "red" or "white."
7. Explain the functioning of the muscle spindle.
8. Explain what is meant by *tonus* and tell why it is important.
9. Describe the anatomy and operation of a striated muscle.
10. Outline the sliding-filament theory of muscle contraction.
11. Explain the role of calcium in muscle contraction.
12. Explain the "all-or-none" response of muscles.
13. Describe the anatomy and operation of a smooth-muscle cell.
14. Describe the anatomy and operation of a cardiac-muscle cell.
15. Discuss the myogenic beat of vertebrate cardiac muscle.

# CONCEPTS IN REVIEW

## *Section I*    The Animal Skeleton: A Living Scaffold

Muscles and bones are effectors whose activities enable an animal to breathe, eat, walk, and run. The muscles accomplish their work by pulling, not pushing, on bones or other kinds of tissue.

There are two basic types of skeletons: hydroskeletons (consisting of the compressible fluid inside the body cavities of various invertebrates) and hard skeletons (composed of crystallized mineral salts, materials such as chitin, or cartilage and bone). Hard skeletons are located either at the body surface (exoskeletons) or internally (endoskeletons), and they are made of separate pieces that are hinged for movement.

In animals with an exoskeleton (clams, snails, lobsters, spiders, beetles), the muscles operate from inside the skeleton. One set of muscles—the flexors—bends each skeletal part, while another set—the extensors—straightens it (text pp. 958–959).

Endoskeletons, composed of cartilage or of a combination of cartilage and bone, are more complex than exoskeletons and contain living cells. Bone emerges from two developmental pathways that result in dermal membrane bones (such as those of the outer skull), and endochondial bones (such as those in the limbs and pelvic girdle). The axial skeleton consists of skull, vertebral column, and ribs; the appendicular skeleton includes the pectoral and pelvic girdles plus forelimb and hind limb bones. Bones meet at joints, where they are held together by ligaments, cushioned by cartilage, and protected from shock by synovial fluid. Compact bone is constructed of Haversian systems that have a single blood capillary, whereas spongy bone has many blood vessels in its bone marrow interior. Fat cells and blood cells arise in the marrow (text pp. 959–962).

Among the interacting muscles that move bones, the agonist, or prime mover, contracts as the antagonist relaxes. Complementary muscles called synergists augment or modify the movement resulting from the effect of the prime mover (text pp. 963–964). Muscles may connect directly to bone or grade into tendons. Tendons can concentrate contractile force or change the direction of muscle force.

## Section II    Regulating Movement: Feedback Control of Muscle Action

To coordinate the activity of hundreds of muscles, the body relies on proprioceptors, which report on the position of the body's main parts. In vertebrates, muscle spindles (or stretch receptors) incorporate sensory neurons that form the first part of reflex arcs and also connect to the brain. These arcs help an animal maintain its stance. The brain-connected neurons help hone fine muscular control.

## Section III    Muscle Structures and Functions

All vertebrate muscle cells are one of three types: smooth, cardiac, or skeletal (striated). Each skeletal muscle is a set of fibers composed of many fused cells. The red component of striated muscle is slow muscle that contains a large amount of myoglobin, many blood capillaries, and many mitochondria. Skeletal muscle can contract repeatedly and resist muscle fatigue. White, or fast, muscle has little or no myoglobin, few capillaries, and few mitochondria. As a result, it becomes fatigued quickly. Many skeletal muscles are a combination of the two colors (text pp. 966–967).

The giant fused cells of skeletal muscle contain many nuclei, and their cytoplasm is packed with myofibrils built of sarcomeres. The sarcomeres are repeating units of actin and myosin proteins and are the actual sites of contraction (text pp. 967–968).

## Sections IV and V    The Molecular Basis of Muscle Contraction and Refinements of Muscle Action: Graded Responses and Muscle Tone

In each sarcomere, a myosin thick filament is surrounded by actin thin filaments (text pp. 968–969). The sliding-filament theory states that myosin heads apply power strokes that cause actin filaments to slide and shorten the sarcomere. The same process occurring simultaneously in thousands of sarcomeres produces muscle contraction. ATP fuels the process, and a flow of calcium ions triggers it via tubules and a network of sarcoplasmic reticulum, which serves as a calcium reservoir. Events inside muscle cells that lead to contraction are set in motion by the release of acetylcholine at the neuromuscular junction (text pp. 970–974).

Most individual striated muscle cells in vertebrates contract in an all-or-none fashion when an incoming nerve impulse exceeds a certain threshold. But whole vertebrate muscles exhibit graded responses to a stimulus (text pp. 974–975). These graded responses are based in part on summation, which can culminate in the state of sustained maximum contraction known as tetanus. In contrast, tonus, the sustained partial contraction of muscles, enables terrestrial animals to maintain normal posture in the face of gravity. Some muscles do not need the stimulus of action potentials to contract; they display intrinsic rhythmicity.

## *Section VI*   Smooth Muscle

The most abundant type of vertebrate muscle after striated is the smooth muscle found in the gut wall, the walls of blood vessels, the iris of the eye, reproductive organs, and glandular ducts. Smooth muscle carries out slow, sustained contractions that are not under voluntary control.

The cells of smooth muscle have a single nucleus and no striations, and they are linked by surface "pegs and sockets," collagen fibers, and gap junctions. Actin and a special kind of myosin are the basis of smooth-muscle contractions (text pp. 975–976). Sympathetic nerves stimulate those contractions and parasympathetic nerves inhibit them.

## *Section VII*   Cardiac Muscle: Striated Tissue with Smooth-Muscle Characteristics

Cardiac, or heart muscle is striated (like skeletal muscle) and consists of cells with a single nucleus (like smooth muscle). These cells, which have an abundant supply of mitochondria, branch and interlock while their ends are bound firmly to each other by intercalated disks. Composed of folded, reinforced cell membrane, the disks transmit electrical impulses from one cell to the next (text pp. 976–977).

Cardiac-muscle cells beat independently of any nerve stimulation and thus generate the myogenic heartbeat of vertebrates. But impulses from sympathetic or parasympathetic nerves, acting through pacemaker cells in the heart wall, can speed or slow the intrinsic contraction of cardiac-muscle cells.

## *Section VIII*   Muscles in Evolution: The Contractile Systems of Invertebrates

Nearly all contractile systems in eukaryotic cells contain actin and myosin, and rely on a changeable flow of calcium ions to induce the myosin power stroke. Evolution, however, has produced differences in the molecular apparatus that underlies contractions.

Smooth-muscle cells were probably the first kind to emerge. In invertebrates, the smooth muscles teem with myosin thick filaments. As a result, they can initiate powerful contractions and sustain these contractions for a long time. Among other variations produced by evolution is the stretch activation of flight muscles in some insects, a kind of cycling between elevators and depressors that allows very high wing-beat frequencies.

By generating the force to move blood, bone, the contents of the gastric cavity or alimentary canal, and whole animals, the molecular mechanisms of contraction have, through evolution, helped shape animal life styles.

---

## KEY TERMS

appendicular skeleton   *text page 959*

axial skeleton   959

bone   959

cardiac muscle   966

cartilage   959

creatine phosphate   967

Haversian system   960

intercalated disk   976

joint   960

ligament   960

muscle fiber   966

muscle spindle   964

myofibril   967

sarcomere   967

sarcoplasmic reticulum   972

skeletal muscle   966

sliding-filament theory   970

smooth muscle   966

T tubule   972

tendon   964

tetanus   975

tonus   975

tropomyosin   971

troponin   971

# SELF-QUIZ: TESTING WHAT YOU HAVE LEARNED

## Matching Key Terms

Match each term on the left with the most appropriate description on the right.

| | | | |
|---|---|---|---|
| 1. joint | a. | contains large amounts of apatite |
| 2. tendon | b. | the actual site of muscle contraction |
| 3. troponin | c. | built from sarcomeres |
| 4. tetanus | d. | resides on the actin thin filament |
| 5. cartilage | e. | joins bones together |
| 6. tonus | f. | areas where two or more bones meet |
| 7. bone | g. | muscle tone |
| 8. myofibril | h. | maximum contraction |
| 9. sarcomere | i. | joins muscle to bone |
| 10. ligaments | j. | composed of collagen and polysaccharides |

## True or False?

1. _____ Bones contain living cells.

2. _____ The pelvis is part of the appendicular skeleton.

3. _____ Agonist muscles are also called prime muscles.

4. _____ Tendons connect bones to bones.

5. _____ Bodybuilders get big muscles by exercising to increase the number of cells in each muscle.

6. _____ "White" muscles have more endurance than "red" muscles.

7. _____ Muscle cell number remains constant; exercise simply increases the size of the cells.

8. _____ Muscle soreness is caused by build-up of waste products such as lactic acid.

## Completion

1. The thin filaments of a sarcomere are composed mostly of the protein _____.

2. A tissue that provides support when filled with a fluid is a _____.

3. The individual cells of cardiac muscle are attached to each other by means of _____.

4. _____ muscle may be either "red" or "white."

5. The major components of cartilage are _____ and complex _____.

6. Haversian systems are found within _____ bones.

## Short Answer

1. What are the major differences between tendons and ligaments?

2. What is the function of the exoskeleton in invertebrate animals?

3. What are the major elements of the sliding-filament theory?

4. What are the anatomical and chemical differences among the three basic types of muscle?

5. What controls the myogenic beat of cardiac muscle?

6. What do the terms *tonus* and *tetanus* mean?

# Multiple-Choice Review

In the following sentences, fill in any blanks. Complete each statement by circling the correct response.

1. In compact bone, a _____ system surrounds a(n):

   a. vein.
   b. artery.
   c. capillary.
   d. joint.
   e. none of the above

2. Ligaments serve to:

   a. bind muscle fibers together.
   b. bind muscles to bones.
   c. bind collagen fibers together.
   d. join synergists.
   e. bind bones to bones at joints.

3. A muscle that is fast to contract, has few capillaries, little myoglobin and fuel, and only a few

   _____ is probably:

   a. a red muscle.
   b. a white muscle.
   c. in the heart of a vertebrate.
   d. in the wing of an insect.
   e. none of the above

4. Zones of contact between bones of the vertebrate skull are called:

   a. sutures.
   b. joints.
   c. gaps.
   d. spindles.
   e. none of the above

5. "Red" muscles are marked by:

   a. slow contractions.
   b. lots of mitochondria.
   c. lots of glycogen.
   d. many capillaries.
   e. all of the above

## Exercise

In the diagram below, show where ATP first binds to the myosin head. Then use labels to show what happens with ATP and myosin at each stage of the power stroke.

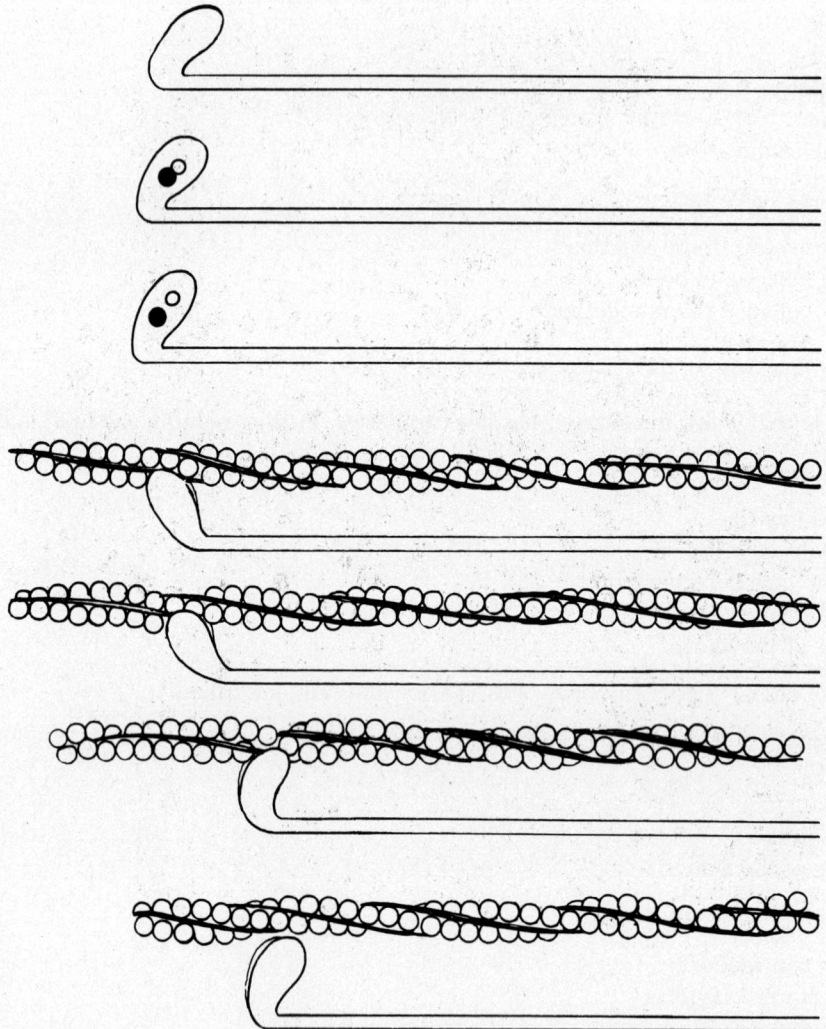

Now, describe each event.

# 40
# THE VERTEBRATE BRAIN

---

## CHAPTER AT A GLANCE

---

## CHAPTER PREVIEW

Human brains are often called the most complex structures in the universe. The ultimate truth of this claim may be hard to discern, but there is no doubt that our brains are the most complex of vertebrate organs. Chapter 40 underscores this fact as it begins by describing the nerve nets, ganglia, and simple nerve cords of the invertebrates, and then proceeds to trace the development and origin of the various parts of the brain.

The next section outlines the methods through which researchers study living brains. These include studying the effects of the accidental loss of brain tissue, stimulating particular regions with electrical probes, and studying neuron anatomy. The section also looks at the phenomenon of sleep and its various stages.

The chapter's final sections explore the three categories of memory, and take a look at the brain's chemicals—hormones and neurotransmitters. In all, Chapter 40 is a fascinating journey into a biological world that science is only beginning to comprehend.

# LEARNING OBJECTIVES

When you have mastered the concepts of this chapter, you will be able to:

1. Make a simple sketch of a vertebrate brain, label the three main regions, and list the structures found in each.
2. Explain the function of cerebrospinal fluid.
3. Describe which areas of the brain are the most primitive or "oldest" (in an evolutionary sense), and which are the most advanced, or "newest."
4. Explain the functioning of Purkinje cells.
5. Name the three components of the limbic system, and explain the functioning of each.
6. Describe the structure of the reticular formation, and list the functions it controls.
7. Describe the structure of the cortex, and list the functions it controls.
8. State and discuss the two basic methods for studying living brains.
9. Describe the findings of "split-brain" studies.
10. Name the areas of the brain that control speech.
11. List the stages of sleep, and explain how sleep may be related to the creation of long-term memory.
12. Discuss the hierarchical neuron relationships found in the visual cortex of mammals.
13. Name the three categories of memory, and tell which part of the brain is used for each.
14. Explain how central nervous system neurons communicate with each other chemically.

# CONCEPTS IN REVIEW

## *Section I*   Structure and Function of Vertebrate Brains

The human brain contains billions of neurons and an even larger number of neuronal interconnections organized into distinctive minisystems that carry out a variety of functions including learning and remembering. These functions arise from changes in the number and activities of synapses, and these changes in turn result from modifications in the electrical and chemical properties of brain cells.

While biologists know little about how the vertebrate brain evolved, they have observed its embryonic development. Three large swellings eventually give rise to a forebrain, midbrain, and hindbrain, each subdivided into regions (text pp. 983–984). Four ventricles (hollow cavities) contain cerebrospinal fluid, which also fills spaces between the protective meninges and between cells of the brain. The blood–brain barrier, formed by the continuous tight junctions and channel-less cytoplasm of cells in the walls of brain capillaries, ensures a constant, nourishing fluid environment for brain cells.

The most posterior part of the brain is the medulla oblongata, the site at which the spinal cord and the brain merge. Moving from back to front, you next encounter the pons and cerebellum, parts of the hindbrain that, together with the midbrain, make up the brain stem. The three parts of the hindbrain control respiration, circulation, swallowing, vomiting, balance, stance, and some locomotor movements. Purkinje cells in the cerebellum help carry out complicated tasks through an estimated 200,000 synapses per cell (text p. 986).

The midbrain of terrestrial vertebrates relays auditory and visual information to the forebrain and controls some auditory and visual reflexes. Activities of various parts of the forebrain regulate instinctive behavior and learning. They also coordinate physiology, sensory information, emotions, and processes that lead to short-term memory (text pp. 987–988). A diffuse network of neurons from various parts of the brain makes up the reticular formation. Together with tracts that lead to the thalamus, it forms the reticular activating system—the gateway to consciousness.

In mammals, the highly folded neocortex (also called simply the cortex) is composed of six layers of neurons organized in columns. Separate regions of the cortex receive input from different senses or act in motor control, and within each region, groups of neurons tie in to different parts of the body (text pp. 989–991). Vast "unassigned" areas of the cortex are the associative areas where memory, learning, language, and personality arise.

## Section II    Exploring the Living Brain

By analyzing the behavioral deficiencies that resulted from damage to various parts of the brain, physicians and researchers helped sort out the functions of different brain areas. Split-brain studies showed that the right brain controls the left side of the body and vice versa. They also established that the left brain is in charge of analytical and language abilities, whereas the right brain specializes in spatial relations, and musical and facial recognition. Areas that control speech are located in either the left or the right brain, and sometimes in both (text pp. 993–994).

Finely controlled electrical stimulation of the brains of humans and other animals has revealed that within the brain are prepackaged neural bases for all sorts of behaviors. Electrical stimulation studies on rats helped map the limbic system. From brain-wave recordings, known as electroencephalograms (EEGs), experimenters have learned of the alpha and beta waves that are related to the absence or presence of visual stimulation, and of the altered brain state we call sleep. In one stage of sleep, rapid eye movement signals a high rate of brain activity concurrent with paralysis of the body's muscles (text pp. 995–997).

The parallel processing of information and the hierarchical organization of neuronal networks are features of the mammalian brain that simplify studies at the cellular level. An interactive combinatorial alternative to the visual cortex column system may be the key to smell as well as to memory and learning (text pp. 997–1000).

## Section III    Memory: Multiple Circuits at Work?

Learning is a hallmark of mammalian behavior, and memory is the basis of learning. There are three categories of memory: immediate, short-term, and long-term. Immediate memory lasts for seconds; short-term for minutes; and long-term for months or years. Many of the long-term memories stored in our brain are inaccessible, while others can be brought to consciousness voluntarily.

Evidence suggests the hippocampus is one site of memory and learning. Stimulation of the hippocampus may cause alterations in the number and properties of synapses that could represent the means of storing information. The cerebellum is another site of memory formation. But most memories seem to be distributed, not localized. Consolidation—the transfer to long-term memory—of an event improves if the event is accompanied by significant physiological consequences (text pp. 1000–1001). In prosopagnastics, head injury or strokes in the occipital and temporal lobes have destroyed the ability to recognize faces. Experiments indicate the memories are there but cannot be retrieved.

## Section IV    The Chemical Messengers of the Brain

There are two classes of chemical messengers in the brain: neuroactive peptides and neurotransmitters. Neuroactive peptides are manufactured in the body of brain cells, from which they travel down the axon to a synaptic terminal. Subsequently, a nerve impulse traveling down the axon causes the secretion of the stored peptide. Many neurons synthesize and secrete both a neuropeptide and a neurotransmitter. Neurotransmitters are manufactured at the synaptic terminal and, after secretion into a synapse, are quickly degraded. Neuroactive peptides resist degradation and act on target cells over a long time. Endorphins and enkephalins are neuroactive peptides that function as pain killers. Some neuroactive peptides are also found outside the brain in various parts of the body, where they have different functions.

Like neurotransmitters, different neuroactive peptides are found in different areas of the brain. Unlike neurotransmitters, the peptides may inhibit the activity of target cells (text pp. 1002–1005). Although researchers are only beginning to understand how neuroactive peptides work, they think one of the substance's main functions is to modulate and refine the nerve activities that govern much of life. Thus, the distinction between the activities of the nervous and the endocrine systems becomes increasingly blurred.

---

## KEY TERMS

blood–brain barrier  *text page 985*

cerebellum  *985*

cerebrospinal fluid  *985*

cerebrum  *987*

column  *989*

corpus callosum  *992*

endorphin  *1003*

enkephalin  *1003*

forebrain  *984*

hindbrain  *984*

limbic system  *988*

medulla oblongata  *985*

midbrain  *984*

neocortex  *989*

neuroactive peptide  *1002*

orientation slab  *997*

pons  *985*

Purkinje cell  *986*

rapid eye movement (REM) sleep  *995*

reticular activating system (RAS)  *988*

reticular formation  *988*

substance P  *1004*

---

# SELF-QUIZ: TESTING WHAT YOU HAVE LEARNED

## Matching Key Terms

Match each term on the left with the most appropriate description on the right.

| | |
|---|---|
| 1. cerebrum | a. a neuroactive peptide |
| 2. limbic system | b. links the brain's hemispheres |
| 3. substance P | c. receives information from sensory systems |
| 4. endorphin | d. mesencephalon |
| 5. corpus callosum | e. regulates complex behavior |
| 6. reticular formation | f. much enlarged in humans |
| 7. Purkinje cell | g. the chemical messenger of pain |
| 8. midbrain | h. involved with emotions |
| 9. pons | i. giant, many-branched cell |
| 10. neocortex | j. part of the brain stem |

## True or False?

1. _____ The number of glial cells in a brain is less than the number of neurons.

2. _____ The telencephalon is found in the forebrain.

3. _____ The autonomic nervous system is regulated by the reticular formation.

4. _____ The neocortex of a cat is larger than the neocortex of a lizard.

5. _____ Gastrointestinal movement increases during sleep.

6. _____ The medulla oblongata evolved before the cortex.

7. _____ Purkinje cells help regulate balance.

8. _____ The medulla oblongata resides in the telencephalon.

## Completion

1. The hindbrain of vertebrates contains the two major subdivisions, the _____ and the _____.

2. The vertebrate midbrain also is called simply the _____.

3. The special connective tissues that line the inside of the skull and wrap the spinal cord are the _____.

4. The _____ is formed by the midbrain, and the medulla and pons.

5. The part of the brain that shows the most phylogenetic modification is the _____.

6. The inability to recognize the faces of familiar people is called _____.

## Short Answer

1. What are the classes and functions of neurotransmitters?

2. What constitutes the limbic system, and what are its various functions?

3. What two areas of the brain are involved in speech, and what are their precise functions?

4. What are the four basic ways through which biologists study living brain tissue?

5. What are the stages of sleep, and what happens in each stage?

6. What are the basic ways in which central nervous system neurons can communicate with each other?

## Multiple-Choice Review

In the following sentences, fill in any blanks. Complete each statement by circling the correct response.

1. The site of long-term _____ is most likely the:

   a. cerebellum.
   b. cortex.
   c. pons.
   d. brain stem.
   e. medulla oblongata.

2. The site of human consciousness is most likely the:

   a. pons.
   b. cerebellum.
   c. hypothalamus.
   d. brain stem.
   e. cerebrum.

3. A person who is right-handed most likely has his or her _____ center located in the:

   a. median fissure.
   b. right hemisphere.
   c. left hemisphere.
   d. pons.
   e. brain stem.

4. If the hippocampus is destroyed, the most serious effect will probably be:

   a. learning impairment.
   b. loss of sex drive.
   c. frequent urination.
   d. increased hunger.
   e. poor speech.

5. Substance P is a(n):

    a. enzyme.
    b. type of cerebrospinal fluid.
    c. depressant.
    d. peptide.
    e. general neurotransmitter.

# Exercise

In the chart below, supply the functions of each main component of the vertebrate brain.

**Parts of the Vertebrate Brain**

| Component | Functions |
| --- | --- |
| Medulla oblongata | |
| Pons | |
| Cerebellum | |
| Midbrain | |
| Thalamus | |
| Hypothalamus | |
| Cerebrum | |

# 41
# EVOLUTION AND THE GENETICS OF POPULATIONS

---

## CHAPTER AT A GLANCE

The Origins of Evolutionary Thought
  *Lamarck and the inheritance of acquired characteristics*
  *Darwin and natural selection*
Variations in Genes: The Raw Material of Natural Selection
  *Looking for genetic variation: Protein electrophoresis*
Population Genetics: The Links Between Genetics and Evolution
  *The gene pool and gene frequencies*
  *Genetic equilibrium and the Hardy-Weinberg law*
Mechanisms of Evolution: Upsetting the Gene-pool Equilibrium
  *Nonrandom mating*
  *Genetic drift*
  *Mutation*
  *Migration and gene flow*

---

## CHAPTER PREVIEW

Chapter 41 is the first in a series of chapters that take a close look at current evolutionary thinking. Here, you start at the most basic level: the genes upon which evolutionary processes ultimately act.

The chapter begins with brief discussions of evolutionary theory, from Lamarck's concepts of acquired characteristics to Darwin's theory of evolution by natural selection. Darwin's ideas, especially natural selection, are presented in the context of historical and current notions of adaptation. The chapter then goes on to examine the case for the existence of extensive variation in populations of sexually reproducing species. It is this variation, in the form of multiple alleles of genes, that provides the raw material upon which natural selection acts.

Next the chapter takes up the Hardy-Weinberg law. The five requirements of this biological "null hypothesis" are straightforward and easy to understand. (But to *really* understand the mathematical basis of the law, you will find that there is no substitute for working through the Hardy-Weinberg problems at the end of the text chapter.) Once the law is explained, the final section of the chapter shows how each of its five requirements is violated in nature. Such "violations" produce the changes in allele frequencies that constitute evolution.

# LEARNING OBJECTIVES

When you have mastered the concepts of this chapter, you will be able to:

1. Describe the theory of acquired characteristics, tell who was its most famous champion, and explain why it is rejected today.
2. Outline the life of Charles Darwin, and briefly explain his theories of evolution by natural selection.
3. Define adaptation, explain how organisms become adapted, and describe the evolutionary benefits adaptation yields.
4. Explain why high levels of genetic variation are thought beneficial to a species.
5. Briefly outline the method of protein electrophoresis, and explain why it is assumed to be a good statistical estimate of variation.
6. Define the terms *gene pool* and *population*, and give a workable definition of evolution in terms of population genetics.
7. Write the Hardy-Weinberg equation and show how it can be used to calculate allele frequencies in a population.
8. Define the five assumptions of the Hardy-Weinberg law (no mutations, no migration, no natural selection, random mating, and large population size), and explain what will happen if each assumption is violated.
9. List and define the two categories of nonrandom mating.
10. Name and define the two extreme cases of genetic drift.
11. Explain why mutation is the *only* source of wholly new genetic material.

# CONCEPTS IN REVIEW

## *Section I*   The Origins of Evolutionary Thought

Crucial to the concept of evolution is the idea that living things change over time. In the late eighteenth century, evidence consistent with this view—especially fossil finds in many locations—began to accumulate. Drawing on this and other evidence, the French zoologist Jean Baptiste de Lamarck presented, in 1809, one of the important early theories of evolution: the theory of the inheritance of acquired characteristics. Although Lamarck's conception of evolution was erroneous—inherited traits are not affected by an animal's use or disuse of its body parts—he was correct in proposing that evolution proceeds according to natural laws.

Charles Darwin's (and Alfred Russell Wallace's) overarching contribution to evolutionary thinking was the concept of natural selection—that organisms best adapted to their environment will have an advantage in the "battle of life." An adaptation is any genetically based feature that results in an individual or species being better suited to some aspect of its environment. (Keep in mind, though, that Darwin knew nothing of genes when he formulated his great work.) And adaptations that improve an organism's ability to function in its environment improve that organism's chances to survive to reproduce. In addition to the relationship between adaptations and natural selection, Darwin also recognized that geographical isolation is a key factor in the development of differences in the adaptations possessed by different populations of a species. Over time, such differences may become so great that the populations diverge into separate (new) species. Speciation is covered in detail in Chapter 42.

## Section II    Variations in Genes: The Raw Material of Natural Selection

As Gregor Mendel discovered, variations in phenotype often reflect variations in the underlying genotypes of organisms. Such genetic variation—implicit in Darwin's theory—can be measured experimentally through the technique of protein electrophoresis, which enables the researcher to distinguish among proteins with characteristic combinations of amino acids. Because a protein's structure is determined by the nucleotide sequence of its coding gene, the variations revealed by electrophoresis serve as a window on the genetic makeup of populations.

One way of expressing genetic variation in a population is to calculate the average heterozygosity—the average frequency of individuals that are heterozygous at each gene locus surveyed (text p. 1017). Another way of expressing the same variation is to figure percent polymorphism (text p. 1018).

## Section III    Population Genetics: The Links Between Genetics and Evolution

The study of population genetics involves using mathematical descriptions of genetic phenomena to trace evolutionary trends within populations. All the various alleles of all the genes carried by individuals in a population constitute that population's gene pool, and one of the most important evolutionary properties of gene pools is their ability to change as a function of space and of time. In fact, population geneticists define evolution as a change in the relative frequencies of different alleles (genes) in a population. The Hardy-Weinberg law approaches change in gene frequencies from the other side—it sets out a series of conditions, stated mathematically, that together serve as a test of whether gene frequencies in a population of sexually reproducing organisms remain stable (text p. 1020). If the conditions for Hardy-Weinberg equilibrium are met, evolution does not occur. As the text points out, though, in real populations those conditions are almost never met. Evolution is a fact of life (text p. 1022).

## Section IV    Mechanisms of Evolution: Upsetting the Gene-pool Equilibrium

Numerous factors upset Hardy-Weinberg equilibrium in real populations. Gene mutations and migration bring new alleles into populations, and natural selection, acting on variant phenotypes, shapes gene pools as well. In addition, there are other phenomena that serve to alter gene frequencies in populations. These include nonrandom mating (inbreeding and assortative mating) and genetic drift. The founder effect, one type of genetic drift, can occur when a few members of a population leave and take up residence away from the original group. It can produce some of the most dramatic changes in gene frequencies. A classic example is the high incidence of a genetic disease, Ellis-van Creveld disease, among the Amish.

Mutations are heritable changes in the chemical structures of genes (text p. 1026). Although they are rare events, they create the new alleles that are a major source of genetic variation in populations. Types of mutations include chromosomal rearrangements of genes and DNA, duplications of genes, point mutations (changes in DNA base sequences), and others.

A final source of new genetic variation in a population is migration. When individuals (or other gametes) migrate from one population to another and interbreed with the existing population, gene flow takes place (text p. 1027). As you might expect, gene flow tends to reduce the genetic differences that exist between populations of a species.

## KEY TERMS

adaptation *text page 1013*

assortative mating *1024*

average heterozygosity *1017*

bottleneck effect *1026*

electrophoresis *1016*

evolution *1011*

founder effect *1025*

gene flow *1027*

gene pool *1019*

genetic drift *1024*

Hardy-Weinberg law *1020*

inbreeding *1023*

inheritance of acquired
characteristics *1012*

migration *1027*

natural selection *1015*

nonrandom mating *1023*

percent polymorphism *1018*

population *1019*

population genetics *1019*

# SELF-QUIZ: TESTING WHAT YOU HAVE LEARNED

## Matching Key Terms

Match each term on the left with the most appropriate description on the right.

1. gene flow
2. natural selection
3. adaptations
4. founder effect
5. bottleneck effect
6. electrophoresis
7. Lamarck
8. genetic drift
9. evolution
10. gene pool

a. descent with change
b. pioneers
c. population crash
d. inheritance of acquired characteristics
e. Darwin's mechanism
f. changing populations
g. chance
h. total alleles in a population
i. protein travel rates
j. specializations

## True or False?

1. _____ A gene pool is simply a population's genotype.

2. _____ Isozymes are different forms of proteins.

3. _____ The Hardy-Weinberg law holds true in small populations.

4. _____ Assortative mating is a random process.

5. _____ The founder effect requires large population size.

6. _____ A bottlenecked population shows less diversity.

7. _____ Average heterozygosity and percent polymorphism are terms associated with electro-phoresis.

8. _____ The founder and bottleneck effects are terms associated with drift.

## Completion

1. When gene frequency changes simply as a result of random sampling errors, the result is

    _____.

2. When organisms travel into or out of a population by migration, the result is

    _____.

3. A population _____ occurs when the population is reduced to only a few survivors.

4. Electrophoresis works by detecting the different _____, or alternate forms of proteins.

5. The mechanism behind Darwin's theory of organic evolution is _____.

6. Electrophoresis measures the _____ of populations.

## Short Answer

1. How do population geneticists measure the variation in a population?

2. What is the driving force of evolution according to the rejected theory of the inheritance of acquired characteristics?

3. What is the mathematical basis for the Hardy-Weinberg law?

4. What are the assumptions of the Hardy-Weinberg law?

5. How does genetic drift change the genetic composition of populations?

6. What is gene flow, and how does it change the gene pool of populations?

## Multiple-Choice Review

In the following sentences, fill in any blanks. Complete each statement by circling the correct response.

1. The theory of _____ by inheritance of acquired characteristics was advanced by:

   a. Darwin.
   b. Hutton.
   c. Smith.
   d. Lyell.
   e. Lamarck.

2. Self-fertilization is a form of:

   a. assortative mating.
   b. inbreeding.
   c. genetic drift.
   d. the founder effect.
   e. the bottleneck effect.

3. According to _____ genetics, the actual units of evolution are:

   a. genes.
   b. individuals.
   c. populations.
   d. species.
   e. higher taxa.

4. Which of the following is not a violation of the Hardy-Weinberg law?

   a. having a small population
   b. lots of drift
   c. the absence of selection
   d. lots of mutation
   e. high migration rates

5. Adaptations:

   a. suit an organism to its environment.
   b. are genetically based.
   c. increase fitness.
   d. enhance survival.
   e. all of the above

# 42
# NATURAL SELECTION

---

## CHAPTER AT A GLANCE

---

## CHAPTER PREVIEW

Biological theories often consist of two parts, and Darwin's theory of evolution was no exception. First, he presented a huge body of factual evidence that evolutionary change had taken place in particular organisms; and second, he postulated a *mechanism* by which he believed that change had occurred. Natural selection was Darwin's mechanism, and Chapter 42 takes what we've briefly considered in previous chapters, and builds upon it to describe the various types of natural selection.

The concept of *fitness* is introduced early on. As you will discover, fitness is measured in terms of *reproductive success*. In this context, a frail person with five children is far more fit—from the point of view of evolution—than a bulging body builder who has no children at all. This idea can sometimes be difficult to grasp, but it is thoroughly and clearly stated in the first section. The next section introduces the idea of selection in a discussion of *artificial selection* of fruit flies (in the laboratory) for a particular trait. Then comes the heart of the chapter—the various patterns or modes of natural selection.

The four basic selection modes (normalizing, directional, diversifying, and balancing) are fairly simple and you should have little trouble understanding what is going on in each. (Those who have a broader background in statistical math may want to sketch frequency distributions for each mode and see what happens to the mean and variance in each.) The chapter's final section presents some of the more recent ideas on evolutionary mechanisms, such as concepts of conservative adaptations, the neutralist–selectionist debate, molecular clocks, and units of selection that are more extensive than single alleles.

# LEARNING OBJECTIVES

When you have mastered the concepts of this chapter, you will be able to:

1. Describe the outcome of natural selection, and explain in general terms how it comes about.
2. Define the term *fitness* as it is used in evolutionary biology.
3. Define the term *fixed allele*, and describe the effect of fixation on genetic variation.
4. Define normalizing selection, and explain its effect on populations.
5. Define directional selection, explain its effect on populations, and give an example.
6. Define diversifying selection, and explain its effect on populations.
7. Define balancing selection, and explain its effect on populations.
8. Explain hybrid vigor and give an example.
9. Explain what is meant by genetic load.
10. Describe the issues in the neutralist–selectionist debate, and present the evidence that supports each view.
11. List the various levels at which natural selection is believed to operate.

# CONCEPTS IN REVIEW

## *Section I*   The Darwinian and Genetic Meanings of Fitness

Chapter 42 focuses on the ways different processes of natural selection affect the various phenotypes in populations and, over time, either alter underlying gene frequencies or maintain the status quo.

A modern statement of the mechanism of natural selection—that is, one that takes genes into account—emphasizes reproductive success. The term *fitness* includes the relative ability of an organism to survive, mate successfully, and reproduce a new organism (text p. 1033). Today, then, Darwin's "survival of the fittest" refers to the fact that individuals in a population carrying variant alleles of a gene or genes will have differential reproductive success.

If an allele has a high degree of fitness—for example, the allele for white fur in an arctic hare—selection pressures may act against variants of that allele and cause them to be lost from the population. Or, one of a pair of alleles may be lost through chance. In either case the remaining allele is then said to be fixed, and automatically has a high degree of fitness. It will remain so until new variants arise. At the opposite end of the fitness spectrum are lethal alleles or alleles for sterility. Their fitness is essentially zero, because affected individuals have no chance to reproduce. In thinking about this topic, remember that every individual inherits in its genotype alleles that are more fit and less fit, and that the total "package," rather than a particular allele, may be what determines the degree of reproductive success.

## *Section II*   Natural Selection: How Phenotype Affects Genotype

Since natural selection tends to eliminate less fit alleles from a population, selective processes tend to lead to the increased adaptation of a population to its environment (text pp. 1034–1036). Selection influences the genetic composition of a population indirectly, by operating on phenotypes. It is the phenotype that survives or perishes, reproduces or leaves no offspring.

## Section III    Types of Selective Processes

There are four major selective processes. As you study this section, keep in mind that selection does not mold an ideal, "best" organism for a particular environment. It operates only on the available options—the existing phenotypes (and genotypes) in a population.

Normalizing selection preserves the genotypic status quo (text p. 1037). It involves selection against atypical phenotypes, and is especially common in environments that have remained stable through long periods. Directional selection involves a trend toward a shift in a phenotype, such as a shift toward disease resistance in bacteria. A classic example is the change in body color in the peppered moths living in industrial areas of Great Britain (text p. 1038).

In diversifying selection the outcome is diversity—two or more phenotypes coexist in the same population. Each phenotype is adapted to some specialized features of a particular portion of the total environment (text p. 1041). An extreme form, known as disruptive selection, may gradually result in the disappearance of intermediate phenotypes.

Balancing selection maintains genetic variability in a population. One form is heterozygote advantage (hybrid vigor), in which organisms with many heterozygous alleles are particularly vigorous and successful in reproducing. Such heterozygosity also has a cost, in that certain alleles are maintained that are deleterious or lethal in the homozygous condition. This cost is termed genetic load. A form of balancing selection called frequency-dependent selection occurs when the relative fitnesses of the genotypes in a population vary according to their frequency (text p. 1043). An example is rare-mate advantage in fruit flies.

## Section IV    New Views on Evolutionary Mechanisms

A current area of research in biology involves the neutralist–selectionist debate (text p. 1044). According to selectionists, all variations in protein molecules have evolved as the result of natural selection; according to neutralists, many variations may be the result of random mutations that are neutral with respect to survival and reproductive success.

Recent research suggests that the gene may not be the exclusive unit of natural selection (text p. 1047). Instead, selection may operate on groups of genes: for example, in genes coding for structural proteins that act together, such as actin and myosin. In the case of higher-order phenomena such as animal behavior, selection may act at the level of the organism, or even on groups of organisms—so-called group selection.

## KEY TERMS

Allen's rule   *text page 1034*

balancing selection   *1041*

Bergmann's rule   *1034*

directional selection   *1038*

disruptive selection   *1041*

diversifying selection   *1041*

fitness   *1033*

fixed allele   *1034*

frequency-dependent
   selection   *1043*

genetic load   *1043*

group selection   *1048*

heterozygote
   advantage   *1041*

hybrid vigor   *1041*

normalizing selection   *1037*

rare-mate advantage   *1044*

# SELF-QUIZ: TESTING WHAT YOU HAVE LEARNED

## Matching Key Terms

Match each term on the left with the most appropriate description on the right.

1. fixed allele
2. directional selection
3. Allen's rule
4. fitness
5. group selection
6. genetic load
7. normalizing selection
8. disruptive selection
9. Bergmann's rule
10. hybrid vigor

a. mono-allelic trait
b. maintains the status quo
c. shift in phenotype
d. mules are stronger than horses or donkeys
e. deleterious homozygotes
f. selection acting on populations
g. members of a species are larger in colder habitats
h. reproductive efficiency
i. against the average phenotype
j. polar bears have shorter legs than brown bears

## True or False?

1. _____ Human incest taboos are moral laws with no biological basis.

2. _____ *Heterozygote advantage* means the heterozygote has a higher fitness than either homozygote.

3. _____ Selectionists utilize evolutionary clock arguments.

4. _____ Normalizing selection maintains the status quo.

5. _____ Directional selection produces new phenotypes from old ones.

6. _____ Populations with a high genetic load have an advantage over those who do not.

7. _____ Random DNA changes occur at different rates in different species.

8. _____ Hybrid vigor is simply a synonym for heterozygote advantage.

## Completion

1. When a set of alleles at a given locus is reduced, by selection or random process, so that only one allele remains, then that allele is said to be _____ in the population.

2. _____ says that when species inhabit a range of great latitude variability, the individuals in high latitudes (colder environments) will be larger than those in warmer areas.

3. Color shifts in peppered moth populations have been due to _____ selection.

4. _____ is the selective favoring of genotypes that enhance the survival of whole populations relative to other populations, rather than of individuals with respect to each other.

5. Populations that have a number of different phenotypes are probably produced by _____ selection.

6. Rare-mate advantage is another term for _____ selection.

## Short Answer

1. What was Charles Darwin's basic argument for evolution by natural selection?

2. What is the probable explanation for the observations of Allen's rule?

3. How does gene amplification work to the advantage of *Culex* mosquito populations?

4. What type of selection allows bent grass plants to thrive on mine dump–contaminated soils? How does it work?

5. What arguments are used for the neutralist position in the neutralist–selectionist debate?

## Multiple-Choice Review

In the following sentences, fill in any blanks. Complete each statement by circling the correct response.

1. Populations may be broken into a number of phenotypically different groups by the process of:

   a. genetic load.
   b. directional selection.
   c. disruptive selection.

   d. normalizing selection.
   e. random drift.

2. A progressive, or collective, build-up of harmful _____ in a population is known as:

   a. Bergmann's rule.
   b. genetic load.
   c. allele fixation.
   d. Allen's rule.
   e. amplification.

3. Populations can undergo phenotypic shifts to keep up with a changing environment as a result of:

   a. disruptive selection.
   b. normalizing selection.
   c. frequency-dependent selection.
   d. diversifying selection.
   e. directional selection.

4. Hybrid _____ is really a form of:

   a. normalizing selection.
   b. balancing selection.
   c. genetic load.
   d. directional selection.
   e. random drift.

5. In evolutionary terms, the individual with the highest _____ is one who:

   a. has the least genetic load.
   b. shows hybrid vigor.
   c. has the most offspring.
   d. has heterozygote advantage.
   e. has few deleterious alleles.

# 43

# THE ORIGIN OF SPECIES

---

## CHAPTER AT A GLANCE

---

## CHAPTER PREVIEW

Chapter 43 begins with a survey of the criteria biologists use to judge whether a biological population is a species. Unlike the early work of Carolus Linnaeus, the modern approach involves the concept of a shared gene pool that is reproductively isolated from all other gene pools.

Using this modern definition, the text reviews the various intrinsic and extrinsic mechanisms that function to produce reproductive isolation. Of these, geographical isolation (the allopatric model of speciation) is the easiest to understand and the most widely accepted explanation. In the following section on the genetic basis of speciation, you will also explore the process of polyploidy—a well-understood mechanism that can produce almost "instant" reproductive isolation without geographical separation.

The next section deals with the concepts of microevolution (the small-scale changes that may lead to speciation) and of macroevolution (the major phenotypic changes that occur over evolutionary time). Three important phenomena related to a discussion of speciation—convergent, parallel, and divergent evolution or radiation—are examined here, as is the fate that inevitably befalls most species: extinction. Organisms on Earth have also experienced several episodes of mass extinction, and you will learn toward the end of this chapter how such biological catastrophes play a role in the punctuated-equilibrium theory—a recent, controversial proposal that posits a sort of stop-and-go pace to evolutionary change. Lastly, a brief final section reviews the basic features of micro- and macroevolution, and describes the kinds of new information that may shed light on the overall process that shaped life on our planet. This finale to the evolution section is a fitting prelude to the chapters you will consider next, on the ecology of organisms.

# LEARNING OBJECTIVES

When you have mastered the concepts of this chapter, you will be able to:

1. Give the modern biological definition of a species, and explain the problems inherent in this definition.
2. List and define four categories of prezygotic isolating mechanisms, and three categories of postzygotic isolating mechanisms.
3. Define a cline, and explain how reproductive isolation may vary along a cline.
4. Explain how speciation is thought to come about, referring to the allopatric speciation model.
5. Explain what is meant by the term *genetic identity.*
6. Explain how small changes in regulatory genes may produce large morphologic changes in organisms.
7. Explain how major changes in chromosome number or structure may result in rapid phenotypic change, even in the absence of geographical isolation.
8. Explain what is meant by the term *macroevolution,* and discuss how macroevolutionary events are identified.
9. Explain how radioactive isotopes are used to determine the age of fossils.
10. Explain what a phylogeny is, and tell how one is constructed.
11. Compare and contrast the evolutionary patterns of divergent, parallel, and convergent evolution, and adaptive radiation.
12. Explain the difference between analogous and homologous body structures in different species, and give examples of each case.
13. Explain what is meant by "mass extinction," and tell when in Earth's history five major mass extinctions are believed to have taken place.
14. Discuss the theory of punctuated equilibria, and compare punctuation with gradualism.

# CONCEPTS IN REVIEW

## *Section I*   How Biologists Define a Species

Modern biologists generally define a species as a group of actually or potentially interbreeding populations that is reproductively isolated from other such groups. Thus, members of a species can interbreed with each other, but they cannot breed with organisms belonging to another species.

One advantage of the standard of reproductive isolation is that it is very precise. Notice, however, it can be applied only to organisms that reproduce sexually. Asexual reproducers, including most prokaryotes, many plants, and some animals, must be classified into species based on physical (that is, biochemical and morphological) traits.

## Section II    Preventing Gene Exchange

Two general types of mechanisms operate to block the exchange of genes between individuals of related groups. The first general type is made up of prezygotic isolating mechanisms—mechanisms that prevent the formation of zygotes (text p. 1053). Prezygotic isolation falls into two categories: ecological and behavioral. In the first case, two related groups may become adapted to slightly different environments—perhaps varying soil types or food sources. Over time, these genetic differences become so great that successful cross-fertilization can no longer take place. In behavioral isolation, related groups evolve differing behaviors—such as specific mating rituals—that restrict the exchange of genes to members of the same group.

Sometimes the differences that result in prezygotic isolation involve mechanical isolation (text p. 1055). For example, mating is physically impossible between members of different species because the genitals of males and females are structurally incompatible, or molecules on the surfaces of sperm and egg fail to bind. A final type of prezygotic mechanism is temporal isolation, in which time-related environmental cues that trigger reproductive processes are different for related species (text p. 1056).

In postzygotic isolating mechanisms mating occurs, but the resulting hybrid organism is inviable or sterile. In a special case of hybrid sterility, termed hybrid breakdown, the second and subsequent generations after a cross show reduced reproductive success. Contrast this situation with the very different outcome of crossbreeding between two genetically distant members of the same species. As you learned in Chapter 42, the result is often heterozygote advantage (hybrid vigor).

Populations of a species that are spread out over a broad geographical range are often arrayed in a cline, a gradient along which gradual change in a characteristic or characteristics occurs as each population evolves adaptations to its own local environment. Along a cline, subspecies with distinct characteristics may arise. Often, individuals at either end of a cline are reproductively isolated.

## Section III    Becoming a Species: How Gene Pools Become Isolated

Ernst Mayr's model of allopatric speciation proposes that species can originate in a two-stage process (text p. 1059). First, populations of existing species are separated by a physical or geographical barrier. As a result, over time, genetic differences leading to pre- or postzygotic isolation arise between the two groups. In the second stage the diverged populations may again come into contact. If this happens, speciation becomes complete through the action of natural selection.

## Section IV    The Genetic Bases of Speciation

The extent of differences between populations that are diverging into separate species, or between species that have already diverged, is represented by a statistic called genetic identity—the relative proportion of the same structural genes present in members of groups being compared (text p. 1059). In general, biologists believe that the genetic events leading to speciation take place gradually. Once a new species has arisen, it tends to diverge genetically from related species at a more rapid pace. In some cases, such as in the primate order, major differences in body form are not reflected by corresponding divergence in structural genes. This has led biologists to hypothesize that small changes in regulatory genes may account for many of the large-scale changes responsible for speciation and the origin of higher taxonomic groups (text p. 1061).

One mechanism that may rapidly split populations genetically is polyploidization—the sudden multiplication of an entire complement of chromosomes (text p. 1062). It can result in sympatric speciation, in which new species arise even though no geographical isolation has taken place. A phenomenon similar to polyploidization, the rearrangement of chromosomes, has been proposed to explain the evolutionary origin of giant pandas (text p. 1063). Clearly, species can originate in a variety of ways.

## *Section V*    Explaining Macroevolution: Higher-order Changes

The changes that generate species are sometimes termed microevolution; those that produce the major phenotypic differences that separate genera, classes, orders, and so on are termed macroevolution. Some lines of descent can be traced by studying the fossil record. In other cases relationships must be inferred by comparison of related living organisms. When lines of descent over evolutionary time are constructed, the result is a phylogeny (text p. 1065).

The rationale for building a phylogeny is simple: one assumes that similarities in body structure, biochemistry, reproductive strategies, and other features of organisms can be used to trace lines of common descent. The process, however, is complex because evolution proceeds in different patterns. In cases of parallel evolution, two or more lineages evolve along similar lines. In convergent evolution, very distantly related lineages become more alike as similar adaptations take hold in response to demands of the environment (text p. 1066). Thus similar structures in different organisms may reflect homology (derivation from a common ancestor) or analogy (independent origin of structures used for similar purposes).

One of the most common evolutionary patterns that can be constructed from the fossil record is divergent evolution, or radiation. It is represented by the branching and rebranching of a single line. Another common feature of evolution is extinction—the complete loss of a species or group of species. Mass extinctions have occurred at least five times in Earth's history.

Gaps in the fossil record have led some paleontologists to propose the punctuated-equilibrium theory of evolution. The theory holds that evolution proceeds by spurts—radical changes over short (in the geologic time scale) periods of time, with intervening periods of equilibrium. The theory is controversial and tends not to be supported when an abundant fossil record is available.

## KEY TERMS

allopatric speciation    *text page 1059*

analogy    *1067*

behavioral isolation    *1054*

cline    *1058*

convergent evolution    *1066*

divergent evolution (radiation)    *1069*

ecological isolation    *1053*

extinction    *1069*

genetic identity    *1059*

homology    *1067*

hybrid breakdown    *1057*

hybrid inviability    *1057*

hybrid sterility    *1057*

macroevolution    *1064*

mechanical isolation    *1055*

microevolution    *1064*

parallel evolution    *1066*

phylogeny    *1065*

polyploidization    *1062*

postzygotic isolating mechanism    *1053*

prezygotic isolating mechanism    *1053*

punctuated equilibrium    *1073*

reproductive isolation    *1052*

species    *1052*

subspecies    *1058*

sympatric speciation    *1063*

# SELF-QUIZ: TESTING WHAT YOU HAVE LEARNED

## Matching Key Terms

Match each term on the left with the most appropriate description on the right.

| | | |
|---|---|---|
| 1. analogy | a. | reproductively isolated |
| 2. homology | b. | multiple chromosome sets |
| 3. microevolution | c. | family tree |
| 4. extinction | d. | minor allele frequency changes |
| 5. hybrid sterility | e. | large phenotype changes |
| 6. macroevolution | f. | geographic separation |
| 7. species | g. | same ancestral origins |
| 8. allopatric speciation | h. | a functional comparison |
| 9. polyploidization | i. | no one breeds mules |
| 10. phylogeny | j. | total species death |

## True or False?

1. _____ Temporal isolation is a postzygotic mechanism.

2. _____ Hybrid sterility is a postzygotic mechanism.

3. _____ Mules exhibit heterozygote advantage.

4. _____ Subspecies are genetically inferior to full species.

5. _____ The allopatric speciation model is based on geography.

6. _____ Allopolyploids arise through hybridization.

7. _____ Carbon-14 dates are not reliable beyond 40,000 years.

8. _____ A phylogeny traces lines of descent.

## Completion

1. _____ occurs when two distantly related lineages evolve structures that are super-ficially similar.

2. _____ describes the situation when two closely related lineages evolve structures that are both superficially and actually similar.

3. The wings of grasshoppers and birds are both used for flight, and are thus _____ .

4. The wings of birds and bats are both used for flight; since these organisms spring from a common ancestor, the structures are _____.

5. The permanent loss of a species is termed _____.

6. The notion that evolution proceeds by rapid change followed by long periods of stasis is called _____.

## Short Answer

1. How are species defined in terms of reproductive isolation?

2. How are phylogenies constructed?

3. What is genetic identity, and how is this statistic calculated?

4. What are the differences between punctuated equilibrium and gradualism?

5. What is meant by the term *macroevolution*, and what is its relationship (if any) to punctuated equilibrium?

6. When did the five mass extinctions occur in Earth's history?

# Multiple-Choice Review

In the following sentences, fill in any blanks. Complete each statement by circling the correct response.

1. The theory of punctuated equilibrium was proposed by:

    a. David Raup.
    b. Eldridge and Gould.
    c. David Jablonski.
    d. M. J. D. White.
    e. Ernst Mayr.

2. The most common evolutionary pattern is that of:

    a. divergence.
    b. convergence.
    c. parallelism.
    d. relay.
    e. microevolution.

3. The notion of accidental _____ isolation (by divergence over time) is:

    a. punctuated equilibrium.
    b. allopatric speciation.
    c. sympatric speciation.
    d. the selection hypothesis.
    e. the by-product hypothesis.

4. Which of the following is *not* an example of a prezygotic _____?

    a. ecological isolation
    b. hybrid sterility
    c. behavioral isolation
    d. mechanical isolation
    e. temporal isolation

5. Most species of crop plants have come about through:

    a. allopatric speciation.
    b. sympatric speciation.
    c. polyploidization.
    d. subspeciation.
    e. punctuation.

# 44
# ECOSYSTEMS AND THE BIOSPHERE

## CHAPTER PREVIEW

Biology texts are generally organized in one of two ways for the study of ecology and evolution: they either start with small units such as species and populations of organisms, and work up to an overall world view, or they reverse the sequence and begin with an overview. Your text takes this second option. Chapter 44 begins with the entire living world—the biosphere—and following chapters work down to the smaller divisions of ecosystems, communities, and populations. This chapter introduces many terms and concepts, but none of them are difficult. In fact, much of what you will learn has wide application outside of biology. For example, you will encounter the ideas that formed the basis of the "ecology movement" of the 1960s and 1970s and that continue to influence society in important ways.

The chapter begins with a definition of ecosystem, and then moves into the biosphere, or entire life-supporting area of the Earth. First, the text surveys the physical factors such as sunlight and rainfall that create a physical environment for life. Note that the sun does much more than supply the energy for photosynthesis. It also is the major factor—by way of the temperature gradient created by solar radiation falling differentially upon a round world—involved in producing Earth's climates.

Next, the chapter explores the world's biomes—the major environmental divisions based on the overall plant composition of large regions. This survey, with its beautiful photographs, is followed by a look at the habitats of life: the air, land, and water. The section on energy flow through ecosystems includes a number of explanatory charts and graphs, and the familiar "pyramids" of ecology. Then the chapter discusses the cycling of finite materials, and contrasts these cycles with the one-way flow of energy from sunlight. Finally, the focus narrows to aquatic ecosystems, and the ecological life of a lake.

# LEARNING OBJECTIVES

When you have mastered the concepts of this chapter, you will be able to:

1. Define the term *ecology*, and state the basic goals of ecological science.
2. Provide a definition for the term *ecosystem*.
3. List the characteristics that distinguish a terrestrial biome.
4. Explain what is meant by the "biosphere," and what its physical limits are.
5. Explain what is meant by solar constant.
6. Explain how the sun drives global climate, and how the physical parameters of our planet (rotation rate, angle of tilt, and so on) also shape climatic conditions.
7. List the nine terrestrial biomes, describe their physical conditions, state where they are found, and list some of their typical plant and animal species.
8. Explain what lentic and lotic communities are.
9. List the five marine biomes.
10. Explain what an organism's "habitat" is.
11. Explain how soil is formed, and list its major components.
12. Explain what food chains and food webs are, and provide examples of each.
13. List and define the major trophic levels.
14. Explain what kind of information is conveyed by pyramids of numbers and biomass.
15. Distinguish between primary and secondary production, and between gross and net production.
16. Explain how biological magnification works, and give an example.
17. Explain why materials, unlike energy, must be recycled on Earth, and outline the cycles of water, carbon, nitrogen, and phosphorus.
18. Describe the serious environmental problems that are increasingly associated with the carbon and phosphorus cycles.
19. Sketch a cross section of a fresh-water lake, and explain how and why ecologists divide lakes into three major zones.
20. Explain what takes place in lake eutrophication.

# CONCEPTS IN REVIEW

## *Section I*   Ecosystems

Ecology is the study of the interplay of organisms and their home environments. The ecosystem is the most complex level of biological organization, and the most complex ecosystem of all is the biosphere: the life-supporting soils, seas, and atmosphere of the Earth, and all the organisms in them.

## *Sections II and III*   The Biosphere *and* Biomes

The biosphere has many parts and subparts that together provide its myriad habitats, or actual places where organisms live. Ultimately, all this life depends on the continual influx of energy from the sun. The solar constant is a value assigned to the amount of solar energy per square meter that reaches the outer edge of the atmosphere. Only about 1–4 percent of this energy is available to do work at the planet's land and water surfaces. Plants convert to sugars 1.2 percent of the impinging solar energy, using a portion of those sugars as metabolic fuel. Ultimately, only about 0.5 percent of the solar constant is available to support the life processes of all other organisms in the biosphere.

The Earth's relationship to the sun also establishes our planet's seasons and the climates that determine both the nature and distribution of living things. The dynamics of air and ocean water movements due to the Earth's rotation, tilt, and orbit around the sun result in global climate and weather patterns. Climatic factors, including regional temperature ranges, precipitation, and wind patterns, determine the predominant vegetation zones of the Earth. These zones, or biomes, are the Earth's major ecosystems (text pp. 1079–1084).

There are nine terrestrial biomes: tropical rain forest, tropical seasonal forest, savanna, desert, temperate grassland, temperate shrubland, temperate forest, taiga, and tundra. Fresh-water aquatic biomes include lentic (lake and pond) communities and lotic (stream) communities. Marine biomes encompass sandy beaches, intertidal zones, coral reefs, pelagic (open ocean) communities, and benthic (bottom) communities.

## *Section IV*   The Habitats of Life: Air, Land, and Water

At a fundamental level, the different states of matter—solid, liquid, and gas—that make up habitats impose constraints on organisms and affect their modes of life. Organisms that live in contact with air, for example, generally must tolerate or cope with temperature extremes, and most have hard body parts to provide support against the downward pull of gravity. The gaseous constituents of air (oxygen, nitrogen, carbon dioxide) also impose limits as well as provide essential raw materials (text p. 1089).

On land, many plants and most animals acquire water and mineral nutrients directly or indirectly from the soil. Soils contain varying amounts of silt, sand, clay, and nutrient-rich humus. Soil chemistry and structure determine the amount and availability to organisms of minerals and water (text p. 1090); water is held between soil particles by capillarity and by imbibition. Soils change over time, undergoing the processes of humification and mineralization due to the activity of organisms that inhabit them and in response to climatic conditions (text p. 1090).

Although water provides protection from temperature extremes and physical support for aquatic organisms, it generally contains limited supplies of mineral nutrients and oxygen. Plants can survive only in a top layer of water, the photic zone.

## Section V    Energy Flow in Ecosystems

Energy flows through ecosystems via food chains that are part of larger, interconnected food webs. The trophic levels of a food chain begin with photosynthesizers: producers that capture solar energy and transform it into a carbohydrate storage form. Energy is transferred to consumers at successive trophic levels. At each step, some energy may also be transferred to decomposers and detritivores. Ecosystems often show pyramids of numbers and biomass, which illustrate the inefficiency of energy transfer between trophic levels (text p. 1093). This inefficiency generally limits the number of trophic levels in a food chain to four or five.

In thinking about the efficiency of energy transfer, it is important to understand that there are two kinds of production in an ecosystem: primary and secondary. Primary production is the energy stored in tissues of photosynthetic organisms; it is further broken down into gross production (all the energy captured by a photosynthesizer) and net production (the energy ultimately available to consumers). Secondary production is the energy stored in the tissues of consumers. Ecologists have calculated that between 75 and 90 percent of the calories of net production at one level are lost by the time organisms at the next level have produced their cytoplasm.

## Section VI    Cycles of Materials

The supply of water and minerals available to organisms living on Earth is finite. Thus, these materials cycle into and out of organisms in complex ways. The global water or hydrologic cycle begins with precipitation and ends with evaporation from the surfaces of water bodies or organisms. In the carbon cycle atmospheric carbon taken up as $CO_2$ by plants is returned to the air by respiration, combustion, or as organic wastes. An increase in atmospheric $CO_2$ as the result of heavy burning of fossil fuels may be raising surface temperatures by creating a global greenhouse effect.

Two biogeochemical cycles—involving nitrogen and phosphorus—also provide organisms with critical nutrients. In the nitrogen cycle atmospheric nitrogen is first fixed by bacteria and cyanobacteria. After use by plants and animals it is then returned to the atmosphere through a complex series of chemical steps (text pp. 1100–1101). Phosphates enter the food chain when they are taken up from soil or water by producer organisms. And although bird droppings are a major avenue for the return of phosphates to the land, much phosphate is lost for long periods to deep marine sediments.

## Section VII    Close-up of a Lake Ecosystem

Because of its clear boundaries, a lake is a good example of a self-contained ecosystem. Of all the physical and chemical features of water that affect the organisms within it, light penetration is the most important. Especially rich in life forms is the shallow edge of a lake where light reaches the bottom—the littoral zone. Beyond this zone, in the central limnetic zone, levels of light, oxygen, and photosynthesis by water plants or phytoplankton are significantly reduced. The profundal zone is the area below the depth of effective light penetration. It is the least productive part of a lake. Eutrophication is the natural process by which lakes become increasingly productive as they fill with silt and organic debris (text pp. 1103–1104).

## KEY TERMS

benthic community   *text page 1089*

biological magnification   *1097*

biomass   *1093*

biome   *1085*

biosphere   *1079*

carbon cycle   *1099*

consumer   *1092*

Coriolis effect   *1082*

decomposer  *1092*

denitrification  *1101*

desert  *1087*

detritivore  *1092*

ecological efficiency  *1097*

ecology  *1078*

ecosystem  *1078*

eutrophication  *1104*

food chain  *1092*

food web  *1092*

gross production  *1096*

habitat  *1079*

humification  *1090*

humus *1090*

hydrologic cycle  *1099*

imbibition  *1090*

lentic community  *1088*

limnetic zone  *1104*

littoral zone  *1103*

loam  *1090*

lotic community  *1088*

mineralization  *1090*

net production  *1096*

nitrification  *1101*

nitrogen cycle  *1100*

nitrogen fixation  *1101*

pelagic community  *1089*

phosphorus cycle  *1102*

photic zone  *1091*

primary production  *1093*

producers  *1092*

profundal zone  *1104*

pyramid of biomass  *1093*

pyramid of numbers  *1093*

savanna  *1086*

secondary production  *1096*

silt  *1090*

solar constant  *1079*

standing crop  *1093*

taiga  *1088*

temperate forest  *1087*

temperate grassland  *1087*

temperate shrubland  *1087*

trophic level  *1092*

tropical rain forest  *1086*

tropical seasonal forest  *1086*

tundra  *1088*

# SELF-QUIZ: TESTING WHAT YOU HAVE LEARNED

## Matching Key Terms

Match each term on the left with the most appropriate description on the right.

| | | | |
|---|---|---|---|
| 1. | littoral zone | a. | based on plants |
| 2. | consumer | b. | where you live |
| 3. | net production | c. | rotting organic material |
| 4. | trophic level | d. | on the beach |
| 5. | biome | e. | heterotrophic |
| 6. | biomass | f. | after metabolic "taxes" |
| 7. | loam | g. | optimal for plant growth |
| 8. | habitat | h. | a weight |
| 9. | humus | i. | ends a food chain |
| 10. | detritivore | j. | based on who eats what |

## True or False?

1. _____ The Coriolis effect causes liquids to rotate clockwise in the southern hemisphere.

2. _____ Tropical rain forests occur at high latitudes.

3. _____ Biomes are based on the dominant vegetation.

4. _____ Lotic "biomes" are those associated with streams.

5. _____ Humification is the process that yields soil.

6. _____ Secondary production is accomplished by herbivores.

7. _____ Ecological efficiency ranges from 10 to 50 percent.

8. _____ The primary site of lake production is the profundal zone.

## Completion

1. Photosynthetic areas around the edges of lakes make up the _____ zone.

2. The amount of solar energy that reaches Earth is known as the _____.

3. The upper, photosynthetic zone in aquatic ecosystems is called the _____ zone.

4. The accumulation of minute amounts of trace materials as one travels up trophic levels is

   _____.

5. _____ is a successional process of lakes.

6. Decomposers are mostly _____ and _____.

## Short Answer

1. How much of the Earth's solar constant actually is incorporated into plants?

2. In what sense does energy from the sun drive the Earth's climate?

3. How does the Coriolis effect influence Earth's climate?

4. How do savannas differ from temperate grasslands?

5. What factors contribute to the development of deserts?

6. How does energy flow through ecosystems?

# Multiple-Choice Review

In the following sentences, fill in any blanks. Complete each statement by circling the correct response.

1. Detritivores include:

   a. photosynthetic plants.
   b. carnivores.
   c. herbivores.
   d. bacteria.
   e. omnivores.

2. Secondary production refers to the _____ stored in:

   a. herbivores.
   b. carnivores.
   c. plants.
   d. fungi.
   e. bacteria.

3. The _____ of water into regions of different temperatures is termed:

   a. zonation.
   b. stratification.
   c. profundal zonation.
   d. eutrophication.
   e. none of the above

4. Steppes, veldt, and pampas are all examples of:

   a. forests.
   b. shrublands.
   c. deserts.
   d. tundras.
   e. grasslands.

5. Permafrost is a feature of the:

   a. taiga.
   b. Antarctic.
   c. tundra.
   d. temperate forest.
   e. desert.

## Exercise

In the chart below, briefly describe the stages of the water, carbon, and nitrogen cycles.

| Cycle | Stage |
|-------|-------|
| Water |  |
| Carbon |  |
| Nitrogen |  |

# 45

# THE ECOLOGY OF COMMUNITIES

---

## CHAPTER AT A GLANCE

---

## CHAPTER PREVIEW

This chapter considers the actual subdivisions of ecological communities (the major plant biomes of Chapter 44), and begins by defining a community as the interacting populations of a species within a geographical area. Then, in the first section, it introduces the concept of biological succession—the ecological process by which communities come into existence and change through time. As you read, keep in mind that succession is a continuum, and that the different stages grade into one another: there are no clear-cut boundaries. Also, note that the two major types of succession— primary and secondary—are differentiated solely on the basis of the substrate on which they begin.

   The next section explores community structure, including the specific meanings ecologists ascribe to the terms *species richness* and *species abundance*. As you will see, *richness* refers to the number of different species (the diversity of species) in a communuty, while *abundance* refers to the number of individuals within a given species in the community. The species equilibrium model of MacArthur and Wilson is the centerpiece of this section, and entails several of the most important principles of modern ecology.

Chapter 45 closes with a discussion of the ecological meaning of the term *niche,* and of the various types of niches that organisms may occupy.

---

# LEARNING OBJECTIVES

When you have mastered the concepts of this chapter, you will be able to:

1. Define an ecological community.
2. Explain the concept of ecological succession, and distinguish between the two types of succession.
3. Explain and discuss the concept of ecological climax.
4. List the factors that influence particular species involved in successional stages.
5. List the factors that play a role in successional species replacement.
6. Define an ecocline, and explain the effect ecoclines have on a community's composition.
7. Explain what is meant by species richness and tell how it may vary among different communities.
8. Explain how species richness differs in different stages of succession.
9. Outline the species equilibrium model of MacArthur and Wilson.
10. Describe the various factors that are included in the niche concept.
11. Compare and contrast the concepts of a fundamental and realized niche.
12. Name the characteristics that determine whether a species is considered generalist or specialist.
13. Outline the concept of competitive exclusion.

---

# CONCEPTS IN REVIEW

## *Section I*   Ecological Succession: A Basic Feature of Life

A community is an association of the interacting members of populations of different species in a particular area. Over time, the types of species that make up a community change—the process known as ecological succession. Ecologists describe succession as either primary or secondary. In primary succession, bare rock, sand, or lava slowly becomes populated with organisms such as lichens or mosses (text p. 1110). As these life forms carry out their life processes, rocky surfaces are slowly broken down and pockets of soil develop. Gradually, more "pioneer" or "fugitive" species (grasses or ferns) colonize the area. As succession proceeds, soil accumulates and water and nutrients become more available. The plant community becomes more diverse, inhabited by larger, longer-lived species. Ultimately, the community may stabilize and become dominated by climax species—those that simply reproduce themselves and are not replaced by new arrivals. A form of primary succession called aquatic succession takes place in lakes and ponds (text p. 1111).

Secondary succession is much more common than primary succession. It begins on already developed soil that has been cleared of vegetation by a natural or man-made disturbance (text p. 1112).

Research has shown that succession is a highly complex and often unpredictable series of events. In general, there are three mechanisms of succession: (1) facilitation, in which early species make colonization or growth easier for later species; (2) inhibition by early species of later ones, so that a disturbance must occur for succession to proceed; and (3) tolerance, in which later species tolerate early ones but are unaffected by them (text pp. 1114–1115). Local geography and many chance events, including disturbances and the types of species inhabiting surrounding areas, can also determine which species succeed in a particular area.

## Section II   The Structure of Communities

Community structure is based on four components: (1) physiognomy, or the physical forms of component species (trees, vines, and so on); (2) the relative abundance of different species, or of individuals of various species; (3) the total number of species, termed species richness; and (4) the niche structure of the community (text p. 1115).

Community physiognomy considers the major growth forms of plants. The growth forms typical of Earth's major biomes were described in Chapter 44. These principal community types grade into one another, often along climate gradients known as ecoclines (text p. 1116). In studies of relative species abundance in communities, two patterns emerge. In tropical communities there is equitability: most of the species have roughly equal population sizes. In other biomes there is generally a pattern of dominance: one or a few species have larger populations than others.

Species richness—the total number of species in a community—varies with latitude. Tropical rain forests near the equator are the richest biological areas on Earth (text p. 1118). The species equilibrium model (MacArthur and Wilson) attempts to explain variations in species richness. It assumes that the number of species found on an island is determined by a balance between the immigration rate of new species and the extinction rate of existing species (text p. 1119). Ecologists believe that most environments are composed of habitat "islands," and that each will provide adequate resources for only a certain number of species.

Species richness and relative abundances in communities are determined by competition between species for limited resources, predation (including herbivory, disease pathogens, and parasites), and physical disturbance (text p. 1121). According to the intermediate disturbance hypothesis, when the intensity of disturbances or predation in a community is neither very low nor very great, the community has the most species.

A species' niche is its total way of life (text p. 1121). Subdivisions include the food niches of animals and the habitat niches of plants and animals. A fundamental niche is the full environmental range that a species can occupy if there is not direct competition from another species. A realized niche is the niche that results when the fundamental niche is narrowed by competition. The competitive exclusion principle postulates that no two species can occupy exactly the same niche at the same time in a locale if resources are limiting (text p. 1124).

Competition for resources can have important consequences for a species' versatility and for the niche structure of a community. Resource partitioning occurs in communities in which competition occurs between species (text p. 1124). In general, some species evolve with narrow niches and are termed specialists, while other species are adapted to life over a wide range of conditions and are termed generalists. Overall, relatively few species occur where resources are limited or unpredictable. The most diverse communities are those in which resources are not limited, or are predictable and exploited by specialists (text p. 1125).

## KEY TERMS

climax community  *text page*  1111

community  1109

competitive exclusion principle  1124

ecocline  1116

ecological succession  1109

food niche  1122

fugitive species  1110

fundamental niche  1124

generalist  1124

habitat niche  1122

niche  1121

primary succession  1109

realized niche  1124

secondary succession  1109

specialist  1124

species equilibrium model  1119

species richness  1118

# SELF-QUIZ: TESTING WHAT YOU HAVE LEARNED

## Matching Key Terms

Match each term on the left with the most appropriate description on the right.

| | |
|---|---|
| 1. generalist | a. the interacting species of an area |
| 2. primary succession | b. what an animal/plant does "for a living" |
| 3. ecocline | c. gradual changes in species composition |
| 4. food niche | d. broad niche |
| 5. ecological succession | e. requires a developed soil |
| 6. niche | f. what a species eats and when |
| 7. community | g. narrow niche |
| 8. specialist | h. number of species in a given area |
| 9. species richness | i. a gradient |
| 10. secondary succession | j. on bare rock |

## True or False?

1. _____ Secondary succession may take place on rock or sand.

2. _____ Aquatic succession is a form of primary succession.

3. _____ Fugitive species are mostly K-selected.

4. _____ Old field succession is more likely in Kansas than in Arizona.

5. _____ The abundance of individuals per species determines species richness.

6. _____ A niche can be likened to an animal's "address."

7. _____ Ecoclines are climatic gradients.

8. _____ A realized niche is larger than a fundamental niche.

## Completion

1. _____ refers to the types of species occurring in an area.
2. When a community reaches a state where its constituent species simply reproduce themselves,

   but are not replaced by new arrivals, the community is said to be in its _____ stage.

3. The number of species present in a community refers to _____.

4. The _____ is an organism's "occupation," while its habitat is the _____.

5. The _____, or physical appearance of a community, is simply the physical form the component species take.

6. The species equilibrium model was developed by _____ and _____.

## Short Answer

1. What is thought to determine the number of species on an island?

2. How have ecologists tested the species equilibrium model?

3. What characteristics are likely to make a plant a successful colonizer?

4. What kinds of species interactions affect equilibrium?

5. What is meant when ecologists say some species are generalists and some are specialists?

6. What experiments are the basis for the competitive exclusion model?

# Multiple-Choice Review

Complete the following statements by circling the correct response.

1. Species that specialize in precarious existences are styled as:
   a. diverse.
   b. generalists.
   c. fugitive.
   d. pioneer.
   e. climax.

2. The complete range of environmental factors a species can occupy make up its:

   a. fundamental niche.
   b. wide-range niche.
   c. generalist niche.
   d. realized niche.
   e. specialist niche.

3. Another name for species richness is:

   a. relative abundance.
   b. density.
   c. species dominance.
   d. species diversity.
   e. species equilibrium.

4. Factors that may affect community equilibrium include:

   a. competition.
   b. natural disasters.
   c. predation.
   d. the weather.
   e. all of the above

5. A species is usually said to occupy a realized niche when the niche is:

   a. the only one available because of physical characteristics of the environment.
   b. equally divided among several subspecies.
   c. narrowed by competition.
   d. shared equally with another species.
   e. broader than normal.

# 46

# THE ECOLOGY OF POPULATIONS

---

## CHAPTER AT A GLANCE

---

## CHAPTER PREVIEW

Chapter 46 surveys special characteristics of populations, with a focus on growth. It begins by exploring the meaning of statistical features such as mortality and natality, and then describes exponential growth in a population—the kind of growth that happens when there are no limits on food, space, or other resources. In the real world, of course, no population can grow exponentially for long. Hence, the chapter next describes a more realistic phenomenon, logistic growth, and introduces the concept of an environmental carrying capacity that ultimately limits the numbers of individuals of a population that an ecosystem can support.

Because growth in biological populations results from reproduction, the chapter next discusses two very different reproductive strategies, "r selection" and "K selection," respectively. These concepts are widely used in many areas of biology, and you will encounter them in later chapters and courses. The focus then shifts to the interrelated factors that place limits on a population's size within a community. The concept of density plays a central role in this discussion—as you will see, factors that limit the numbers of populations are broadly classified into those such as competition that are tied to density, and those such as natural disasters that are density-independent.

A final section looks at another feature that is determined by interacting factors—that is, how a population is distributed within its potential range. Competition for resources is important here, too, and has led to the evolution of phenomena such as allelopathy in plants, resource partitioning, and character displacement. As you learn more about these concepts, you will gain a fuller understanding of how interacting species have a profound impact on the communities in which they live.

# LEARNING OBJECTIVES

When you have mastered the concepts of this chapter, you will be able to:

1. Explain what biologists mean when they use the term *population*, and name at least three fundamental characteristics that all populations exhibit.
2. Draw and explain an exponential growth curve, and describe the conditions that must be met in order for exponential growth to take place.
3. Explain the ecological meaning of intrinsic rate of increase and environmental carrying capacity.
4. Draw and label a logistic growth curve, and discuss the factors that determine its shape.
5. Discuss the concept of a reproductive time lag, and explain how such a lag affects a population's growth curve.
6. Explain what a survivorship curve is, and what information it provides about population growth rates.
7. Define, and then compare and contrast, the reproductive strategies of *r* and *K* selection.
8. Explain the difference between density-dependent and density-independent growth factors, and give four examples of each.
9. Explain the cyclic growth and decline patterns found in interacting predator–prey populations.
10. Explain resource partitioning.
11. Discuss the phenomenon of character displacement, and explain how it could lead to the evolution of new species.
12. Discuss the hypothesis that very complex, diverse communities are more stable than simpler communities, and then present a summary of arguments for and against this position.

# CONCEPTS IN REVIEW

## Section I    Population Growth

Populations—groups of individuals belonging to the same species—all have certain statistical characteristics. These are the per capita birth rate, or natality; the per capita death rate, or mortality; and density—the number of individuals per unit area.

As first described by Malthus, a population can, in theory, grow exponentially (geometrically) if there are no limits on resources such as food or hiding places and no predation or competition (text p. 1130). This kind of population growth is represented by an exponential growth curve. The conditions necessary for exponential growth rarely occur in nature, however. The finite levels of resources in any environment set an upper limit to population size—termed carrying capacity (*K*)—that can be reached but never long exceeded. The leveling-off of growth when population size reaches equilibrium with available resources is plotted on a logistic growth curve (text p. 1131).

When resource limits are approached or exceeded, time is required for the birth rate to fall and the death rate to rise. This response time is known as reproductive time lag. It is one reason for the fluctuations in size that are seen in every population. In many natural populations, carrying capacity, and hence population size, fluctuates seasonally. If a population (temporarily) drastically exceeds the carrying capacity of its environment, damage may occur that permanently lowers the carrying capacity (text p. 1132).

In addition to environmental carrying capacity, a population's age structure and reproductive strategy also affect the rate at which the population grows. Age structure reflects the relative numbers of young, middle-aged, and older individuals in a given population. In a population having many members at or nearing reproductive age, significant growth may occur. Age structure may also be represented by a survivorship curve (text p. 1134).

The reproducing members of a population follow a reproductive strategy that is a complex adaptation evolved over millenia. Reproductive strategies generally fit into one of two categories: those of *r-selected* species and those of *K-selected* species. In *r* selected species, individuals reach reproductive age quickly and produce many offspring. Each offspring is small and enters the world with relatively few resources. Out of the many produced, a few will survive. In *K* selection, a strategy related to environmental carrying capacity and the need to compete for resources evolves, and results in individuals that mature slowly and produce few offspring. However, parents invest a great amount of resources in each offspring, so that, after a long period of growth to large size, its chances of survival are high (text p. 1135).

## Section II    Limits on Population Size

The size of a population is measured in terms of its density. Whether density is high or low, distribution of individuals is usually uneven. Common patterns are clumped, uniform, and random (text p. 1136). There are often negative consequences attached to high (or rising) population density. These density-dependent factors include increased predation, parasitism, disease, and intraspecific and interspecific competition. Population size may also be reduced by density-independent factors, a category that includes natural catastrophes (text pp. 1138–1139).

The interactions of predators and their prey affect population size in complex ways. Such populations sometimes cycle regularly between growth and decline, in part due to the effects of reproductive time lags. In general, predation may slow or stop the growth of a prey population only when many reproducing individuals are eliminated. If only weak, sick, or very young prey are taken, the effect of predation on the density of the population as a whole may be slight (text p. 1141).

An area of controversy among ecologists is the question of whether species diversity in a community generates stability, or whether the reverse is true. One aspect of this argument is that a complex food web is more stable than a simple one. However, in nature many stable, highly diverse communities are characterized by the presence of many simple food webs. It may be that stable environments beget diversity because they allow rare species to persist (text p. 1142).

## Section III    How Populations Are Distributed

Just as competition, predation, and other elements interact to determine the size of a population within a community, other factors interrelate to determine its distribution. Overall, distribution of a population in its potential range depends on the location of food and suitable habitat, interspecific competition for resources, and other variables (text p. 1144). Among plants, one of the most effective forms of interspecific competition for resources is allelopathy. Among species that share similar or identical habitat niches, resource partitioning is often seen. In character displacement, closely related species have evolved physical differences in the body structures used for exploiting a limited resource (text pp. 1144–1145). Eventually, such solutions to the need for dividing up a scarce resource may lead to speciation.

## KEY TERMS

age structure    *text page 1133*

allelopathy    *1144*

carrying capacity    *1131*

character displacement    *1144*

density-dependent factor    *1138*

density-independent factor    *1137*

exponential growth curve    *1130*

interspecific competition    *1139*

intraspecific competition    *1139*

K-selected species    *1135*

logistic growth curve    *1131*

mortality    *1129*

natality    *1129*

population density    *1135*

r-selected species    *1135*

reproductive strategy    *1135*

reproductive time lag    *1132*

resource partitioning    *1144*

survivorship curve    *1134*

---

# SELF-QUIZ: TESTING WHAT YOU HAVE LEARNED

## Matching Key Terms

Match each term on the left with the most appropriate description on the right.

1. natality
2. K-selected species
3. interspecific competition
4. allelopathy
5. population density
6. exponential growth
7. r-selected species
8. mortality
9. carrying capacity
10. intraspecific competition

a. K
b. per capita birth rate
c. reproduce many young
d. between species
e. reproduction tied to K
f. within species
g. chemical warfare
h. individuals per area
i. "no limits"
j. death rate

## True or False?

1. _____ The carrying capacity of an environment cannot be permanently exceeded.

2. _____ Logistic growth produces a sigmoid curve pattern.

3. _____ Large yolked eggs are characteristic of r-selected species.

4. _____ A herd of horses shows a random distributional pattern.

5. _____ Competition and predation are examples of density-independent factors.

6. _____ Allelopathy involves the production of toxic chemicals.

7. _____ Galápagos finches exhibit character displacement.

8. _____ Natural populations do not grow exponentially for long.

## Completion

1. The number of individuals per unit of area is known as the _____ of a population.
2. A graphic display that shows the numbers of survivors in different age groups of a population

   is called a _____.

3. _____ competition involves only members of the same species.
4. The amount of time it takes for birth rates to fall and death rates to rise is the

   _____ .

5. _____-selected species mature slowly, and produce fewer young.

6. Territorial animals show a _____ dispersion pattern.

## Short Answer

1. Name and give examples of the three types of population dispersion patterns.

2. List the characteristics of $r$- and $K$-selected species.

3. List at least five density-dependent factors.

4. List five density-independent factors.

5. Compare and contrast intraspecific and interspecific competition.

6. What is the theoretical basis of the logistic growth curve?

## Multiple-Choice Review

In the following sentences, fill in any blanks. Complete each statement by circling the correct response.

1. In the _____ growth equation the symbol $dN/dt$ represents:

   a. population size.
   b. rate.
   c. time.
   d. change in population size.
   e. none of the above

2. Parasites and pathogens spread most rapidly in a population when:

   a. the host population is randomly distributed.
   b. the host is adapted to them.
   c. the parasite is highly virulent.
   d. the host population is dense.
   e. the host is weakened.

3. Competition among different species for a _____ resource is termed:

   a. interspecific competition.
   b. scramble competition.
   c. diffuse competition.
   d. intraspecific competition.
   e. density-independent competition.

4. Closely related species, exploiting the same limited resources, often acquire different features in a process of:

   a. allopatric speciation.
   b. sympatric speciation.
   c. character displacement.
   d. allelopathy.
   e. resource partitioning.

5. Density-independent factors may include:

   a. seasonal changes.
   b. accidents.
   c. bad weather.
   d. natural catastrophes.
   e. all of the above

# 47

# ADAPTATION: ORGANISMS EVOLVING TOGETHER

---

## CHAPTER AT A GLANCE

---

## CHAPTER PREVIEW

This chapter presents some of the living world's most unusual and striking inhabitants as it explores organisms' adaptations for defense and escape, and the close relationships of symbiosis. It begins by introducing the concept of coevolution—the source of biological relationships between two species so close that the fitness of one is dependent on the other. A fascinating first section presents an array of mechanical and chemical defense adaptations of plants and animals, and the phenomena of Batesian and Müllerian mimicry. Be sure you have a clear idea of the difference between these two forms of mimicry.

Next the chapter considers the intriguing kinds of adaptations that enable organisms to more successfully escape predation—including different forms of camouflage, protective grouping, running away, or avoidance of predators in time and space. As with the previous section, be sure to note the subtle differences between phenomena such as countershading and disruptive and cryptic coloration.

The final section of Chapter 47 describes the various types of symbiotic relationships that occur among different species. Here you may find it helpful to construct a study chart that lists the type of symbiosis (parasitism, and so on) on one axis, and the effect of the relationship on the two species on the other axis.

# LEARNING OBJECTIVES

When you have mastered the concepts of this chapter, you will be able to:

1. Define the term *coevolution*.
2. List three mechanical defenses seen in animals, and three mechanical defenses seen in plants.
3. List three examples of animal chemical defenses, and give the two sources of animal defensive chemicals.
4. List and discuss three types of plant chemical defenses.
5. Discuss how predator–prey coevolution has produced a type of biological "arms race."
6. Compare and contrast Müllerian and Batesian mimicry.
7. Discuss the different kinds of escape strategies that have evolved in animals and plants, and provide examples of each.
8. Outline the various methods of camouflage used by animals, including cryptic coloration, disruptive coloration, and countershading.
9. Distinguish between the symbiotic relationships of parasitism, commensalism, and mutualism.
10. Explain how coevolution of parasites and their hosts can actually act against the more virulent parasites.
11. Distinguish between obligatory and facultative commensalism.

# CONCEPTS IN REVIEW

## *Section I*   Adaptations for Defense

Adaptations evolve over generations as natural selection acts directly on variation in phenotypes and indirectly on their underlying genotypes. Two general categories of adaptations for defense evolved by virtually all eukaryotes, animals and plants, are mechanical and chemical defenses.

Very large animals may escape predation on the basis of size alone. Smaller creatures commonly rely on mechanical defenses, including hard exoskeletons, claws, teeth, and horns. Among plants, analogous structures are spines, thorns, and tough outer tissues (text p. 1151). In general, mechanical and other mechanisms for defense exact a biological price, one that is "paid" in the energy, nutrients, and time required to develop and maintain them.

Chemical defenses encompass a wide variety of chemicals manufactured in the body cells of animals or derived from foods and stored in animals' tissues for later use. Snake venoms, bee stings, and toxins in skin glands in amphibians are examples in animals (text p. 1152). Among plants, chemical defenses are extremely common. A passive type of defense is nutrient exclusion—plant tissues that contain such low amounts of nutrients that, for animal predators, they are not worth eating. In the case of cellulose-rich plant parts, such as tree trunks and mature leaves, the parts are indigestible to the vast majority of herbivores.

Plants also produce metabolically a range of secondary compounds that serve as chemical defenses. These include terpenes, steroids, alkaloids, tannins, cyanogenic glycosides, and quinones (text p. 1155). Most often these compounds are produced by short metabolic pathways starting with acetate or amino acids. Among the poisonous alkaloids produced by members of the legume family are caffeine, opium, nicotine, and strychnine. As one might expect, many herbivores have adaptations for evading or metabolizing plant toxins, sometimes using plant chemicals for their own defense (text p. 1156). However, diverse plant communities present herbivores with a battery of chemicals to which no single herbivore is likely to be adapted.

Many animal predators recognize prey on the basis of visual cues. As a result, some prey have evolved warning (aposematic) coloration. In Müllerian mimicry, identical or similar warning colors and patterns may serve different aposematic species (text p. 1158). A related adaptation is Batesian mimicry, in which palatable species mimic the warning coloration of noxious ones.

## Section II    Adaptations for Escape

Escape, in various forms, is another common adaptation for defense. Animal prey may flee or hide, or both. Among certain species, schooling, flocking, and herding are believed to serve as effective forms of "escape." A species can also escape its predators by avoiding them in space or time. Among animals, palatable insects (such as periodic cicadas) that spend much of their life cycle underground are prime examples. Specialized plant life cycles, such as flowering and seeding only at lengthy intervals, are also adaptations that thwart predation (text pp. 1160–1161).

Camouflage (crypsis) is an often seen adaptation for escape. Cryptic coloration hides body outlines and bulk; variations include disruptive coloration and countershading (text p. 1162).

## Section III Symbiosis

Biologists recognize varying degrees of symbiosis, an association of more than one species living together. These divisions reflect the degree of benefit and harm stemming from the association. In parasitism the predator lives in or on its host. From an evolutionary standpoint, the most successful parasites are those that, while clearly doing some harm to the host, do not kill it. Among vertebrates a major defense against parasites is attack by antibodies or T lymphocytes. In response, some parasites release chemicals that suppress the host's immune system; others are adapted in other ways to circumvent immune system defenses (text p. 1164).

In the association known as commensalism, an organism of one species lives in or on another while neither harming nor benefiting its host. In mutualism, typified by the classic example of ants and acacias, the association clearly benefits both partners (text pp. 1165–66).

## KEY TERMS

aposematic coloration  *text page 1158*

Batesian mimicry  *1159*

camouflage  *1161*

coevolution  *1150*

commensalism  *1165*

countershading  *1162*

cryptic coloration  *1161*

disruptive coloration  *1161*

Müllerian mimicry  *1158*

mutualism  *1166*

nutrient exclusion  *1153*

symbiosis  *1163*

warning coloration  *1158*

# SELF-QUIZ: TESTING WHAT YOU HAVE LEARNED

## Matching Key Terms

Match each term on the left with the most appropriate description on the right.

1. Batesian mimicry
2. countershading
3. Müllerian mimicry
4. warning coloration
5. camouflage
6. mutualism
7. coevolution
8. symbiosis
9. cryptic coloration
10. commensalism

a. reciprocal evolution
b. aposematic coloration
c. a barnacle on a whale
d. a relationship that benefits both
e. edible prey that resemble noxious prey
f. conceals body outlines
g. similar color patterns among truly noxious species
h. living together
i. dark-colored on top, light-colored on bottom
j. blend into background

## True or False?

1. _____ The term *coevolution* was coined by Ehrlich and Rauen.

2. _____ Parasitism is an example of symbiosis.

3. _____ Nutrient exclusion is a defense mechanism seen in fish.

4. _____ Tannins are chemical defenses used by insects.

5. _____ Defense chemicals are produced as secondary compounds.

6. _____ Skunks show Müllerian mimicry.

7. _____ Barnacles on whales are an example of parasitism.

8. _____ A skunk's fur markings represent warning coloration.

## Completion

1. _____ is any color or pattern that allows an organism to blend in with its background.

2. _____ are incorporated into the actual physical structure of an organism.

3. An organism that is brightly colored, and also has a potent stinger, is said to be _____ colored.

4. A fish that is dark-colored on the top and light-colored on the bottom has _____.

5. Both species are benefited by the symbiotic relationship of _____.

6. The cardiac glycosides of the monarch butterfly are derived from _____ and not from the butterfly itself.

## Short Answer

1. In what type of community do plants have the most concentrated toxic weapons?

2. What are the advantages of warning coloration?

3. What kinds of behavioral defenses have predators evolved to compensate for the chemical defenses of their prey?

4. How do tannins serve as defenses?

5. What defensive compounds are produced by legumes, and how do they work?

6. Why do plants often have more than one self-defense chemical?

## Multiple-Choice Review

In the following sentences, fill in any blanks. Complete each statement by circling the correct response.

1. Batesian _____ was first suggested by work on:
   a. South Sea fruit flies.
   b. blue jays.
   c. South American insects.
   d. milkweeds.
   e. monarch butterflies.

2. A bug that looks like bird droppings on a leaf is an example of:

   a. warning coloration.
   b. Müllerian mimicry.
   c. simple camouflage.
   d. Batesian mimicry.
   e. aposematic coloration.

3. A facultative relationship is one in which:

   a. the relationship is required for the survival of both species.
   b. the relationship requires thought.
   c. one species needs the relationship in order to survive.
   d. the organisms can survive independently.
   e. None of the above describes a facultative relationship.

4. A mutualistic relationship such as the ant-*Acacia* example could have started out as:

   a. commensalism.
   b. parasitism.
   c. separation in space and time.
   d. aposematic coloration.
   e. a predator–prey relationship.

5. What percentage of _____ drugs contains plant-derived substances?

   a. 10
   b. 25
   c. 50
   d. 60
   e. 70

# 48
# BEHAVIORAL ADAPTATIONS TO THE ENVIRONMENT

## CHAPTER AT A GLANCE

## CHAPTER PREVIEW

This chapter introduces the biology of behavior, beginning with the most simple types of behaviors: the tropisms, taxes, kineses, and reflexes. As you will see, these in turn set the stage for more complex behaviors. Next the text focuses on the once-shadowy world of instinct. Research by the brilliant ethologist Konrad Lorenz and others has revealed some of the basic elements of such "inborn" behavior patterns, including fixed motor patterns (once started they must be fully completed), closed programs that cannot be changed, and open programs that to some extent can be changed by simple experience.

Learning, the modification of behavior by experience, is the next chapter topic. As with more rudimentary behaviors, it, too, occurs at varying levels of complexity. Once again, research by pioneers such as Lorenz, Niko Tinbergen, Ivan Pavlov, and Karl von Frisch has revealed types of learning as diverse as imprinting, latent learning, trial and error, and associative learning (classical conditioning) in animals from bees to geese to dogs.

The subsequent discussion of animal migration and navigation draws on previous material to show how innate and learned behaviors, as well as mechanisms such as a "magnetic sense," may explain the remarkable journeys of birds, caribou, and other animals that travel long distances. A final section places behavior in perspective, pointing out that even those species we consider the most intelligent (such as ourselves) probably have significant components of instinct and reflex among their learned behaviors.

# LEARNING OBJECTIVES

When you have mastered the concepts of this chapter, you will be able to:

1. Provide a concise and functional definition of the term *ethology*, and explain how ethology differs from comparative psychology.
2. List the three broad categories of animal behavior.
3. List the three types of reflexes, and explain why reflexes are considered the building blocks of other, more complex, behaviors.
4. Discuss the nature of instinct, and give an example of a fixed motor pattern.
5. Distinguish between closed and open behavioral programs.
6. Explain how a sign stimulus can trigger a behavioral response, and provide an example.
7. Explain what is meant by programmed learning, and provide an example.
8. Outline the work of Konrad Lorenz, and explain what is meant by imprinting.
9. Discuss the category of latent learning, and explain how song birds are able to reproduce complicated species-specific song patterns.
10. Explain what is meant by the learning mechanisms of trial and error, classical conditioning, and instinct learning.
11. Provide a definition of insight learning, or reasoning, and tell which animals use it.
12. Compare and contrast navigation and migration.

# CONCEPTS IN REVIEW

## *Section I*    Reflex Behavior

Ethologists—scientists who study animals' behavioral interactions with their environment—divide animal behavior into several categories. The most basic category, innate behaviors, includes two subgroups: reflexes and instincts.

Reflexes are automatic, involuntary responses to external stimuli. There are two fundamental types: tropisms and kineses. Animal tropisms—orientations to light, moisture, or some other stimulus—are termed taxes. In a kinesis the intensity of the animal's simple reflexive behavior is proportional to the intensity of the stimulus.

## Section II     Instinct: Inherited Behavioral Program

An instinct is defined as a stereotyped, inherited pattern of behavior that takes place in response to a specific environmental stimulus. Konrad Lorenz performed some of the classic studies of instinct, particularly in the greylag goose. Among the phenomena he described are fixed motor patterns—unvarying, genetically determined series of precise physical movements (text p. 1174). Such stereotyped behavior is readily observed in the courtship of birds.

There are actually two types of fixed motor patterns. The first, termed closed programs, are entirely and permanently set at birth. By contrast, open programs are innate motor patterns in which certain elements can be modified by learning (text pp. 1174–1175). Most instinctive behaviors are triggered by an environmental event. Such an event is called a sign stimulus; this, in turn, sets in motion an innate releasing mechanism that results in a particular behavior (text p. 1176). When a sign stimulus is produced by another member of an animal's own species, it is called a releaser. In some instances animals will preferentially respond to a more intense or exaggerated stimulus—a so-called supernormal stimulus (text p. 1177).

In order for a sign stimulus to exert its effect, an animal must be motivated; that is, it must be in a physiological state that renders it receptive to the stimulus. When an animal is physiologically primed to execute a behavior but does not encounter the appropriate stimulus, it may exhibit appetitive behavior—activity that increases its likelihood of being exposed to the necessary stimulus (text p. 1178). Some animals become increasingly sensitized to a particular stimulus when they are deprived of it, responding to ever cruder approximations of the stimulus as time passes.

The opposite of sensitization is habituation, in which repeated exposure to a sign stimulus lessens an animal's responsiveness to it. Apparently, habituation works by lowering the sensitivity of the nervous system to sign stimuli. It is also thought to be involved in a simple type of learning (text p. 1179).

## Section III     Learning

Learning is defined as behavior that can be modified on the basis of experience, although it can also have reflexive and instinctive components. In some animals, learning may be possible only in what are known as sensitive periods—times, usually early in an animal's life, when permanent behavioral repertoires can be built into the nervous system (text p. 1179). Such periods play a role in two of the simplest types of learning: programmed learning and imprinting.

Programmed learning is initiated by instinct during a sensitive period. An excellent example is the learning of flower types and other information by foraging honeybees (text p. 1180). Imprinting, a term coined by Lorenz, refers to the type of programmed learning in which a young animal learns to recognize its parent (or some other stimulus) during a short-lived sensitive period. Goslings, swans, and other baby birds follow the first object they see that satisfies the crudest sign stimulus requirements of size and pattern of movement (text p. 1182).

More complex than either imprinting or other programmed learning is latent learning, in which there is a delay between the time when a behavior is first learned and when it is used. The learning of complex songs by young birds is a common example (text pp. 1182–1183). Like other simple forms of learning, this type can be definitely linked to hormonal and developmental processes.

There are numerous complex learning patterns. Many species exhibit trial-and-error learning, which is also called feedback learning (text p. 1185). In feedback learning, a behavior is said to be reinforced if it elicits a favorable outcome, and the animal tends to repeat it. Such learning is the basis for operant conditioning, the training method in which an accidental behavior that approximates a desired action is rewarded.

A different kind of learning, known as classical or Pavlovian conditioning (or associative learning), depends on an animal learning to associate receiving a reward with a particular stimulus. Then, the animal evinces the response associated with the reward whenever the stimulus is present (text p. 1185).

The most complex form of learning, insight learning or reasoning, is the ability to use abstractions based on past learning in novel ways, combinations, or situations. It is apparently available only to animals such as humans and other primates that have highly complex nervous systems.

Learned behavior can be transmitted from one member of a population or species to other members of the group. This sort of cultural inheritance is typical of humans and is sometimes seen in other primates and even in birds (text p. 1185).

## *Section IV* Complex Behavior: Navigation and Migration

Navigation and migration are examples of complex learning phenomena. Each is a highly adaptive mix of innate and learned behaviors. Modes of navigation in which learning plays a role include use of solar, stellar, or magnetic compasses, and recognition of physical landmarks. The periodic journeys known as migrations also rely on the learning of solar or stellar cues, a magnetic sense, landmark recognition, and possibly even low-frequency sound vibrations (text p. 1188).

## KEY TERMS

appetitive behavior   *text page 1178*

associative learning   *1185*

classical (Pavlovian) conditioning   *1185*

closed program   *1174*

ethology   *1171*

feedback learning   *1185*

fixed motor pattern   *1174*

habituation   *1178*

imprinting   *1182*

innate   *1171*

innate releasing mechanism   *1176*

insight learning   *1185*

instinct   *1173*

kinesis   *1173*

latent learning   *1182*

learning   *1179*

migration   *1187*

navigation   *1187*

open program   *1174*

operant conditioning   *1185*

programmed learning   *1180*

reasoning   *1185*

reflex   *1172*

reinforcement   *1185*

releaser   *1176*

sensitive period   *1179*

sign stimulus   *1176*

supernormal stimulus   *1177*

taxis   *1172*

trial-and-error learning   *1185*

## SELF-QUIZ: TESTING WHAT YOU HAVE LEARNED

## Matching Key Terms

Match each term on the left with the most appropriate description on the right.

1. habituation
2. releaser
3. reflex
4. imprinting
5. taxis
6. migration

a. instincts
b. animal tropisms
c. unvarying series' of movements
d. sign stimulus
e. "behavioral boredom"
f. Konrad Lorenz

7.  supernormal stimulus    g.  finding your way from point *a* to point *b*
8.  navigation              h.  periodic journey
9.  fixed motor pattern     i.  exaggerated
10. innate                  j.  involuntary response

# True or False?

1. _____ Reflex behaviors are all automatic.

2. _____ Fixed motor patterns are examples of learning.

3. _____ Open program behavior is found among insects.

4. _____ Releasers must always come from members of the same species.

5. _____ Programmed learning is never seen among insects.

6. _____ Latent learning is only slightly more complex than imprinting.

7. _____ Insight learning is found onl· among primates.

8. _____ True navigation requires a map sense.

# Completion

1. Navigation and _____ are examples of complex learning.

2. Attribution of human qualities to nonhuman species is known as _____.

3. Automatic and involuntary responses to an external stimulus are _____.

4. Egg laying and rolling behavior of greylag geese are examples of _____.

5. A more exaggerated stimulus is called a _____ stimulus.

6. The direct opposite of sensitization is _____.

7. In many animals there is a critical time, or _____, during which a particular kind of learning readily takes place.

# Short Answer

1. What causes animals to imprint?

2. What role does reinforcement play in trial-and-error learning?

3. How does insight learning differ from feedback learning?

4. What topics are encompassed by the field of ethology?

5. How does a reflex work?

6. Why is appetitive behavior an important adaptation?

## Multiple-Choice Review

In the following sentences, fill in any blanks. Complete each statement by circling the correct response.

1. Latent learning involves:

   a. a supernormal stimulus.
   b. reinforcement.
   c. delay between stimulus and behavior.
   d. programmed insight.
   e. feedback.

2. _____ conditioning was developed by:

   a. B. F. Skinner.
   b. Peter Marler.
   c. Max Planck.
   d. Konrad Lorenz.
   e. Ivan Pavlov.

3. An animal's response is proportional to the _____ of the stimulus:

    a. in instinct.
    b. in operant conditioning.
    c. in a kinesis.
    d. in programmed learning.
    e. in sensitization.

4. A planarian that orients itself to a current of water shows:

    a. a kinesis.
    b. a taxis.
    c. a reflex.
    d. conditioning.
    e. a fixed motor pattern.

5. An open program is most likely to be found among:

    a. plants.
    b. bacteria.
    c. planaria.
    d. insects.
    e. dogs.

# 49
# SOCIAL BEHAVIOR

---

## CHAPTER AT A GLANCE

Behavior and Ecology
Communication
Mating Behavior
Altruistic Behavior
Insect Societies
Social Systems of Vertebrates
Societies of Mammals

---

## CHAPTER PREVIEW

If you are like most people, you have at least a general idea of the subject matter of animal behavior. Some of the classical studies in this field—Pavlov's work with dogs, Jane Goodall's research on chimps, and so on—have received wide publicity. The entire field of ethology, however, is a very young science compared to most other areas of biology. Much of the work ethologists carry out—especially in studies of social behavior—still involves months and years of collecting data in the field.

Chapter 49 begins with a look at the possible adaptive advantages of sociality, and then focuses on a fundamental aspect of social interaction—communication. Among other examples, this section describes the remarkable "waggle dancing" of honeybees that enables these tiny social insects to communicate complex information about the locations and quality of food sources.

In succeeding sections the chapter surveys mating behavior, including its concomitants territoriality and aggression, and the little-understood phenomena of altruism and kin selection. It then explores in turn animal societies as different as those of insects and, among the vertebrates, wolves and primates.

Many students have difficulty in understanding the concepts of altruism, kin selection, and the like. Simply remember—evolution by natural selection is the cornerstone of all biological phenomena. Although much remains to be learned of why animals behave as they do, the answers almost surely relate to mechanisms that increase the chances that an organism—or its genes—will survive to reproduce.

# LEARNING OBJECTIVES

When you have mastered the concepts of this chapter, you will be able to:

1. Explain why behavior is considered part of an organism's ecological role.
2. Explain the importance of the concepts of r and K selection to behavioral science.
3. Outline the "waggle dance" communication of honeybees, and explain why it is considered a type of language.
4. State the evidence for and against the notion that some species of nonhuman primates have true language capabilities.
5. Define the terms *polygyny, polyandry, sexual dimorphism, threat display,* and *territoriality,* and explain how they relate to mating behavior.
6. Define altruism, and tell why it is such a difficult behavior to explain in terms of natural selection theory.
7. Outline the concept of kin selection, and use it to explain the seemingly altruistic behavior of honeybees.
8. Explain the concept of dominance in vertebrate social structure.
9. Outline the social organization of a wolf pack, and compare it to other vertebrate social behaviors.
10. Compare the social organization found among the nonhuman primates, using gibbons and baboons as examples.
11. List the evidence that many human behaviors have a genetic basis, and may be acted upon by natural selection.

# CONCEPTS IN REVIEW

## *Section I*    Behavior and Ecology

A number of animal species organize themselves into societies; the subfield of biology that studies such social behavior is known as sociobiology. A society is defined as a group of animals of the same species in which the members communicate and interact in cooperative ways.

Social behavior can be adaptive, giving the members of a group a competitive advantage so that they reproduce more successfully than solitary individuals. Thus, it can be shaped by natural selection. Ecological factors and reproductive strategy also affect behavior in major ways (text pp. 1194–1195). For example, the sorts of behavior built around parent–offspring interaction tend not to be seen in r-selected species. By contrast, social behavior, especially parental care of young, is often characteristic of K-selected species. Parental care and survival of young to reproductive age both relate to the concept of inclusive fitness, which is a measure of the likelihood that an individual will contribute to the successful passing on of genes it shares with related animals. Sociality can also be an advantage when group membership offers the strength of numbers.

## Section II    Communication

Social behavior depends on communication—the transmission of information from one organism to another that alters the behavior of that individual or individuals (text p. 1195). This flow of information takes place as stimuli (signals) are perceived by senses. Sometimes the communication channel is visual, as in the mating display of a peacock; in other cases it may be chemical (pheromones), acoustic (human speech, bird calls), or tactile (text p. 1196). Communication stimuli often have highly specific meanings, and research has shown that specific properties of sensory systems and brain circuits are essential components of the specificity of communication (text p. 1197).

Many communicative behaviors originally arose as adaptations that served strictly biological functions. The modification of a functional behavior to include a communicative role is termed ritualization. Examples include courtship patterns of birds, fish, and mammals that are derived from activities related to feeding, cleaning, or nurturing (text pp. 1197–1198).

Many insects communicate through sounds, pheromones, and other chemical substances. One of the most remarkable instances of insect communication is the waggle dancing of honeybees, which employs tactile stimuli to convey information about food sources (text pp. 1198–1199). An important implication of research on honeybee communication is that it appears to involve learning, in spite of the relatively simple makeup of the insect brain.

Biologists refer to waggle dancing as a language because it is symbolic and refers to events or objects distant in time or space, and depends on arbitrary conventions, such as using the sun as a reference point. Even so, there is little scientific basis for attributing true language capabilities to animals other than humans.

## Section III    Mating Behavior

Mating behavior is a major component of inclusive fitness. It tends to be simpler in monogamous species than in polygamous ones, in which the two sexes tend to display sexual dimorphism and competition for mates is often keen (text p. 1200). For both monogamous and polygamous species, however, territoriality and aggression are components of reproductive success. Aggression includes behaviors such as fighting and threat displays; usually it is stylized, avoiding real damage to either party (text pp. 1201–1202).

## Section IV    Altruistic Behavior

Like aggression and territoriality, altruism is a special type of behavior that may influence reproductive success. It is usually defined as self-sacrifice by one member of an animal species that brings benefit to others of the species. At least some altruistic acts, such as a parent endangering or sacrificing itself for its offspring, can be thought of as resulting from "selfish genes" (text p. 1203), since such altruistic behavior toward relatives contributes to inclusive fitness and increases the statistical chances of having genes identical to one's own passed on to subsequent generations. Other altruistic acts may result from kin selection, a poorly understood phenomenon in which cooperative behavior promotes the survival of children and other kin, and, hence, genes like one's own (text p. 1204).

## Section V    Insect Societies

Bees, ants, and other hymenopterans may have quite complex social organizations. Although fascinating, the behavioral repertoires of these societies are highly mechanical, including a great deal of innate, closed-program behavior. Frequent physical contacts, interdependence of members, and division of labor are the hallmarks of such societies, providing benefits such as efficient food production or harvesting and effective group defense against predation (text p. 1206).

## *Section VI*   Social Systems of Vertebrates

Vertebrate societies have evolved primarily because of the advantages conferred by group living. A typical feature of vertebrate social groups is the social hierarchy (or dominance hierarchy) in which members possess ranks with varying responsibilities and privileges. An example is the pecking order of chickens. Often, the dominant individual is a male whose privileges include preferential access to food and mates. Dominance usually also confers the responsibility to act as primary defender of the group (text p. 1207).

## *Section VII*   Societies of Mammals

Mammals display a vast array of social organizations, from the enormous herds of African ungulates to the packs of wolves and troops of baboons. As usual, each of these structures is ecologically adaptive, conferring benefits relating to food acquisition, defense, and other aspects of life. The social organization of wolves is closely meshed with their ecological roles as pursuers of large game. Both economic and behavioral limits apply. Packs grow, divide, and dissolve in relation to leadership changes and as the wolf population fluctuates in response to availability of prey (text p. 1209). Such events reflect the ability to alter behavior as environmental circumstances change.

Primate societies include several types of stable, complex social systems, from small family units to large assemblies organized in multilevel social structures. Hamadryas baboons typify the latter: the largest social unit, the troop, is subdivided into smaller bands, which are the units for food gathering and defense. Bands, in turn, consist of harems made up of a male, several females, and their offspring (text p. 1210). A feature of many primate societies, including that of humans, is the cultural transmission of information and behaviors.

Studies of development in human infants underscores our place in the scheme of primate evolution. Clearly, early behaviors such as supported "walking," breath holding and paddling when submerged, the sucking reflex, and smiling and crying are programmed genetically. Much of what humans do may ultimately be controlled by brain circuits, chemicals such as hormones, and by our genetic heritage. At the same time, it is important to remember that the human brain has an enormous capacity for learning and generating versatile actions in the face of novel circumstances (text p. 1212).

## KEY TERMS

| | | |
|---|---|---|
| aggression   *text page 1202* | kin selection   *1203* | social hierarchy   *1207* |
| altruism   *1203* | pecking order   *1207* | society   *1194* |
| band   *1210* | polyandry   *1200* | sociobiology   *1194* |
| communication   *1195* | polygyny   *1200* | territoriality   *1201* |
| dominance hierarchy   *1207* | ritualization   *1197* | threat display   *1202* |
| harem   *1210* | sexual dimorphism   *1200* | troop   *1210* |
| inclusive fitness   *1194* | | |

# SELF-QUIZ: TESTING WHAT YOU HAVE LEARNED

## Matching Key Terms

Match each term on the left with the most appropriate description on the right.

| | | | |
|---|---|---|---|
| 1. | sociobiology | a. | serves a communicative role |
| 2. | polyandry | b. | E. O. Wilson |
| 3. | polygyny | c. | W. D. Hamilton |
| 4. | altruism | d. | male lions are maned; females are not |
| 5. | kin selection | e. | primate society |
| 6. | sexual dimorphism | f. | self-sacrifice |
| 7. | troop | g. | many wives (or female mates) |
| 8. | pecking order | h. | many husbands (or male mates) |
| 9. | ritualization | i. | chickens |
| 10. | territoriality | j. | definition and defense of a feeding/breeding area |

## True or False?

1. _____ A collie saving a baby chick is an example of altruism.

2. _____ Societies are interacting groups of the same species.

3. _____ The discipline of sociobiology was founded by Charles Darwin.

4. _____ Mating behavior is simplest in polygamous species.

5. _____ Aggression may include posturing.

6. _____ The term *reciprocal altruism* was coined by R. L. Trivers.

7. _____ Pecking order is an example of dominance.

8. _____ E. O. Wilson described the waggle dance of honeybees.

## Completion

1. _____ is the modification of a behavior so that it plays a communicative rather than a strictly biological role.

2. _____ fitness is increased by altruism.

3. Some of the most convincing evidence that many human behaviors are innate comes from the behaviors seen in _____ humans.

4. The basic social unit of the hamadrayas baboon is the _____.

5. Aggressive territoriality behavior is often between _____ of the same species.

6. _____ species are those with one female and many males involved in mating.

7. Species with _____-selected reproductive strategies are most likely to involve complex social behaviors.

8. Lions, which have maned males and nonmaned females, are sexually _____.

## Short Answer

1. What are the advantages and disadvantages of living in a society?

2. How does the concept of inclusive fitness complement altruism?

3. Why don't we see complex parent–offspring social behavior in *r*-selected species?

4. What information is conveyed by the waggle dance of honeybees?

5. What are the "reasons" for (or benefits of) territoriality?

6. Why is kin selection difficult to account for in evolutionary theory?

# Multiple-Choice Review

In the following sentences, fill in any blanks. Complete each statement by circling the correct response.

1. The _____ of chickens is a clear example of:

    a. kin selection.
    b. a dominance hierarchy.
    c. territoriality.
    d. pair bonding.
    e. altruism.

2. The waggle dance of honeybees was first described by:

    a. E. O. Wilson.
    b. South American Indians.
    c. R. L. Trivers.
    d. Charles Darwin.
    e. Karl von Frisch.

3. _____ courts are examples of:

    a. a symbolic territory.
    b. kin selection.
    c. polyandry.
    d. ritualization.
    e. a dominance hierarchy.

4. The theory of reciprocal _____ is the work of:

    a. E. O. Wilson.
    b. Charles Darwin.
    c. R. L. Trivers.
    d. Karl von Frisch.
    e. W. D. Hamilton.

5. The theory of kin selection is the work of:

    a. Charles Darwin.
    b. Karl von Frisch.
    c. R. L. Trivers.
    d. W. D. Hamilton.
    e. E. O. Wilson.

# 50
# HUMAN ORIGINS

---

## CHAPTER AT A GLANCE

---

## CHAPTER PREVIEW

Chapter 50 covers exciting ground—the origins of our own species, *Homo sapiens*. The chapter begins with a survey of the various primates and their geographical distributions—material that is summarized in Table 50-1. You will need to know the primate characteristics, and their distribution among the taxa, in order to understand the transition from apes to hominids described in the next section.

The stage for this transition is set by a discussion of the recently discovered molecular data that provides evidence of the closeness of relationships between humans and the great apes. In part, it is from this evidence that biologists now know that chimps and humans share a more recent common ancestor than do chimps and gorillas.

The next section, on hominid evolution, presents a number of taxa, their associated geological ages, and their physical characteristics. As you study, make use of the phylogenetic chart (Figure 50-13), and the summary in Table 50-2. Chapter 50 closes with the origins and diversification of recent human subspecies, including the remarkably rich cultures of the popularly misunderstood Neandertals and Cro-Magnon peoples.

# LEARNING OBJECTIVES

When you have mastered the concepts of this chapter, you will be able to:

1.  Name the two major groups of living primates, list the anatomical features that distinguish them, and tell where they are found and the major habitats to which they are adapted.
2.  Outline early primate evolution, and set out the time frame (in millions of years ago) during which each stage took place.
3.  List the major characteristics that biologists use to define the order Primates, and name the features that distinguish apes from the hominids.
4.  Discuss the recent biochemical evidence that suggests that humans are more closely related to apes than previously thought.
5.  Discuss the importance of a bipedal gait in separating apes from hominids, and cite evidence for the emergence of bipedalism in human ancestors.
6.  Explain why biologists believe that reproductive behavior has played an important role in human evolution.
7.  List the major genera and species of fossil hominids, telling when and where each group lived.
8.  State the evidence for an African origin of the genus *Homo*.
9.  Describe the Oldowan, Acheulian, and Mousterian industries.
10. Name the biological factors that produced the phenotypic differences among modern human races.

# CONCEPTS IN REVIEW

## *Section I*   The Primates

The evolutionary history of humans and our primate relatives is a fascinating topic. All primates have bodies that display adaptations for an arboreal (tree-dwelling) life style. In the lineage leading to humans, erect bipedalism, the dextrous hand with opposable thumb, and an interplay of biological and cultural evolution probably led to the rapid expansion in size and complexity of the brain.

The primates are divided into prosimians—lemurs, lorises, and tarsiers—and anthropoids—New World monkeys, Old World Monkeys, apes (gibbons, orangutans, gorillas, and chimpanzees), and the human lineage, of which *Homo sapiens* is the only living representative (text pp. 1216–1217). The first primates with prosimian features appeared about 54 million years ago, at the beginning of the Eocene epoch, and the earliest anthropoids appeared in Africa about 30 million years ago. By the beginning of the Miocene, about 26 million years ago, the ancestors of New World monkeys inhabited South America, while those of Old World monkeys, apes, and humans existed in the tropical and subtropical Old World (text p. 1220). Fossils indicate that during the first half of the Miocene in the Old World, a genus of apelike creatures named *Dryopithecus* flourished.

By about 5 million years ago dryopithecines had died out in many areas (according to the fossil record). And between 7 and 5 million years ago there evolved new primate forms, including the early members of the family Hominidae, to which humans belong. All modern primates have body characteristics that reflect the arboreal ecological niche to which primate ancestors were adapted. These characteristics include highly mobile arms, shoulders, and legs adapted to locomotion through trees; thumbs and big toes separate from other digits and adapted for grasping branches; binocular, stereoscopic vision; and other features (text p. 1221).

## Section II    The Transition from Ape to Hominid

To study the evolutionary transition from apes to the hominid lineage, biologists use numerous types of evidence. These include anatomical comparisons, genetics, biochemistry, and behavior. Humans and chimpanzees share many anatomical and biochemical traits, indicating the groups' close evolutionary relationship (text p. 1224). The diagnostic feature judged best for identifying early hominids is evidence of an erect, bipedal stance. Bipedalism is thought to have allowed early hominids to make use of the hands for carrying various types of objects. Hand use in turn is believed to have contributed in major ways to expansion of the cerebral cortex in human evolution (text p. 1225).

## Section III    The Evolution of the Hominids

The oldest individuals clearly of human type probably lived 10 to 5 million years ago, not long after humans and chimpanzees diverged. There are two overlapping periods of human evolution, with the first—from 4 million to 1.3 million years ago—encompassing members of the genus *Australopithecus*. These earliest known humans, including *A. afarensis*, *A. africanus*, and *A. boisei* and *robustus*, were bipedal in stance and gait, but had small brains (text pp. 1226–1228). The genus *Homo* arose about 2 million years ago. It includes *H. habilis*, *H. erectus*, and *H. sapiens*. It is in *H. habilis* that evidence of a major increase in brain size appears. It is also possible that members of this species made the earliest stone tools (text p. 1230).

Homo habilis is credited with two characteristically human activities—toolmaking and butchering of large animals for meat. The crude toolmaking practice of this group has been termed the Oldowan industry (text p. 1231). *H. erectus*, the successor to *H. habilis*, began the spread of humans outside Africa. Along with a spectacular increase in brain size, members of this group also experienced changes in behavior patterns, as evidenced by their relatively sophisticated toolmaking, known as the Acheulian industry (text p. 1232). They probably used fire, and may have had social interactions that included the use of language.

The last 300,000 years of human evolution have been occupied by two groups: archaic *Homo sapiens*, including Neandertals, and a later group who were physically indistinguishable from modern humans. The Neandertals lived from about 130,000 to about 35,000 years ago. The variety and sophistication of their stone tools, known as the Mousterian industry, and other evidence testify to the wide range of learned behaviors typical of this group. The Neandertals built shelters, hunted large game, created body ornaments, and buried their dead (text p. 1234).

## Section IV    The Origin and Diversification of Recent Humans

Early forms of modern humans, sometimes called *Cro-Magnon Man*, appear to have originated in Africa, spreading rapidly across the inhabitable Earth. In many areas they may have coexisted with late Neandertal populations. Like many of their ancestors, these people had the ability to adapt to varied and often demanding environments. They differed physically from Neandertals in that their bodies were less massive overall, and they lacked the prominent Neandertal brow ridge (text p. 1235). Physical changes continued as early modern humans developed agriculture and adopted a less nomadic life style.

Modern humans experienced a number of important behavioral changes. These included an elaboration of technology (toolmaking and tool use), the use of symbolic communication and the development of art, and the building of complicated structures and large communal living sites. These cultural changes took place along with a population explosion and the development of slight genetic differences that gave rise to the subspecies that we today characterize as races.

## KEY TERMS

Acheulian industry *text page 1232*

anthropoid *1217*

ape *1217*

*Australopithecus* *1226*

brachiation *1217*

*Dryopithecus* *1220*

Hominidae *1221*

*Homo erectus* *1232*

*Homo habilis* *1230*

knuckle walking *1223*

Mousterian industry *1234*

Neandertal *1234*

New World monkey *1217*

Old World monkey *1217*

Oldowan industry *1231*

primate *1216*

prosimian *1216*

# SELF-QUIZ: TESTING WHAT YOU HAVE LEARNED

## Matching Key Terms

Match each term on the left with the most appropriate description on the right.

1. brachiation
2. *Dryopithecus*
3. primates
4. prosimian
5. New World monkey
6. Neandertal
7. *Homo erectus*
8. *Homo habilis*
9. Hominidae
10. *Australopithecus*
11. Old World monkey

a. quadrupeds with prehensile tails
b. "before apes"
c. human family
d. the earliest known humans
e. branch-to-branch swinging
f. found in Africa and Asia
g. buried their dead
h. "handyman"
i. first hominid to leave Africa
j. forest ape
k. a mammalian order

## True or False?

1. _____ *Homo erectus* lived about 3 million years ago.

2. _____ *Australopithecus* produced the Mousterian industry.

3. _____ Tarsiers and lemurs are species of New World apes.

4. _____ Primates have existed for roughly 60 million years.

5. _____ Bipedalism requires a large brain.

6. _____ The earliest campfires date to about 500,000 years ago.

7. _____ Humans and chimps share a more recent common ancestor than do humans and gorillas.

## Completion

1. Tool use and butchering of animals for food first appears in the hominid species

   _____ .

2. All New World (South American) monkeys are _____, and live in the trees.

3. The name *Dryopithecus* means _____, which would imply that this species was very good at brachiation.

4. Modern apes are all quadrupeds, and walk on their rear feet and front _____.

5. According to molecular data, the split between chimps and humans occurred about _____ years ago.

6. The earliest known prehumans are members of the genus _____.

## Short Answer

1. What is the evidence that our ancestors were full bipeds by 3.5 million years ago?

2. What are four sexual characteristics that are unique to humans?

3. Where are the *Australopithecus* fossils found, and what was the habitat of these individuals like?

4. What was the Mousterian industry, and what species of fossil human is it associated with?

5. What happened in human evolution between 7 and 2 million years ago?

6. When did humans first spread outside of Africa?

## Multiple-Choice Review

In the following sentences, fill in any blanks. Complete each statement by circling the correct response.

1. The primate suborder Catarrhini contains:

   a. lemurs.
   b. the aye-aye.
   c. hominids.
   d. New World monkeys.
   e. Old World monkeys.

2. The earliest known _____ are represented by fossils of:

   a. *Dryopithecus*.
   b. *Purgatorius*.
   c. Hominidae.
   d. *Siuapithecus*.
   e. *Ramapithecus*.

3. The earliest anthropoids appear in _____ during the:

   a. Oligocene.
   b. Eocene.
   c. Miocene.
   d. Pliocene.
   e. Pleistocene.

4. Fossil remains of *Australopithecus afarensis* have been found:

   a. commonly, since Darwin's time.
   b. only in Europe.
   c. during the last century.
   d. only in the past ten years.
   e. only in Asia.

5. Actual evidence of bipedalism in *Australopithecus,* in the form of fossil footprints, was discovered by:

   a. Donald Johanson.
   b. Charles Darwin.
   c. Mary Leakey.
   d. Raymond Dart.
   e. Tim White.

# 51
# HUMANKIND AND THE FUTURE OF THE BIOSPHERE: AN EPILOGUE

---

## CHAPTER AT A GLANCE

From Hunter-Gatherer to Farmer
The Population Explosion
Food and Agriculture: Feeding the Burgeoning Masses
   *Climate*
   *Land use*
   *Energy*
   *Food distribution*
   *Human diet*
   *The green revolutions*
Special Problems of the Tropics
Biological Science, Social Responsibility, and Our Future

---

## CHAPTER PREVIEW

This chapter, an epilogue, seeks to show why your education in biology is relevant, timely, and possibly even critical to the survival of your world. Although many instructors will not test you on this chapter, you should think of it as one of the most important in the text. The first section proceeds logically from Chapter 50, outlining *Homo sapiens*'s transition from hunters–gatherers to societies that developed agriculture and began extensive domestication of animals. The contributions (and consequences) that may be attributed to agriculture are listed here, and deserve some reflection.

Next the chapter considers the explosive increase in world human population that the agricultural revolution—together with discoveries and advances related to technology, medicine, and public sanitation—made possible. In this section you will trace the historical growth of the human population and study the projections of future growth and the problems your generation will encounter in simply producing enough food to keep pace with that growth into the twenty-first century. As the text suggests, human diets will have to change in major ways if everyone is to have food to eat. As an example of the devastation a burgeoning population can wreak, Chapter 51 then considers the destruction of tropical forests and the growing industrialization of Third World countries.

Even as we write this, people in many parts of the globe are destroying habitats and committing other acts that threaten the future of our Earth—all in the name of "progress." And yet, as the text notes, there is hope. You now know much more than you did about biological principles, and about the delicate balance that is the framework of every ecosystem on our planet. In the end, it is that kind of new understanding that will provide the key to a livable future.

## LEARNING OBJECTIVES

In this epilogue chapter, there are no *Concepts in Review* and *Self-Quiz* sections. Even so, you owe it to yourself and your future to read and think about the material presented. When you have done so, you will be able to:

1. Explain what is meant by the term *nomadic hunter-gatherer*, and describe the way of life it entails.
2. List five changes that resulted from the shift of hunter-gatherer peoples to agriculture.
3. Trace the exponential growth of the human population from about 10,000 years ago to the present day, and explain what is now happening to population doubling times.
4. State, in round figures, the projected human population levels through the end of the twenty-first century.
5. List and discuss the problems that humans must confront in order to support the population growth projected through the twenty-first century.
6. List five factors that determine the amount of food that humans can produce through agriculture.
7. Explain what human dietary changes will take place as the world population grows.
8. Explain what is meant by the terms *first green revolution* and *second green revolution*.
9. Explain what problems a growing human population is creating for the tropical areas of our planet, and what slash-and-burn agriculture may, in turn, mean for the world's climate.

# ANSWERS TO SELF-QUIZ

---

## CHAPTER 1

**Matching Key Terms**

1. f   2. m   3. h   4. d   5. k   6. e   7. g   8. l   9. i   10. j   11. b   12. a
13. c

**True or False?**

1. T   2. T   3. F   4. F   5. T   6. F   7. F   8. T   9. F   10. F

**Completion**

1. cells   2. Jean Baptiste Lamarck   3. hypothesis   4. coevolution   5. natural law
6. deductive reasoning   7. natural selection

**Short Answer**   *Page numbers listed refer to the location of answers in your text*

1. pp. 3–4   2. pp. 8–9   3. p. 13   4. p. 16   5. p. 16   6. pp. 17–18

**Multiple-Choice Review**

1. natural selection, d   2. disprove, b   3. microscope, b   4. water, liquid, a   5. theory, c

---

## CHAPTER 2

**Matching Key Terms**

1. k   2. l   3. n   4. a   5. o   6. d   7. f   8. c   9. m   10. h   11. g   12. b
13. e   14. i   15. j

**True or False?**

1. T   2. F   3. F   4. F   5. T   6. T   7. F   8. T   9. T   10. T

## Completion

1. atomic weights, isotopes    2. positive, no charge, negative    3. polar bonds    4. two kilocalories
5. easily broken, kilocalories, mole    6. heat of vaporization, specific heat    7. tensile strength
8. acid

## Short Answer

1. p.30    2. p. 31    3. pp. 33–34    4. p. 40    5. p. 35    6. pp. 37–39

## Multiple-Choice Review

1. electrons, e    2. valence, b    3. electronegativity, c    4. weak, a    5. atoms, e

# CHAPTER 3

## Matching Key Terms

1. l    2. i    3. e    4. k    5. b    6. d    7. j    8. f    9. a    10. h    11. n    12. o
13. c    14. g    15. m

## True or False?

1. F    2. F    3. T    4. T    5. T    6. F    7. T    8. F    9. F    10. T

## Completion

1. structural isomers    2. monosaccharides, disaccharides, polysaccharides    3. fructose    4. chitin
5. glycerol, fatty acids    6. polar, hydrophobic    7. peptide bonds    8. β-pleated sheet

## Short Answer

1. p. 47    2. p. 48    3. pp. 49–50    4. pp. 53–55    5. p. 56    6. p. 57

## Multiple-Choice Review

1. properties, d    2. triglycerides, b    3. starch, a    4. phosphate, b    5. ester, b

# CHAPTER 4

## Matching Key Terms

1. e    2. i    3. h    4. o    5. b    6. l    7. k    8. a    9. m    10. n    11. g    12. c
13. j    14. f    15. d

## True or False?

1. T    2. F    3. T    4. F    5. T    6. T    7. F    8. F    9. T    10. F

Completion

1. chemical, potential; kinetic    2. coupled    3. anabolism, catabolism    4. competitive inhibition
5. endergonic    6. exergonic    7. catalysts    8. negative feedback

Short Answer

1. p. 76    2. pp. 76–77    3. p. 77    4. p. 80    5. p. 81    6. pp. 83–84

Multiple-Choice Review

1. entropy, b    2. endergonic, b    3. c    4. d    5. biosynthetic, a

# CHAPTER 5

Matching Key Terms

1. j    2. g    3. h    4. b    5. a    6. d    7. e    8. c    9. f    10. i

True or False?

1. T    2. T    3. F    4. T    5. F    6. F    7. T    8. F    9. F    10. T

Completion

1. co-transport    2. cellulose    3. anchorage dependence    4. plasma membrane, glycocalyx
5. tight junctions, gap junctions    6. active transport, concentration gradient

Short Answer

1. p. 99    2. p. 102    3. p. 100    4. p. 105    5. p. 106    6. p. 107    7. pp. 110–112

Multiple-Choice Review

1. c    2. d    3. e    4. a    5. c

# CHAPTER 6

Matching Key Terms

1. g    2. n    3. k    4. i    5. o    6. m    7. f    8. b    9. d    10. e    11. j    12. l
13. a    14. h    15. c

True or False?

1. T    2. F    3. T    4. T    5. T    6. T    7. F    8. T    9. F    10. F

## Completion

1. engulfed    2. Golgi complex, lysosomes    3. lipofuscin granules, aging    4. mitochondria, chloroplasts    5. microfilaments, intermediate filaments, microtubules    6. anchorage
7. chemotactic    8. basal body

## Short Answer

1. p. 125    2. See text, Chapter 6    3. pp. 128–129    4. p. 132    5. p. 132    6. p. 136    7. p. 138    8. pp. 139–141    9. pp. 142–143

## Multiple-Choice Review

1. c    2. d    3. a    4. d    5. ribosomes, b

# CHAPTER 7

## Matching Key Terms

1. e    2. l    3. f    4. k    5. m    6. b    7. o    8. d    9. a    10. h    11. n    12. j
13. i    14. c    15. g

## True or False?

1. F    2. F    3. F    4. T    5. T    6. T    7. F    8. T    9. T    10. F

## Completion

1. glycolysis, fermentation, cellular respiration    2. heat    3. oxidation, reduction    4. $CO_2$
5. oxidative phosphorylation    6. biosynthesis    7. covalent modification    8. 2, 2

## Short Answer

1. p. 153    2. p. 153    3. p. 154    4. pp. 157–159    5. p. 160    6. p. 166    7. p. 165
8. pp. 170–171

## Multiple-Choice Review

1. ATP, c    2. prokaryotes, a    3. d    4. b    5. b

# CHAPTER 8

## Matching Key Terms

1. e    2. n    3. g    4. a    5. o    6. j    7. l    8. d    9. c    10. m    11. h    12. b
13. k    14. i    15. f

## True or False?

1. F    2. T    3. F    4. F    5. T    6. F    7. F    8. T    9. T    10. T

## Completion

1. water, oxidized, carbon dioxide, reduced    2. carotenoid pigments    3. antenna complexes
4. P700, I    5. zigzag    6. $O_2$, $CO_2$    7. bundle-sheath    8. greenhouse effect

## Short Answer

1. p. 176    2. p. 176    3. p. 176    4. p. 178    5. pp. 180–181    6. p. 182    7. p. 184
8. p. 187

## Multiple-Choice Review

1. a    2. c    3. b    4. b    5. d

# CHAPTER 9

## Matching Key Terms

1. m    2. k    3. o    4. j    5. i    6. l    7. f    8. n    9. a    10. b    11. h    12. g
13. c    14. e    15. d

## True or False?

1. F    2. T    3. F    4. T    5. F    6. T    7. T    8. T    9. F    10. F

## Completion

1. karyotype, autosomes, sex chromosomes    2. synthesis, replicated, synthesized    3. microtubule-organizing    4. synapsis    5. 2, diploid    6. 4, haploid    7. kinetochores    8. centromere

## Short Answer

1. p. 201    2. p. 202    3. p. 206    4. p. 203    5. p. 205    6. p. 207    7. pp. 207–208
8. p. 214

## Multiple-Choice Review

1. a    2. d    3. c    4. b    5. c

# CHAPTER 10

## Matching Key Terms

1. d    2. o    3. e    4. i    5. c    6. b    7. l    8. m    9. g    10. a    11. f    12. n
13. h    14. k    15. j

## True or False?

1. F    2. T    3. T    4. F    5. T    6. F    7. T    8. T    9. T    10. F

## Completion

1. incomplete dominance    2. T. H. Morgan    3. dihybrid crosses    4. Hugo de Vries
5. Punnett square    6. genes    7. self-fertilizing    8. vitalism

## Short Answer

1. p. 219    2. pp. 225, 228    3. pp. 223–225    4. p. 228    5. p. 229    6. p. 230

## Multiple-Choice Review

1. pangenesis, a    2. blending, b    3. c    4. d    5. e

---

# CHAPTER 11

## Matching Key Terms

1. e    2. h    3. g    4. k    5. b    6. i    7. d    8. j    9. a    10. f    11. c

## True or False?

1. T    2. T    3. F    4. F    5. T    6. T    7. T    8. F    9. T    10. F

## Completion

1. wild-type allele, +    2. relative distances    3. linear    4. linkage    5. locus    6. recombinant
7. recombination analysis    8. functional, structural

## Short Answer

1. p. 239    2. p. 243    3. p. 245    4. p. 247    5. p. 249    6. p. 250    7. p. 253

## Multiple-Choice Review

1. c    2. d    3. b    4. a    5. e

---

# CHAPTER 12

## Matching Key Terms

1. g    2. c    3. h    4. j    5. f    6. b    7. i    8. a    9. d    10. e

## True or False?

1. T    2. T    3. F    4. F    5. T    6. T    7. F    8. T    9. T    10. F

## Completion

1. genes, enzymes    2. electrophoresis    3. P. A. Levene    4. nucleoside    5. Chargaff
6. template    7. semiconservative    8. polynucleotide ligase

## Short Answer

1. pp. 267–268    2. pp. 272–273    3. p. 273    4. p. 275    5. pp. 276–277    6. p. 279

## Multiple-Choice Review

1. a    2. a    3. b    4. d    5. c

# CHAPTER 13

## Matching Key Terms

1. j    2. l    3. a    4. k    5. f    6. d    7. o    8. m    9. n    10. b    11. e    12. i
13. g    14. c    15. h

## True or False?

1. T    2. F    3. F    4. T    5. T    6. F    7. T    8. T    9. T    10. F    11. F

## Completion

1. nonoverlapping    2. nondegenerate    3. transcription    4. clover leaf    5. elongation
6. peptidyl transferase    7. polypeptide, structural    8. cytoplasm

## Short Answer

1. p. 300    2. p. 288    3. p. 295    4. p. 295    5. pp. 290–295
6 pp. 290–292

## Multiple-Choice Review

1. b    2. d    3. c    4. a    5. a

# CHAPTER 14

## Matching Key Terms

1. m    2. e    3. l    4. f    5. i    6. j    7. n    8. k    9. c    10. o    11. g    12. b
13. h    14. a    15. d

## True or False?

1. T    2. F    3. T    4. T    5. T    6. F    7. F    8. T    9. T    10. T    11. F

## Completion

1. transmission genetics    2. auxotrophs    3. conjugation    4. recombination    5. insertion sequences    6. operon    7. multiple, families    8. exons

## Short Answer

1. p. 304    2. p. 305    3. p. 304    4. p. 307–310    5. pp. 317–318    6. p. 321

## Multiple-Choice Review

1. c    2. b    3. a    4. d    5. c

---

# CHAPTER 15

## Matching Key Terms

1. m    2. j    3. d    4. c    5. l    6. i    7. o    8. b    9. a    10. n    11. g    12. f
13. e    14. h    15. k

## True or False?

1. T    2. F    3. T    4. T    5. F    6. F    7. T    8. F    9. T    10. T    11. F

## Completion

1. pedigrees    2. metaphase    3. trisomy 21    4. heterokaryons    5. polygenic    6. 45
7. male    8. Barr bodies

## Short Answer

1. p. 341    2. p. 334    3. p. 346    4. p. 348    5. p. 351    6. p. 354

## Multiple-Choice Review

1. b    2. d    3. c    4. d    5. a

---

# CHAPTER 16

## Matching Key Terms

1. i    2. l    3. o    4. f    5. h    6. j    7. n    8. c    9. e    10. a    11. k    12. g
13. m    14. d    15. b

## True or False?

1. T    2. F    3. T    4. T    5. F    6. T    7. F    8. T    9. T    10. T    11. F

## Completion

1. Sertoli    2. Golgi complex    3. albumen    4. jelly coat    5. fertilization membrane
6. morula    7. morphogenesis    8. neurulation

## Short Answer

1. p. 377    2. p. 397    3. p. 380    4. p. 361    5. p. 373    6. p. 366

## Multiple-Choice Review

1. c    2. a    3. c    4. d    5. b

# CHAPTER 17

## Matching Key Terms

1. g    2. c    3. n    4. m    5. e    6. d    7. l    8. k    9. b    10. a    11. o    12. i
13. f    14. h    15. j

## True or False?

1. F    2. F    3. T    4. T    5. F    6. F    7. T    8. T    9. F    10. T    11. T

## Completion

1. tissue interaction    2. proteins    3. division    4. collagen    5. regulatory    6. arteries
7. leukemias, lymphomas    8. cytoplasm

## Short Answer

1. p. 389    2. p. 391    3. p. 399    4. p. 403    5. pp. 401–402    6. p. 399

## Multiple-Choice Review

1. d    2. d    3. c    4. a    5. b

# CHAPTER 18

## Matching Key Terms

1. k    2. l    3. b    4. m    5. o    6. e    7. c    8. a    9. g    10. d    11. f    12. h
13. j    14. i    15. n

## True or False?

1. T    2. T    3. F    4. F    5. F    6. T    7. F    8. T    9. F    10. F    11. F

## Completion

1. cavernosa, spongiosum    2. corpus luteum    3. Bartholin's    4. indifferent    5. Mullerian duct
6. Wolffian duct    7. implantation    8. afterbirth

## Short Answer

1. p. 411    2. p. 425    3. p. 427    4. p. 424    5. pp. 419–420    6. p. 419

## Multiple-Choice Review
1. a    2. d    3. a    4. c
5. vasectomy, c

# CHAPTER 19

## Matching Key Terms

1. c    2. m    3. g    4. l    5. k    6. o    7. f    8. a    9. e    10. h    11. n    12. i
13. j    14. b    15. d

## True or False?

1. F    2. F    3. F    4. T    5. F    6. F    7. F    8. F    9. T    10. T    11. T

## Completion

1. naked genes    2. endosymbiosis    3. Paleozoic, Mesozoic, Cenozoic    4. Gondwana, Laurasia
5. genus, species    6. kingdom    7. reproductive    8. monophyletic

## Short Answer

1. p. 450    2. p. 451    3. p. 440    4. p. 444    5. pp. 438–439    6. p. 450

## Multiple-Choice Review

1. c    2. d    3. life, b    4. clays, a    5. a

# CHAPTER 20

## Matching Key Terms

1. n    2. e    3. o    4. i    5. g    6. d    7. k    8. l    9. b    10. j    11. h    12. c
13. a    14. m    15. f

## True or False?

1. T    2. T    3. T    4. F    5. T    6. F    7. T    8. T    9. T    10. T    11. F

## Completion

1. mycoplasmas, wall, DNA    2. capsule    3. transformation, transduction, conjugation
4. chemoautotrophs    5. spirochetes    6. lysogeny    7. oncogenes    8. binary fission

## Short Answer

1. p. 466    2. p. 475    3. p. 462    4. p. 463    5. p. 459    6. p. 467

## Multiple-Choice Review

1. b    2. c    3. a    4. c    5. d

# CHAPTER 21

## Matching Key Terms

1. l    2. f    3. b    4. a    5. k    6. o    7. i    8. m
9. d    10. n    11. g    12. c    13. e    14. j    15. h

## True or False?

1. F    2. T    3. F    4. F    5. T    6. F    7. F    8. F    9. T    10. T    11. F

## Completion

1. pellicle    2. paramylum    3. trichocysts    4. fucoxanthin    5. myxomycetes, acrasiomycetes
6. tsetse fly    7. foraminiferans, radiolarians    8. ciliates

## Short Answer

1. pp. 485–486    2. p. 487    3. p. 488    4. pp. 491–492    5. p. 494    6. p. 497

## Multiple-Choice Review

1. b    2. animal-like, c    3. cell wall, d    4. a    5. b

# CHAPTER 22

## Matching Key Terms

1. h    2. n    3. d    4. f    5. o    6. k    7. i    8. l
9. m    10. c    11. e    12. j    13. a    14. g    15. b

## True or False?

1. T    2. F    3. T    4. F    5. F    6. T    7. T    8. F    9. F    10. F    11. T

## Completion

1. histones    2. Oomycetes    3. mycorrhizae    4. 80    5. asexual budding    6. basidiocarp
7. pollution    8. chitin, cellulose

## Short Answer

1. p. 508    2. p. 509    3. p. 523    4. pp. 510, 518    5. p. 519

## Multiple-Choice Review

1. variation, c    2. b    3. d    4. lichen, e    5. a

# CHAPTER 23

## Matching Key Terms

1. j    2. h    3. d    4. a    5. n    6. o    7. k    8. e
9. i    10. m    11. f    12. b    13. g    14. c    15. l

## True or False?

1. F    2. T    3. T    4. T    5. T    6. T    7. F    8. T    9. F    10. T    11. T

## Completion

1. alternation, generations    2. gametophyte    3. algae    4. primary    5. Chlorophyta
6. phycobilins    7. Devonian    8. Bryophyta

## Short Answer

1. p. 529    2. p. 531    3. p. 537    4. p. 540    5. p. 542    6. p. 540

## Multiple-Choice Review

1. b    2. pores, e    3. e    4. c    5. a

# CHAPTER 24

## Matching Key Terms

1. l    2. f    3. h    4. b    5. g    6. a    7. m    8. n
9. o    10. d    11. e    12. i    13. j    14. k    15. c

## True or False?

1. T    2. F    3. T    4. F    5. F    6. T    7. F    8. T    9. T    10. T    11. F

## Completion

1. integuments    2. paleobotanists    3. sporophyte    4. deciduous    5. fertilization    6. ovary wall    7. fertilization    8. monoecious

## Short Answer

1. p. 550    2. pp. 559–562    3. pp. 550–551    4. p. 556    5. p. 558    6. p. 556

## Multiple-Choice Review

1. c    2. embryo, e    3. a    4. conifer, e    5. b

# CHAPTER 25

## Matching Key Terms

1. l    2. j    3. d    4. f    5. m    6. c    7. e    8. n
9. o    10. k    11. g    12. h    13. a    14. i    15. b

## True or False?

1. T    2. F    3. T    4. T    5. F    6. F    7. F    8. F    9. T    10. T    11. F

## Completion

1. protostomes, deuterostomes    2. spicules    3. Hydrozoa, Scyphozoa, Anthozoa
4. unidirectional, cephalization    5. pseudocoelom    6. convergent evolution    7. Hemichordata
8. ribbonworms    9. segmentation

## Short Answer

1. p. 587    2. p. 604    3. p. 602    4. p. 591    5. p. 584    6. p. 576    7. p. 569

## Multiple-Choice Review

1. a    2. c    3. e    4. e    5. a

# CHAPTER 26

## Matching Key Terms

1. d    2. k    3. h    4. a    5. f    6. l    7. i    8. o
9. m    10. n    11. g    12. c    13. j    14. e    15. b

## True or False?

1. T    2. F    3. F    4. T    5. F    6. F    7. T    8. F    9. F    10. T    11. F
12. T

## Completion

1. thyroid    2. Ordovician    3. swim bladder    4. Ichthyostega    5. cleidoic    6. marsupial or metatherian    7. Squamata

## Short Answer

1. p. 622    2. p. 619    3. p. 630    4. p. 619    5. pp. 631–632    6. p. 633

## Multiple-Choice Review

1. rhipidistian; c    2. Cephalochordata; a, b, c, d    3. sea squirts; b, d, e    4. ostracoderms; b, c
5. Acanthodii; c

# CHAPTER 27

## Matching Key Terms

1. f    2. i    3. d    4. b    5. j    6. h    7. c    8. e    9. g    10. a

## True or False?

1. F    2. F    3. T    4. T    5. F    6. T    7. T    8. T

## Completion

1. Cretaceous    2. rod, epidermis    3. pith    4. pericycle    5. prop roots    6. ground parenchyma

## Short Answer

1. p. 645    2. p. 648    3. p. 647    4. pp. 655–658    5. p. 643    6. p. 651

## Multiple-Choice Review

1. storage, d    2. b    3. d    4. a    5. c

# CHAPTER 28

## Matching Key Terms

1. e    2. j    3. c    4. i    5. h    6. b    7. d    8. g    9. f    10. a

## True or False?

1. T    2. T    3. F    4. T    5. F    6. F    7. T    8. T

## Completion

1. vegetative reproduction     2. pollination, fertilization     3. microgametogenesis     4. micropyle
5. endosperm     6. ovary

## Short Answer

1. p. 678     2. p. 680     3. pp. 678, 682     4. p. 676     5. p. 687     6. p. 688

## Multiple-Choice Review

1. reproduction, c     2. a     3. pollen tube, b     4. d     5. e     6. a

# CHAPTER 29

## Matching Key Terms

1. e     2. g     3. i     4. h     5. f     6. j     7. a     8. b     9. c     10. d

## True or False?

1. T     2. T     3. F     4. F     5. F     6. T     7. F     8. F

## Completion

1. solute potential     2. hypertonic     3. zero     4. tracheids, vessel elements     5. guard     6. 30
7. turgor pressure     8. macronutrients

## Short Answer

1. pp. 702–703     2. p. 704     3. p. 707     4. p. 707     5. p. 696     6. p. 696

## Multiple-Choice Review

1. d     2. d     3. water, e     4. a     5. abscisic, b

# CHAPTER 30

## Matching Key Terms

1. h     2. g     3. i     4. j     5. a     6. b     7. e     8. d     9. f     10. c

## True or False?

1. T     2. T     3. T     4. F     5. F     6. F     7. T     8. T

## Completion

1. gravitropism, gibberellins    2. auxins, gibberellins, cytokinins    3. vivipary    4. ethylene
5. photoperiod, flower    6. $P_r$; $P_{fr}$

## Short Answer

1. p. 725    2. p. 726    3. p. 712    4. p. 712    5. p. 720    6. p. 723

## Multiple-Choice Review

1. a    2. e    3. c    4. downward, b    5. e

# CHAPTER 31

## Matching Key Terms

1. g    2. f    3. h    4. e    5. b    6. c    7. a    8. j    9. d    10. i

## True or False?

1. T    2. F    3. F    4. F    5. T    6. F    7. T    8. T

## Completion

1. bulk flow    2. peristaltic    3. sphincter    4. plasma    5. Starling's law    6. endothelial

## Short Answer

1. p. 749    2. p. 749    3. p. 735    4. p. 736    5. pp. 753–754    6. p. 742

## Multiple-Choice Review

1. c    2. d    3. e    4. c    5. b

# CHAPTER 32

## Matching Key Terms

1. f    2. g    3. b    4. i    5. c    6. e    7. h    8. j    9. a    10. d

## True or False?

1. T    2. F    3. T    4. T    5. T    6. F    7. F    8. T

## Completion

1. T cells, B cells     2. disulfide bonds     3. Burnet, Jerne     4. killer, helper, suppressor T cells
5. active immunity     6. antihistamine, histamine

## Short Answer

1. pp. 761–762     2. p. 762     3. p. 765     4. pp. 766–767     5. p. 770     6. pp. 770–771

## Multiple-Choice Review

1. d     2. digest, c     3. enzymatic reactions, a     4. b     5. b

# CHAPTER 33

## Matching Key Terms

1. a     2. d     3. i     4. f     5. c     6. e     7. h     8. j     9. g     10. b

## True or False?

1. T     2. F     3. T     4. F     5. T     6. T     7. T     8. F

## Completion

1. watery, innermost     2. surfactant     3. diaphragm, intercostal     4. myoglobin     5. respiratory center     6. mechanoreceptor reflex

## Short Answer

1. p. 787     2. p. 794     3. p. 797     4. pp. 797–798     5. p. 801     6. p. 803

## Multiple-Choice Review

1. a     2. e     3. c     4. b     5. birds, d

# CHAPTER 34

## Matching Key Terms

1. j     2. d     3. f     4. h     5. e     6. g     7. c     8. i     9. b     10. a

## True or False?

1. F     2. T     3. T     4. F     5. T     6. F     7. T     8. F

## Completion

1. villi    2. incisors, canines, premolars, molars    3. peristalsis    4. omnivorous    5. small intestine    6. somatomedin

## Short Answer

1. p. 834    2. p. 836    3. pp. 816; 820–827    4. p. 820    5. p. 826    6. p. 827

## Multiple-Choice Review

1. d    2. b    3. e    4. b    5. a

# CHAPTER 35

## Matching Key Terms

1. e    2. a    3. h    4. i    5. j    6. f    7. b    8. d    9. g    10. c

## True or False?

1. T    2. T    3. T    4. F    5. T    6. T    7. F    8. F

## Completion

1. rectal    2. uric acid    3. contractile vacuole    4. Malpighian tubules    5. mesonephric kidneys    6. vasopressin

## Short Answer

1. p. 858    2. p. 859    3. p. 859    4. pp. 860–862    5. p. 866    6. p. 867–868

## Multiple-Choice Review

1. surface, e    2. c    3. aquatic, a    4. b    5. collecting, d

# CHAPTER 36

## Matching Key Terms

1. d    2. e    3. j    4. f    5. g    6. h    7. i    8. a    9. c    10. b

## True or False?

1. T    2. F    3. F    4. T    5. T    6. T    7. F    8. F

## Completion

1. dendrites     2. myelin     3. Mauthner cells     4. saltatory propagation     5. summation
6. interneurons

## Short Answer

1. p. 890     2. p. 892     3. p. 891     4. pp. 881–882     5. p. 881     6. pp. 879–880

## Multiple-Choice Review

1. d     2. b     3. c     4. e     5. spinal cord, b

# CHAPTER 37

## Matching Key Terms

1. a     2. i     3. b     4. f     5. d     6. g     7. c     8. h     9. j     10. e

## True or False?

1. T     2. F     3. T     4. T     5. F     6. F     7. F     8. F

## Completion

1. exocrine, endocrine     2. corpora allata     3. islets of Langerhans     4. somatostatin     5. neotony
6. dilate, constrict

## Short Answer

1. pp. 912–913     2. pp. 913–914     3. pp. 914–915     4. p. 920     5. pp. 903–904     6. pp. 905–906

## Multiple-Choice Review

1. pupa, e     2. b     3. oxidation, c     4. d     5. a

# CHAPTER 38

## Matching Key Terms

1. c     2. j     3. b     4. f     5. a     6. g     7. i     8. d     9. h     10. e

## True or False?

1. T     2. F     3. T     4. F     5. T     6. T     7. T     8. T

## Completion

1. baseline level   2. lateral line   3. endolymph   4. incus, malleus, stapes   5. rods, cones
6. olfactory epithelium

## Short Answer

1. p. 936   2. pp. 952–953   3. p. 943   4. pp. 946–947   5. p. 949   6. p. 942

## Multiple-Choice Review

1. e   2. b   3. a   4. middle, c   5. d

# CHAPTER 39

## Matching Key Terms

1. f   2. i   3. d   4. h   5. j   6. g   7. a   8. c   9. b   10. e

## True or False?

1. T   2. T   3. T   4. F   5. F   6. F   7. T   8. T

## Completion

1. actin   2. hydroskeleton   3. intercalated disks   4. striated   5. collagen, polysaccharides
6. compact

## Short Answer

1. pp. 960, 964   2. p. 958   3. pp. 968–972   4. pp. 966–978   5. p. 977   6. p. 975

## Multiple-Choice Review

1. Haversian, c   2. e   3. mitochondria, b   4. a   5. e

# CHAPTER 40

## Matching Key Terms

1. e   2. h   3. g   4. a   5. b   6. c   7. i   8. d   9. j   10. f

## True or False?

1. F   2. T   3. F   4. T   5. T   6. T   7. T   8. F

## Completion

1. metencephalon, myelencephalon    2. mesencephalon    3. meninges    4. brain stem
5. cerebrum    6. prosopagnosia

## Short Answer

1. p. 1002    2. p. 988    3. p. 993    4. p. 992    5. p. 995    6. p. 997

## Multiple-Choice Review

1. memory, b    2. e    3. speech, c    4. a    5. d

# CHAPTER 41

## Matching Key Terms

1. f    2. e    3. j    4. b    5. c    6. i    7. d    8. g    9. a    10. h

## True or False?

1. T    2. T    3. F    4. F    5. F    6. T    7. T    8. T

## Completion

1. genetic drift    2. gene flow    3. bottleneck    4. isozymes    5. natural selection    6. average heterozygosity (or percent polymorphism, or genetic variability)

## Short Answer

1. pp. 1016-1018    2. p. 1012    3. p. 1020    4. pp. 1020–1021    5. p. 1024    6. p. 1027

## Multiple-Choice Review

1. evolution, e    2. b    3. population, a    4. c    5. e

# CHAPTER 42

## Matching Key Terms

1. a    2. c    3. j    4. h    5. f    6. e    7. b    8. i    9. g    10. d

## True or False?

1. F    2. T    3. F    4. T    5. T    6. F    7. T    8. T

## Completion

1. fixed    2. Bergmann's rule    3. directional    4. group selection    5. diversifying
6. frequency-dependent

## Short Answer

1. p. 1032
2. p. 1034    3. p. 1039    4. p. 1041    5. pp. 1044–1045

## Multiple-Choice Review

1. c    2. genes, b    3. e    4. vigor, b    5. fitness, c

# CHAPTER 43

## Matching Key Terms

1. h    2. g    3. d    4. j    5. i    6. e    7. a    8. f    9. b    10. c

## True or False?

1. F    2. T    3. T    4. F    5. T    6. T    7. T    8. T

## Completion

1. convergent evolution    2. parallel evolution    3. analogous    4. homologous    5. extinction
6. punctuated equilibrium

## Short Answer

1. p. 1052    2. p. 1065    3. p. 1059    4. pp. 1073–1074    5. pp. 1064; 1073    6. pp. 1070–1071

## Multiple-Choice Review

1. b    2. a    3. gene pool, e    4. isolating mechanism, b    5. c

# CHAPTER 44

## Matching Key Terms

1. d    2. e    3. f    4. j    5. a    6. h    7. g    8. b    9. c    10. i

## True or False?

1. F    2. F    3. T    4. T    5. T    6. T    7. F    8. F

Completion

1. littoral    2. solar constant    3. photic    4. biological magnification    5. eutrophication
6. bacteria, fungi

Short Answer

1. p. 1079    2. p. 1080    3. p. 1082    4. pp. 1086–1087    5. p. 1087    6. pp. 1091–1092

Multiple-Choice Review

1. d    2. energy, a    3. layering, b    4. e    5. c

# CHAPTER 45

Matching Key Terms

1. d    2. j    3. i    4. f    5. c    6. b    7. a    8. g    9. h    10. e

True or False?

1. F    2. T    3. F    4. T    5. F    6. F    7. T    8. F

Completion

1. species composition    2. climax    3. species richness    4. niche, address    5. physiognomy
6. MacArthur and Wilson

Short Answer

1. p. 1120    2. pp. 1120-1121    3. p. 1110    4. p. 1121    5. p. 1124    6. p. 1124

Multiple-Choice Review

1. c    2. a    3. d    4. e    5. c

# CHAPTER 46

Matching Key Terms

1. b    2. e    3. d    4. g    5. h    6. i    7. c    8. j    9. a    10. f

True or False?

1. T    2. T    3. F    4. F    5. F    6. T    7. T    8. T

## Completion

1. density    2. survivorship curve    3. intraspecific    4. reproductive time lag    5. *K*
6. uniformly spaced

## Short Answer

1. p. 1136    2. p. 1135    3. p. 1137    4. p. 1138    5. p. 1139    6. p. 1131

## Multiple-Choice Review

1. exponential, b    2. d    3. limited, a    4. c    5. e

# CHAPTER 47

## Matching Key Terms

1. e    2. i    3. g    4. b    5. j    6. d    7. a    8. h    9. f    10. c

## True or False?

1. T    2. T    3. F    4. F    5. T    6. F    7. F    8. T

## Completion

1. camouflage    2. mechanical defenses    3. aposematically    4. countershading    5. mutualism
6. milkweed

## Short Answer

1. p. 1157    2. pp. 1158–1159    3. p. 1156    4. p. 1153    5. p. 1156    6. p. 1156

## Multiple-Choice Review

1. mimicry, c    2. c    3. d    4. e    5. prescription, c

# CHAPTER 48

## Matching Key Terms

1. e    2. d    3. j    4. f    5. b    6. h    7. i    8. g    9. c    10. a

## True or False?

1. T    2. F    3. F    4. F    5. F    6. T    7. T    8. T

## Completion

1. migration    2. anthropomorphism    3. reflexes    4. fixed motor patterns    5. supernormal
6. habituation    7. sensitive period

## Short Answer

1. p. 1182    2. p. 1185    3. p. 1185    4. p. 1171    5. p. 1172    6. p. 1178

## Multiple-Choice Review

1. c    2. operant, a    3. intensity, c    4. b    5. e

# CHAPTER 49

## Matching Key Terms

1. b    2. h    3. g    4. f    5. c    6. d    7. e    8. i    9. a    10. j

## True or False?

1. F    2. T    3. F    4. F    5. T    6. T    7. T    8. F

## Completion

1. ritualization    2. inclusive    3. newborn    4. harem    5. males    6. polyandrous    7. *K*
8. dimorphic

## Short Answer

1. p. 1194    2. pp. 1203–1204    3. p. 1195    4. p. 1198    5. p. 1201    6. p. 1205

## Multiple-Choice Review

1. pecking order, b    2. e    3. display, a    4. altruism, c    5. d

# CHAPTER 50

## Matching Key Terms

1. e    2. j    3. k    4. b    5. a    6. g    7. i    8. h    9. c    10. d    11. f

## True or False?

1. F    2. F    3. F    4. T    5. F    6. T    7. T

## Completion

1. *Homo habilis*    2. arboreal    3. forest ape    4. knuckles    5. 6 million    6. *Australopithecus*

## Short Answer

1. pp. 1227–1228    2. pp. 1225–1226    3. pp. 1226–1227    4. p. 1234    5. pp. 1226–1227
6. p. 1232

## Multiple-Choice Review

1. e    2. primates, b    3. Africa, a    4. d    5. c

# APPENDIX

---

## LATIN ROOTS FOR BIOLOGICAL TERMS

**a-** not, without
**ab** away from
**acanth** thorn
**acr** point, end, top
**actin** ray, radial
**ad-** to, at, toward
**aden** gland
**allo** apart, different
**ambi** both
**amphi** on both sides; of both kinds
**an** not, without
**andr** male
**angi** enclosed
**anthrop** human
**anti** not, without
**ap** without; from
**arch** first
**arthr** jointed
**asco** sac
**astro, aster** star
**audio** hearing
**auto** self

**benth** bottom
**beta** second
**bi** two
**bio** life
**blasto, blast** sprout; germination; embryo
**brady** slow
**branchi** gills
**bronch** lungs
**bursa** pouch, sac

**caecum** blind
**cardio** heart
**carp** fruit

**caryo** nucleus
**cata** down
**caudad** tail
**cava** hollow
**cent** hundred
**cephalo, ceph** head
**cerebr** brain
**cheli** claw
**chlor** green
**choano** mouth; collar
**chondrio, chondro** cartilage
**chromo, chrome** color
**chron** time
**cide** killing
**circum** around
**cis** same side
**clad** branch
**cloaca** sewer
**cnido** threat
**co, con** together
**coel** cavity
**contra** against
**crypto** hidden
**cyan** blue
**cyst** pouch; bladder
**cyto, cyte** cell

**dactyl** finger; digit
**deca** ten
**dendr, dendron** tree
**dent** teeth
**derm** skin
**dextr** right
**di** two
**dia** through, across
**dino** to whirl; terrible
**diplo** double
**dont** tooth

**duct** leading, drawing off
**duo** two
**dys** bad

**echino** spiny
**ecto** outer
**en** inner
**endo** inner
**entero** intestine
**epi** top
**erythro** red
**eu** true; good
**eury** wide, broad
**ex, extr** outer, away from

**ferre** bearing
**flagellum** whip
**flav** yellow
**folium** leaf

**gamo** joined, united
**gastr** gut, stomach
**genesis** origin; development
**geo** earth
**germ** bud; sprout
**globus** sphere
**gloea** glutinous substance; glue
**gluco, glyco** sugar
**gnath** jaw
**grade** walking, moving
**gram, graph** written; drawn
**gymn** naked
**gynec, gyne** woman

**halo** salt; circle
**hapl** single
**helminth** worm
**hemi** half

**hem**  blood
**hepat**  liver
**hetero**  other; different
**hex**  six
**hist**  tissue
**holo**  entire, whole
**hom, homeo**  same
**Homo**  man
**hyal**  glassy; transparent
**hydro**  water
**hygro**  wet; moist
**hyper**  over; more; higher
**hypo**  under; less; lower

**icthy**  fish
**-id**  state of, condition
**infra**  below
**inter**  between
**intra**  within
**iris**  rainbow
**is-**  same; equal
**itis**  inflammation

**karyo**  nucleus
**kata**  down
**kine, kines**  movement

**labium**  lip
**lact**  milk
**lepido**  scaly
**leuc, leuk**  colorless; white
**loc**  place, room
**luteo**  yellow
**lysis**  freeing; loosening;
 dissolving

**macro**  large
**manu**  hand
**medulla**  marrow
**mega**  large
**meiosis**  reduction
**melan**  black
**mero**  part
**merous**  having parts
**meso**  middle; between
**meta**  after, beyond; among
**micro**  small
**milli**  1,000
**mon**  one
**morph**  form
**mort**  death
**mult**  many
**myc**  fungus
**myel**  marrow; spine
**myo**  muscle
**myxo**  slime; mucus

**narc**  sleep; stupor
**naso**  nose

**nemat**  thread
**neo**  new
**nephr**  kidney
**neur**  nerves
**nodus**  knot
**nomen**  name
**noto**  back

**ob**  toward, for
**oct**  eight
**ocul**  eye
**olfac**  smelling
**olig**  few
**ology**  study
**oo**  egg
**operculum**  lid, cover
**optic**  of the eye
**ornith**  bird
**orth**  straight
**osmo**  impulse, thrust
**oste**  bone
**ostium**  door
**ovi**  egg

**pachy**  thick
**pan**  all
**para**  side by side
**pari**  equal
**parous**  producing
**parthenos**  virgin
**ped**  foot
**pent**  five
**peri**  around
**phagos**  to eat
**phero**  to bear
**philic**  loving
**phobic**  fearing
**phore**  bearing
**photo**  light
**phyco**  algae
**phyll**  leaf
**phylo**  tribe
**physis**  growth
**phyte**  plant
**placo**  flat plate
**plasm, plast**  something molded;
 formative material
**platy**  flat
**ploid**  having chromosomes
**pod**  foot; having a foot
**poly**  many
**post**  after
**posterior**  back, hind part
**pre, pro, proto**  before; first
**pseud**  false
**psilo**  bare
**pterido**  fern
**ptero**  wing
**pyr**  fire

**quad**  four

**rachis**  spine; ridge
**radicula**  small root
**ramous**  branching
**re**  again
**rectus**  straight
**reticulum**  small net
**retro**  back; again
**rhabdo**  rod
**rhage, rhea**  flow
**rhizo**  root
**rhod**  red

**sapr**  rotten, decaying
**sarc**  flesh
**saur**  lizard
**schist, schiz**  splitting
**scler**  hard
**scyph**  cup
**semi**  half, part way
**septic**  toxic, poisonous
**septum**  enclosure, wall
**seta**  bristle
**siphon**  tube
**som**  body
**sperm**  seed
**sphen**  wedge
**sphygm**  pulse
**spiro**  breathing; coil
**squam**  scale; husk
**stasis**  stopping, standing
**stato**  standing
**stego**  roof
**steno**  narrow
**stoma**  mouth
**stratum**  covering
**strobo**  twisting
**sub**  below, under
**super, supra**  above; beyond
**sym, syn, sys**  together;
 resembling

**tachy**  rapid
**tardi**  slow; late
**taxis**  arrangement; movement
**tax**  group, division
**tele**  distant
**teleo, telo**  end, completion
**tetr**  four
**thallus**  twig; young shoot
**theca, thecium**  case, cover
**therm**  temperature; heat
**thrombo**  clot; lump
**thyla**  pouch
**tome**  section
**tonic**  stretching
**trans**  across
**trema**  hole

**tri**  three
**trich**  hair
**troch**  wheel
**trop**  turn
**troph**  nourishment, growth
**trypano**  borer
**tuber, tuberculum**  swelling
  knob
**tunica**  covering
**turba**  stir, turmoil

**ultra**  above, beyond

**un**  not
**ur**  urine
**uro**  tail, rump

**vas**  vessel
**velum**  veil
**verse**  turning
**vis**  face
**viscera**  internal organs
**visc**  cohesive, sticky
**vitellus**  yolk
**vitr**  glass

**vivo**  life
**vore**  feeding

**xanth**  yellow
**xer**  dry
**xylo**  wood

**zo**  animal
**zyg**  joined; yoke
**zym**  leaven; ferment